Teacher Edition

Eureka Math® Grade 1 Module 2

Special thanks go to the Gordon A. Cain Center and to the Department of Mathematics at Louisiana State University for their support in the development of *Eureka Math*.

For a free *Eureka Math* Teacher
Resource Pack, Parent Tip
Sheets, and more please visit
https://eurekamath.greatminds.org/teacher-resource-pack

Published by Great Minds

Printed in the U.S.A.
This book may be purchased from the publisher at eureka-math.org
BAB 10 9 8 7 6 5 4 3
ISBN 978-1-63255-349-2

***Eureka Math: A Story of Units®* Contributors**

Katrina Abdussalaam, Curriculum Writer
Tiah Alphonso, Program Manager—Curriculum Production
Kelly Alsup, Lead Writer / Editor, Grade 4
Catriona Anderson, Program Manager—Implementation Support
Debbie Andorka-Aceves, Curriculum Writer
Eric Angel, Curriculum Writer
Leslie Arceneaux, Lead Writer / Editor, Grade 5
Kate McGill Austin, Lead Writer / Editor, Grades PreK–K
Adam Baker, Lead Writer / Editor, Grade 5
Scott Baldridge, Lead Mathematician and Lead Curriculum Writer
Beth Barnes, Curriculum Writer
Bonnie Bergstresser, Math Auditor
Bill Davidson, Fluency Specialist
Jill Diniz, Program Director
Nancy Diorio, Curriculum Writer
Nancy Doorey, Assessment Advisor
Lacy Endo-Peery, Lead Writer / Editor, Grades PreK–K
Ana Estela, Curriculum Writer
Lessa Faltermann, Math Auditor
Janice Fan, Curriculum Writer
Ellen Fort, Math Auditor
Peggy Golden, Curriculum Writer
Maria Gomes, Pre-Kindergarten Practitioner
Pam Goodner, Curriculum Writer
Greg Gorman, Curriculum Writer
Melanie Gutierrez, Curriculum Writer
Bob Hollister, Math Auditor
Kelley Isinger, Curriculum Writer
Nuhad Jamal, Curriculum Writer
Mary Jones, Lead Writer / Editor, Grade 4
Halle Kananak, Curriculum Writer
Susan Lee, Lead Writer / Editor, Grade 3
Jennifer Loftin, Program Manager—Professional Development
Soo Jin Lu, Curriculum Writer
Nell McAnelly, Project Director

Ben McCarty, Lead Mathematician / Editor, PreK–5
Stacie McClintock, Document Production Manager
Cristina Metcalf, Lead Writer / Editor, Grade 3
Susan Midlarsky, Curriculum Writer
Pat Mohr, Curriculum Writer
Sarah Oyler, Document Coordinator
Victoria Peacock, Curriculum Writer
Jenny Petrosino, Curriculum Writer
Terrie Poehl, Math Auditor
Robin Ramos, Lead Curriculum Writer / Editor, PreK–5
Kristen Riedel, Math Audit Team Lead
Cecilia Rudzitis, Curriculum Writer
Tricia Salerno, Curriculum Writer
Chris Sarlo, Curriculum Writer
Ann Rose Sentoro, Curriculum Writer
Colleen Sheeron, Lead Writer / Editor, Grade 2
Gail Smith, Curriculum Writer
Shelley Snow, Curriculum Writer
Robyn Sorenson, Math Auditor
Kelly Spinks, Curriculum Writer
Marianne Strayton, Lead Writer / Editor, Grade 1
Theresa Streeter, Math Auditor
Lily Talcott, Curriculum Writer
Kevin Tougher, Curriculum Writer
Saffron VanGalder, Lead Writer / Editor, Grade 3
Lisa Watts-Lawton, Lead Writer / Editor, Grade 2
Erin Wheeler, Curriculum Writer
MaryJo Wieland, Curriculum Writer
Allison Witcraft, Math Auditor
Jessa Woods, Curriculum Writer
Hae Jung Yang, Lead Writer / Editor, Grade 1

Board of Trustees

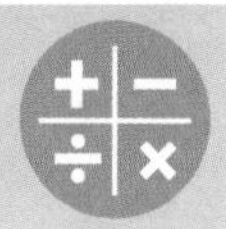

Table of Contents

GRADE 1 • MODULE 2

Introduction to Place Value Through Addition and Subtraction Within 20

G1-M2-TE-BK2-1.3.1-01.2016

Grade 1 • Module 2

Introduction to Place Value Through Addition and Subtraction Within 20

OVERVIEW

Module 2 serves as a bridge from problem solving within 10 to work within 100 as students begin to solve addition and subtraction problems involving teen numbers (**1.NBT.2ab**). In Module 1, students were encouraged to move beyond the Level 1 strategy of counting all to the more efficient counting on. Now, they go beyond Level 2 to learn Level 3 decomposition and composition strategies, informally called make ten or take from ten.[1]

Level 1: Count all *Level 2: Count on* *Level 3: Decompose an addend to compose*

Though many students may continue to count on as their primary means of adding and subtracting, the larger purpose of composing and decomposing ten is to lay the foundation for the role of place value units in addition and subtraction. Meanwhile, from the beginning of the year, fluency activities have focused on the three prerequisite skills for the Level 3 decomposition and composition methods:

1. Partners to ten (**K.OA.4**).
2. Decompositions for all numbers within 10 (**K.OA.3**).
3. Representations of teen numbers as 10 + *n* (**K.NBT.1** and **1.NBT.2b**). For example, students practice counting the Say Ten way (i.e., ten 1, ten 2, ...) from Kindergarten on.

To introduce students to the make ten strategy, in Topic A students solve problems with three addends (**1.OA.2**) and realize it is sometimes possible to use the associative and commutative properties to compose ten, e.g., "Maria made 1 snowball. Tony made 5, and their father made 9. How many snowballs did they make in all?" 1 + 5 + 9 = (9 + 1) + 5 = 10 + 5 = 15. Since we can add in any order, we can pair the 1 with the 9 to make a ten first. Having seen how to use partners to ten to simplify addition, students next decompose a second addend in order to compose a ten from 9 or 8 (e.g., "Maria has 9 snowballs and Tony has 6. How many do they have in all?"). 9 + 6 = 9 + (1 + 5) = (9 + 1) + 5 = 10 + 5 = 15 (**1.OA.3**). Between the intensive work with addends of 8 and 9 is a lesson exploring commutativity so that students realize they can compose ten from the larger addend.

---

[1]See Progressions Document, *"Counting and Cardinality: Operations and Algebraic Thinking,"* p. 6.

Throughout Topic A, students also count on to add. Students begin by modeling the situations with concrete materials, move to representations of 5-groups, and progress to modeling with number bonds. The representations and models make the connection between the two strategies clear. For example, using the 5-groups pictured above, students can simply count on from 9 to 15, tracking the number of counts on their fingers just as they did in Module 1. They repeatedly compare and contrast counting on with making ten, seeing that the latter is a convenient shortcut. Many start to make the important move from counting on, a Level 2 strategy, to make ten, a Level 3 strategy, persuaded by confidence in their increasing skill and the joy of the shortcut. This is a critical step in building flexible part-whole thinking whereby students see numbers as parts and wholes rather than as discrete counts or one part and some ones. Five-groups soon begin to be thought of as ten-frames, focusing on the usefulness of trying to group 10 when possible. This empowers students in later modules and future grade levels to compose and decompose place value units and work adeptly with the four operations. For example, in Grade 1, this is applied in later modules to solve problems such as 18 + 6, 27 + 9, 36 + 6, 49 + 7 (**1.OA.3**).

To introduce students to the take from ten strategy, Topic B opens with questions such as, "Mary has two plates of cookies, one with 10 and one with 2. At the party, 9 cookies were eaten from the plate with 10 cookies. How many cookies were left after the party?" 10 – 9 = 1 and 1 + 2 = 3. Students then reinterpret the story to see its solution can also be written as 12 – 9.

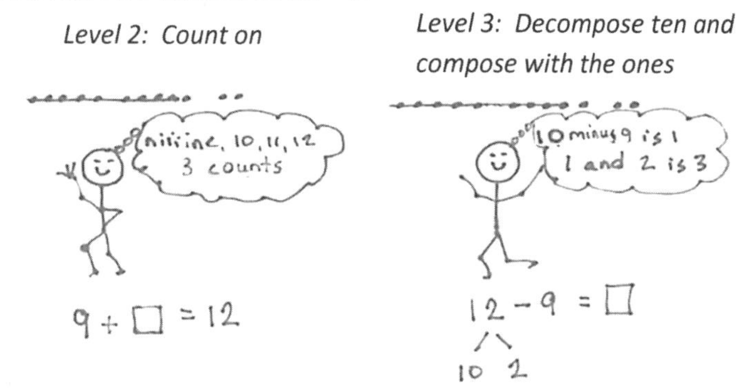

Students relate counting on and subtraction as pictured above. Notice the model is identical, but the thinking is very different.

S: To solve 12 – 9, I count on from 9 to 12, niiiine, 10, 11, 12, three counts. → To solve 12 – 9, I make 12 into 10 and 2 and subtract 9 from ten. 1 + 2 = 3.

Students practice a pattern of action, take from ten and add the ones, as they face different contexts in word problems (MP.8) (e.g., "Maria has 12 snowballs. She threw 8 of them. How many does she have left?"). (**1.OA.3**). This is important foundational work for decomposing in the context of subtraction problem solving in Grade 2 (e.g., "Hmmm. 32 – 17, do I take 7 ones from 2 ones or from a ten?"). Grade 1 students begin using horizontal linear models of 5-groups or ten-frames to begin the transition toward a unit of ten, as shown in the above image.

Topic C presents students with opportunities to solve varied *add to with change unknown, take from with change unknown, put together with addend unknown*, and *take apart with addend unknown* word problems. These situations give ample time for exploring strategies for finding an unknown. The module so far has focused on counting on and subtracting by decomposing and composing (**1.OA.1**). These lessons open up the possibilities to include other Level 3 strategies (e.g., 12 – 3 = 12 – 2 – 1).[2] Teachers can include or adjust such strategy use dependent on whether they feel it enhances understanding or rather undermines or overwhelms. The topic closes with a lesson to further solidify student understanding of the equal sign as it has been applied throughout the module. Students match equivalent expressions to construct true number sentences and explain their reasoning using words, pictures, and numbers (e.g., 12 – 7 = 3 + 2, 10 + 5 = 9 + 6) (**1.OA.7**).

In Topic D, after all the work with 10, the module culminates with naming a ten (**1.NBT.2a**). Familiar representations of teen numbers, such as two 5-groups, the Rekenrek, and 10 fingers, are all renamed as a ten and some ones (**1.NBT2b**), rather than 10 ones and some more ones (**K.NBT.1**). The ten is shifting to being one unit, a structure from which students can compose and decompose teen numbers (**1.NBT.2b**, MP.7). This significant step forward sets the stage for understanding all the numbers within 100 as composed of a number of units of ten and some ones (**1.NBT.2b**). The horizontal linear 5-group modeling of 10 is moved to a vertical representation in preparation for this next stage, in Module 4, as shown in the image on the right. This topic's work is done while solving both abstract equations and contextualized word problems.

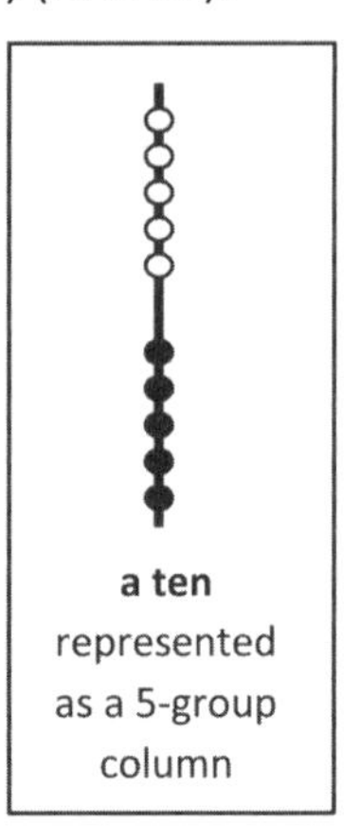
a ten represented as a 5-group column

Notes on Pacing for Differentiation

If pacing is a challenge, embed conversations about efficiency and strategy comparison throughout Module 2. Application Problems and Student Debriefs can provide opportunities to share and compare students' varied strategies. This allows omission of four lessons: 5, 9, 11, and 21. In Lesson 16, consider focusing on the finger work to practice the take from ten strategy rather than focusing on relating counting on to making ten and taking from ten. Consider omitting Lesson 24 if Application Problems are completed daily and if students have completed Lessons 22 and 23, which also focus on solving word problems. Note that it may be useful to extend Lessons 10, 19, 20, or 25 to provide extra practice as students develop their understanding of making ten, taking from ten, and the meaning of the equal sign.

[2]See Progressions Document, *Counting and Cardinality: Operations and Algebraic Thinking*, p. 14.

Distribution of Instructional Minutes

This diagram represents a suggested distribution of instructional minutes based on the emphasis of particular lesson components in different lessons throughout the module.

- Fluency Practice
- Concept Development
- Application Problems
- Student Debrief

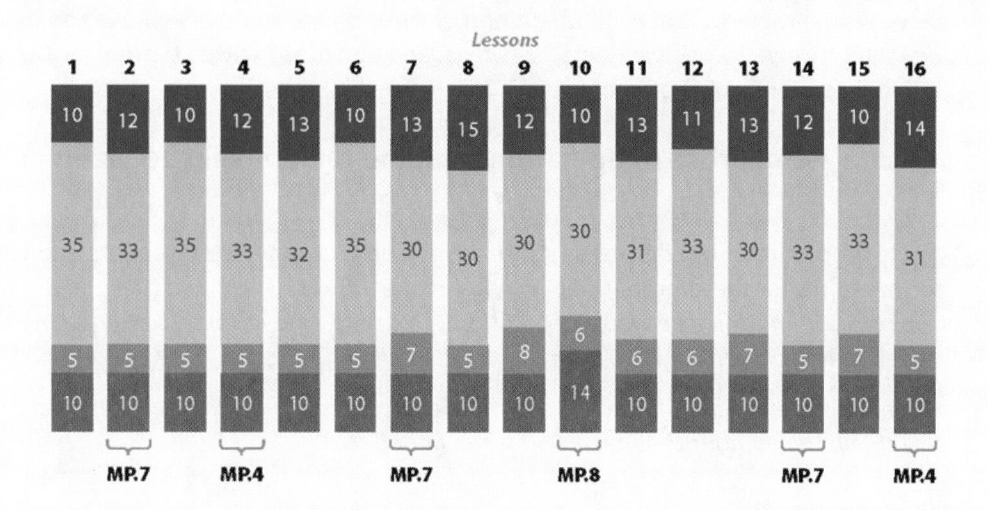

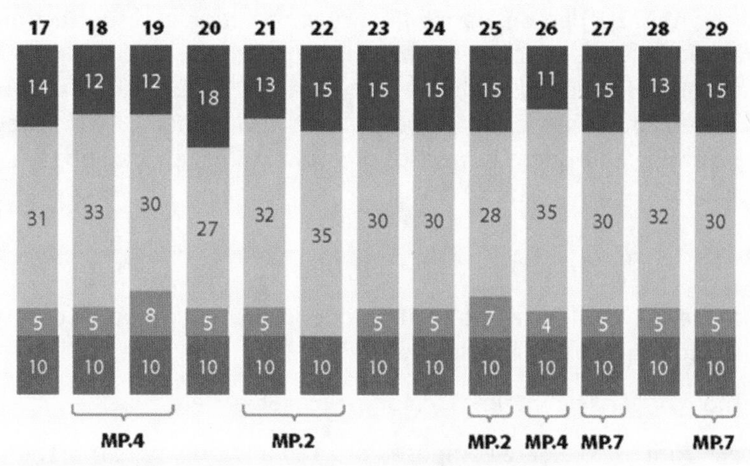

Focus Grade Level Standards

Represent and solve problems involving addition and subtraction.

1.OA.1 Use addition and subtraction within 20 to solve word problems involving situations of adding to, taking from, putting together, taking apart, and comparing, with unknowns in all positions, e.g., by using objects, drawings, and equations with a symbol for the unknown number to represent the problem.

1.OA.2 Solve word problems that call for addition of three whole numbers whose sum is less than or equal to 20, e.g., by using objects, drawings, and equations with a symbol for the unknown number to represent the problem.

Understand and apply properties of operations and the relationship between addition and subtraction.

1.OA.3 Apply properties of operations as strategies to add and subtract. (Students need not use formal terms for these properties.) *Examples: If 8 + 3 = 11 is known, then 3 + 8 = 11 is also known. (Commutative property of addition.) To add 2 + 6 + 4, the second two numbers can be added to make a ten, so 2 + 6 + 4 = 2 + 10 = 12. (Associative property of addition.)*

1.OA.4 Understand subtraction as an unknown-addend problem. *For example, subtract 10 – 8 by finding the number that makes 10 when added to 8.*

Add and subtract within 20.[3]

1.OA.6 Add and subtract within 20, demonstrating fluency for addition and subtraction within 10. Use mental strategies such as counting on; making ten (e.g., 8 + 6 = 8 + 2 + 4 = 10 + 4 = 14); decomposing a number leading to a ten (e.g., 13 – 4 = 13 – 3 – 1 = 10 – 1 = 9); using the relationship between addition and subtraction (e.g., knowing that 8 + 4 = 12, one knows 12 – 8 = 4); and creating equivalent but easier or known sums (e.g., adding 6 + 7 by creating the known equivalent 6 + 6 + 1 = 12 + 1 = 13).

Understand place value.[4]

1.NBT.2 Understand that the two digits of a two-digit number represent amounts of tens and ones. Understand the following as special cases:

a. 10 can be thought of as a bundle of ten ones—called a "ten."

b. The numbers from 11 to 19 are composed of a ten and one, two, three, four, five, six, seven, eight, or nine ones.

[3]The balance of this cluster is addressed in Module 1.

[4]The focus in this module is on numbers to 20. The balance of this cluster is addressed in Modules 4 and 6.

Foundational Standards

K.OA.3 Decompose numbers less than or equal to 10 into pairs in more than one way, e.g., by using objects or drawings, and record each decomposition by a drawing or equation (e.g., 5 = 2 + 3 and 5 = 4 + 1).

K.OA.4 For any number from 1 to 9, find the number that makes 10 when added to the given number, e.g., by using objects or drawings, and record the answer with a drawing or equation.

K.NBT.1 Compose and decompose numbers from 11 to 19 into ten ones and some further ones, e.g., by using objects or drawings, and record each composition or decomposition by a drawing or equation (e.g., 18 = 10 + 8); understand that these numbers are composed of ten ones and one, two, three, four, five, six, seven, eight, or nine ones.

Focus Standards for Mathematical Practice

MP.2 **Reason abstractly and quantitatively**. Students solve *change unknown* problem types such as, "Maria has 8 snowballs. Tony has 15 snowballs. Maria wants to have the same number of snowballs as Tony. How many more snowballs does Maria need to have the same number as Tony?" They write the equation 8 + __ = 15 to describe the situation, make ten or count on to 15 to find the answer of 7, and reason abstractly to make a connection to subtraction, that the same problem can be solved using 15 – 8 = __.

MP.4 **Model with mathematics**. Students use 5-groups, number bonds, and equations to represent decompositions when both subtracting from the teens and adding to make teens when crossing the ten.

MP.7 **Look for and make use of structure**. This module introduces students to the unit *ten*. Students use the structure of the ten to add within the teens, to add to the teens, and to subtract from the teens. For example, 14 + 3 = 10 + 4 + 3 = 17, 8 + 5 = 8 + 2 + 3 = 10 + 3 and conversely, 13 – 5 = 10 – 5 + 3 = 5 + 3.

MP.8 **Look for and make use of repeated reasoning**. Students realize that when adding 9 to a number 1-9, they can complete the ten by decomposing the other addend into "1 and ___." They internalize the commutative and associative properties, looking for ways to make ten within situations and equations.

Overview of Module Topics and Lesson Objectives

Standards		Topics and Objectives		Days
1.OA.1 **1.OA.2** **1.OA.3** **1.OA.6**	A	**Counting On or Making Ten to Solve *Result Unknown* and *Total Unknown* Problems**		11
		Lesson 1:	Solve word problems with three addends, two of which make ten.	
		Lesson 2:	Use the associative and commutative properties to make ten with three addends.	
		Lessons 3–4:	Make ten when one addend is 9.	
		Lesson 5:	Compare efficiency of counting on and making ten when one addend is 9.	
		Lesson 6:	Use the commutative property to make ten.	
		Lessons 7–8:	Make ten when one addend is 8.	
		Lesson 9:	Compare efficiency of counting on and making ten when one addend is 8.	
		Lesson 10:	Solve problems with addends of 7, 8, and 9.	
		Lesson 11:	Share and critique peer solution strategies for *put together with total unknown* word problems.	
		Mid-Module Assessment: Topic A (assessment 1 day, return 1 day, remediation or further applications 1 day)		3
1.OA.1 **1.OA.3** **1.OA.4** **1.OA.6** 1.OA.5 1.OA.7	B	**Counting On or Taking from Ten to Solve *Result Unknown* and *Total Unknown* Problems**		10
		Lessons 12–13:	Solve word problems with subtraction of 9 from 10.	
		Lessons 14–15:	Model subtraction of 9 from teen numbers.	
		Lesson 16:	Relate counting on to making ten and taking from ten.	
		Lessons 17–18:	Model subtraction of 8 from teen numbers.	
		Lesson 19:	Compare efficiency of counting on and taking from ten.	
		Lesson 20:	Subtract 7, 8, and 9 from teen numbers.	
		Lesson 21:	Share and critique peer solution strategies for *take from with result unknown* and *take apart with addend unknown* word problems from the teens.	

Standards		Topics and Objectives		Days
1.OA.1 **1.OA.4** **1.OA.6** 1.OA.5 1.OA.7 1.OA.8	C	**Strategies for Solving *Change* or *Addend Unknown* Problems**		4
		Lesson 22:	Solve *put together/take apart with addend unknown* word problems, and relate counting on to the take from ten strategy.	
		Lesson 23:	Solve *add to with change unknown* problems, relating varied addition and subtraction strategies.	
		Lesson 24:	Strategize to solve *take from with change unknown* problems.	
		Lesson 25:	Strategize and apply understanding of the equal sign to solve equivalent expressions.	
1.OA.1 **1.OA.6** **1.NBT.2a** **1.NBT.2b** 1.NBT.5	D	**Varied Problems with Decompositions of Teen Numbers as 1 Ten and Some Ones**		4
		Lesson 26:	Identify 1 ten as a unit by renaming representations of 10.	
		Lesson 27:	Solve addition and subtraction problems decomposing and composing teen numbers as 1 ten and some ones.	
		Lesson 28:	Solve addition problems using ten as a unit, and write two-step solutions.	
		Lesson 29:	Solve subtraction problems using ten as a unit, and write two-step solutions.	
		End-of-Module Assessment: Topics A–D (assessment 1 day, return 1 day, remediation or further applications 1 day)		3
Total Number of Instructional Days				**35**

Terminology

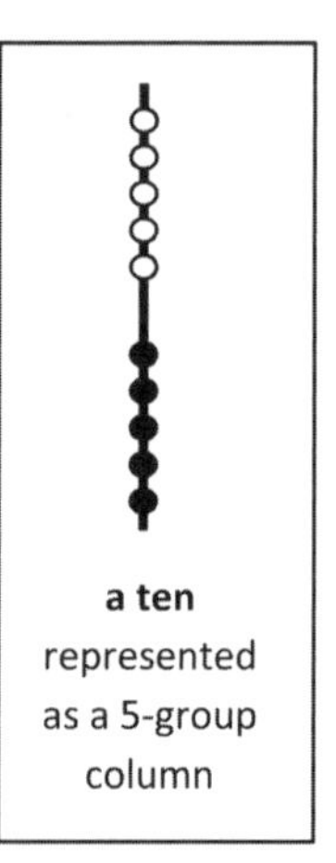
a ten represented as a 5-group column

New or Recently Introduced Terms

- A ten (a group, or unit, consisting of 10 items)
- Ones (individual units, 10 of which become a ten)

Familiar Terms and Symbols[5]

- 5-groups
- Add
- Equals
- Number bonds
- Partners to ten
- Subtract
- Teen numbers

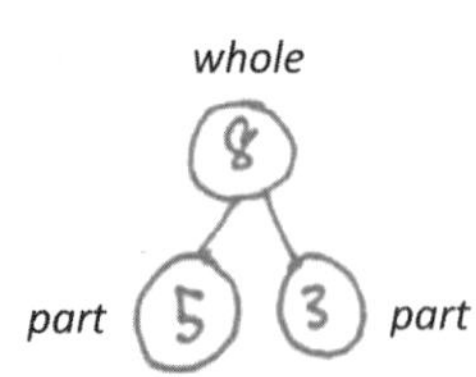

Number Bond

Suggested Tools and Representations

- 5-group formations: 5-groups (and 5-group cards), 5-group rows, 5-group column
- Hide Zero cards
- Number bonds
- Number path
- Rekenrek

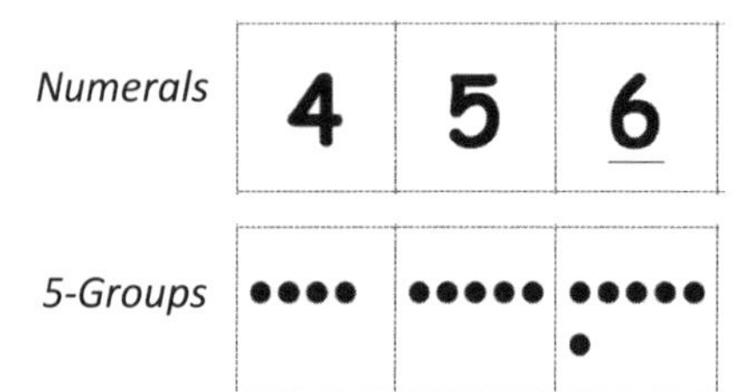

5-Group Cards

ooooo ooooo

5-Group Rows

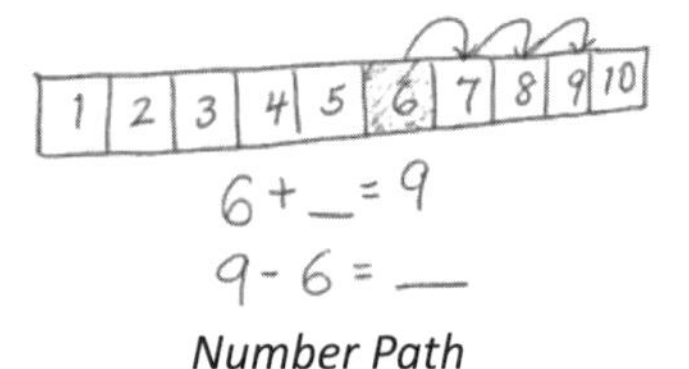

Number Path

Rekenrek

5-Group Column

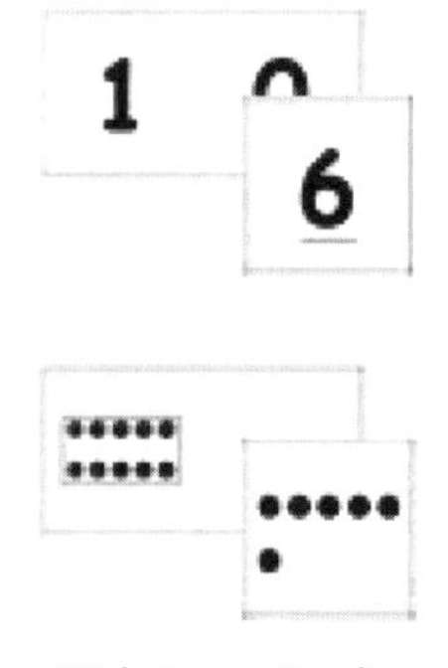

Hide Zero Cards

---

[5]These are terms and symbols students have seen previously.

Homework

Homework at the K–1 level is not a convention in all schools. In this curriculum, homework is an opportunity for additional practice of the content from the day's lesson. The teacher is encouraged, with the support of parents, administrators, and colleagues, to discern the appropriate use of homework for his or her students. Fluency exercises can also be considered as an alternative homework assignment.

Scaffolds[6]

The scaffolds integrated into *A Story of Units®* give alternatives for how students access information as well as express and demonstrate their learning. Strategically placed margin notes are provided within each lesson, elaborating on the use of specific scaffolds at applicable times. They address many needs presented by English language learners, students with disabilities, students performing above grade level, and students performing below grade level. Many of the suggestions are organized by Universal Design for Learning (UDL) principles and are applicable to more than one population. To read more about the approach to differentiated instruction in *A Story of Units*, please refer to "How to Implement *A Story of Units*."

Assessment Summary

Type	Administered	Format	Standards Addressed
Mid-Module Assessment Task	After Topic A	Constructed response with rubric	1.OA.1 1.OA.2 1.OA.3 1.OA.6
End-of-Module Assessment Task	After Topic D	Constructed response with rubric	1.OA.1 1.OA.2 1.OA.3 1.OA.4 1.OA.6 1.NBT.2a 1.NBT.2b

[6]Students with disabilities may require Braille, large print, audio, or special digital files. Please visit the website www.p12.nysed.gov/specialed/aim for specific information on how to obtain student materials that satisfy the National Instructional Materials Accessibility Standard (NIMAS) format.

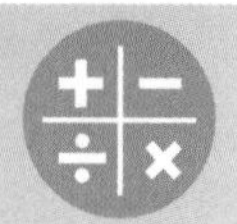

Topic A

Counting On or Making Ten to Solve *Result Unknown* and *Total Unknown* Problems

1.OA.1, 1.OA.2, 1.OA.3, 1.OA.6

Focus Standards:	1.OA.1	Use addition and subtraction within 20 to solve word problems involving situations of adding to, taking from, putting together, taking apart, and comparing, with unknowns in all positions, e.g., by using objects, drawings, and equations with a symbol for the unknown number to represent the problem.
	1.OA.2	Solve word problems that call for addition of three whole numbers whose sum is less than or equal to 20, e.g., by using objects, drawings, and equations with a symbol for the unknown number to represent the problem.
	1.OA.3	Apply properties of operations as strategies to add and subtract. (Students need not use formal terms for these properties.) *Examples: If 8 + 3 = 11 is known, then 3 + 8 = 11 is also known. (Commutative property of addition.) To add 2 + 6 + 4, the second two numbers can be added to make a ten, so 2 + 6 + 4 = 2 + 10 = 12. (Associative property of addition.)*
	1.OA.6	Add and subtract within 20, demonstrating fluency for addition and subtraction within 10. Use mental strategies such as counting on; making ten (e.g., 8 + 6 = 8 + 2 + 4 = 10 + 4 = 14); decomposing a number leading to a ten (e.g., 13 – 4 = 13 – 3 – 1 = 10 – 1 = 9); using the relationship between addition and subtraction (e.g., knowing that 8 + 4 = 12, one knows 12 – 8 = 4); and creating equivalent but easier or known sums (e.g., adding 6 + 7 by creating the known equivalent 6 + 6 + 1 = 12 + 1 = 13).
Instructional Days:	11	
Coherence -Links from:	GK–M4	Number Pairs, Addition and Subtraction to 10
-Links to:	G2–M3	Place Value, Counting, and Comparison of Numbers to 1,000
	G2–M5	Addition and Subtraction Within 1,000 with Word Problems to 100

Topic A begins with students solving word problems with three addends (**1.OA.2**) as a way for them to begin to explore the make ten Level 3 strategy in a meaningful context. With problems that always include at least two numbers that yield 10 when added together, Lesson 1 encourages students to use the associative and commutative properties as they set up and read equations in various ways. The story problem on the right, for instance, can be solved by adding 1 + 9 first and then adding the five (see image below story problem).

We had 1 upper-grade buddy come visit with 9 more buddies following him. Soon after that, 5 more buddies came to our classroom. How many buddies came altogether?

1 + 9 + 5
10 + 5 = 15

9 + 1 = 10
10 + 5 = 15

This leads into Lesson 2's focus of explicitly using the associative and commutative properties[1] to add three addends without the context of story problems (**1.OA.3**). This lesson is where students practice associating the two addends that make ten (**1.OA.6**) and then adding the third addend; they prove to themselves that this simplification of addition is a viable strategy.

Following this introduction, Lessons 3, 4, and 5 afford students ample practice with adding 9 and another single-digit number as they decompose the second addend to make ten with the 9. Students solve problems such as "Maria has 9 snowballs, and Tony has 6. How many do they have in all?" as follows: $9 + 6 = 9 + (1 + 5) = (9 + 1) + 5 = 10 + 5 = 15$. This triad of lessons takes students through a concrete–pictorial–abstract progression as they work with physical 5-groups using objects, 5-group drawings, and finally number bonds.

Lesson 6 reminds students of the commutative property again, by focusing them on when and why they might apply commutativity: to compose ten from the larger addend. Lessons 7, 8, and 9 mirror the earlier set of three lessons, but students decompose one addend to make ten with 8 as the key addend. This extensive practice allows students to internalize both why and how they would compose ten from the larger addend as they come to realize that this is an efficient strategy.

Students use the make ten strategy with 5-group drawings and number bonds to solve a variety of problems involving a mixture of 7, 8, or 9 as addends in Lesson 10. This gives students an opportunity to not only practice their newly discovered strategies, but it also allows them to generalize this make ten strategy to a new number: 7. It is important to note that students can continue to use counting on as a strategy throughout the entirety of Topic A, although many students begin to use the make ten strategy more and more as they continually discuss addition strategies and efficiency with one another.

Topic A ends with Lesson 11 where students solve story problems with two addends (**1.OA.1**) using independently selected methods. By asking questions such as "Why did you solve the problem that way? How did we solve these differently?" students are able to engage in rich dialogue about the mathematical strategies and determine which are most useful.

[1]Just as the Common Core State Standards note, students do not learn or use these formal terms.

A Teaching Sequence Toward Mastery of Counting On or Making Ten to Solve *Result Unknown* and *Total Unknown* Problems

Objective 1: Solve word problems with three addends, two of which make ten.
(Lesson 1)

Objective 2: Use the associative and commutative properties to make ten with three addends.
(Lesson 2)

Objective 3: Make ten when one addend is 9.
(Lessons 3–4)

Objective 4: Compare efficiency of counting on and making ten when one addend is 9.
(Lesson 5)

Objective 5: Use the commutative property to make ten.
(Lesson 6)

Objective 6: Make ten when one addend is 8.
(Lessons 7–8)

Objective 7: Compare efficiency of counting on and making ten when one addend is 8.
(Lesson 9)

Objective 8: Solve problems with addends of 7, 8, and 9.
(Lesson 10)

Objective 9: Share and critique peer solution strategies for *put together with total unknown* word problems.
(Lesson 11)

Lesson 1

Objective: Solve word problems with three addends, two of which make ten.

Suggested Lesson Structure

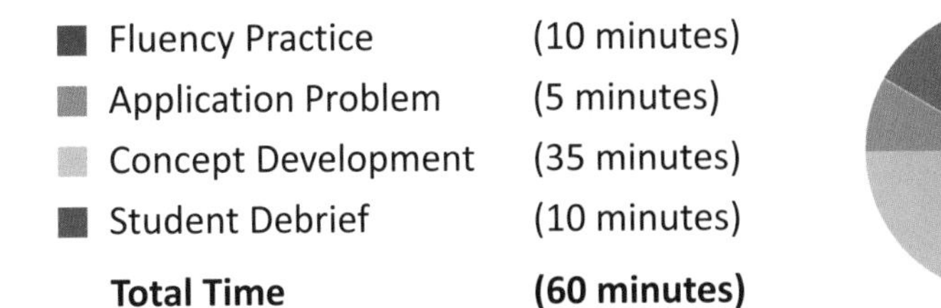

Fluency Practice	(10 minutes)
Application Problem	(5 minutes)
Concept Development	(35 minutes)
Student Debrief	(10 minutes)
Total Time	**(60 minutes)**

Fluency Practice (10 minutes)

- Sparkle: The Say Ten and Regular Way **1.NBT.2** (3 minutes)
- Take Out 1 **1.OA.5** (2 minutes)
- Equal Number Pairs for Ten **1.OA.6** (5 minutes)

Sparkle: The Say Ten and Regular Way (3 minutes)

Note: Say Ten counting reinforces place value and prepares students to add ten and some ones.

Count from 10 to 20, alternating between the regular way and the Say Ten way (e.g., 10, ten 1, 12, ten 3, 14, ten 5). If students are still building fluency with counting within the teen sequence, consider counting the regular way or the Say Ten way without alternating. If time permits, try counting back, too.

Students stand in a circle. Introduce the counting pattern, the start number, and end number: "Today we will count the Say Ten way from 10 to 20." The number range may be adjusted to fit the size of the class. Before the game, practice the counting sequence as a group, and say, "Sparkle!" after the ending number is said aloud: "Let's practice by counting from 10 to 15. Ten, ten 1, 12, ten 3, 14, ten 5, sparkle!"

Begin the game. Students count around the circle, each student saying one number in the counting sequence. After the ending number is said, the next student says, "Sparkle!" and the following player sits. Begin again with the start number, and continue counting in the same direction around the circle until only one player is standing.

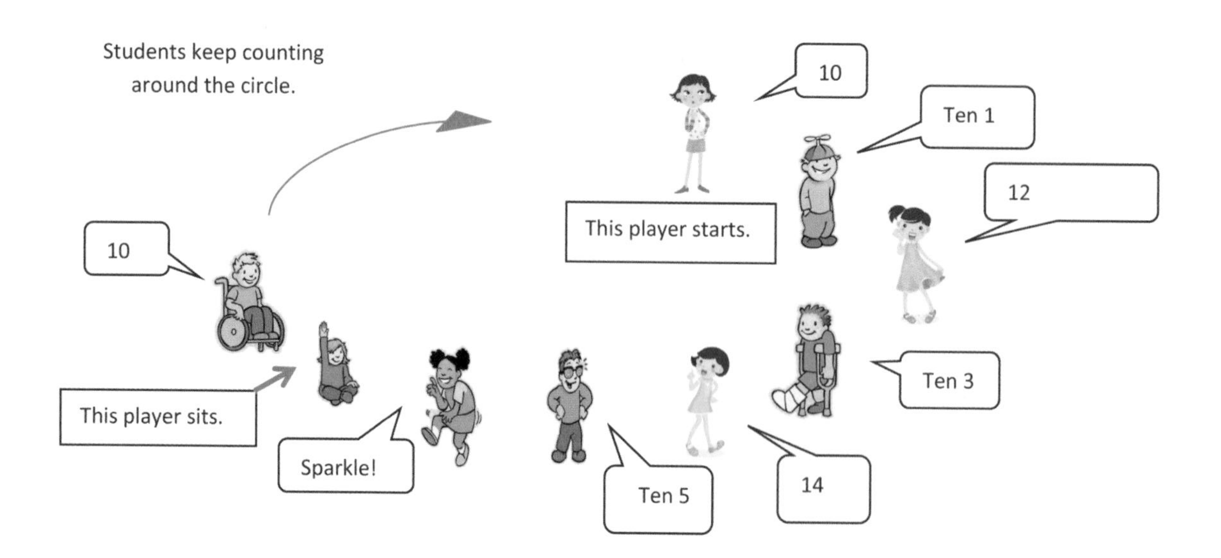

Take Out 1 (2 minutes)

Note: This activity supports fluency with decomposing numbers within 10. This skill is critical for using the upcoming Level 3 addition strategy of make ten. Students need to fluently get 1 out of the second addend when adding to 9.

T: Take out 1 on my signal. For example, if I say "5," you say "1 and 4."
T: 3.
S: 1 and 2.
T: 10.
S: 1 and 9.

Continue with all numbers within 10.

Equal Number Pairs for Ten (5 minutes)

Materials: (S) 5-group cards 0 through 10 with two 5 cards, one "=" card, and two "+" cards per set of partners (Fluency Template)

Note: This activity builds fluency with partners to ten and promotes an understanding of equality.

Assign students partners of equal ability. Students arrange 5-group cards from 0 to 10, including the extra 5, and place the "=" card between them. Write 4 numbers on the board (e.g., 5, 9, 1, and 5). Partners take the 5-group cards that match the numbers written to make two equivalent expressions (e.g., 9 + 1 = 5 + 5).

Suggested sequence: 5, 9, 1, 5; 0, 1, 9, 10; 2, 5, 5, 8; 2, 3, 7, 8; 4, 1, 9, 6; 3, 4, 6, 7.

Application Problem (5 minutes)

John, Emma, and Alice each had 10 raisins. John ate 3 raisins, Emma ate 4 raisins, and Alice ate 5 raisins. How many raisins do they each have now? Write a number bond and a number sentence for each.

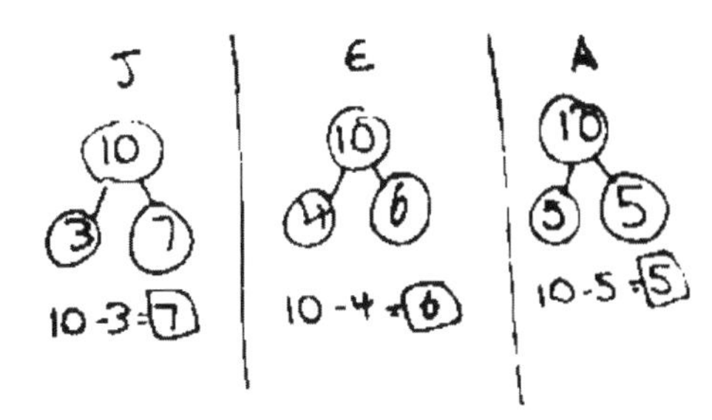

Note: This problem was chosen as an application of the culminating subtraction work from Module 1. All three subtraction sentences and number bonds focus on partners to ten, which are foundational to the first lesson of Module 2.

Concept Development (35 minutes)

Materials: (T) Bin, three different kinds of blocks/pattern blocks, 18-inch length of string tied to form a loop (S) Three different kinds of pattern blocks (10 of each shape, e.g., trapezoid, triangle, and square blocks), personal white board

Have students sit in a semicircle at the meeting area with their personal white boards.

T: The first-grade classrooms each have these special bins with different types of blocks in them. Let's figure out how many we have! (Lay out 9 triangle blocks in a 5-group configuration.) How many triangle blocks do we have?

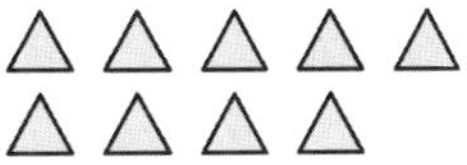

S: 9 triangle blocks!

T: (Lay out 1 square block and 4 trapezoid blocks. Ask students to state the quantity of each group.) We need to figure out how many there are altogether. Help me write the expression.

S: 9 + 1 + 4 = ____.

T: (Write this on the board.)

T: Talk to your partner. What are some ways we could add these blocks together?

S: (As students discuss, the teacher circulates and selects students to share.) We could start with the larger number and count on. → We could add the groups together by counting them all.

T: True! Also, I wonder if we can make ten since it is such a friendly number. Talk with your partner.

S: (Discuss.) 9 and 1. → The 9 triangles and the 1 square.

T: Let's check to be sure your idea is true! (Select a student, one who may particularly benefit from proving this to be true, to move the 1 to the 9 in order to make ten.) Did 9 and 1 make ten?

S: Yes!

T: (Place the square block back in its original position.) I'm going to make the 9 and 1 one group to show this is 10. (Place string around the 9 and 1. Circle 9 and 1 in the equation.) We have 10 (gesture to the 10) and 4 more (gesture to the 4). How many blocks?

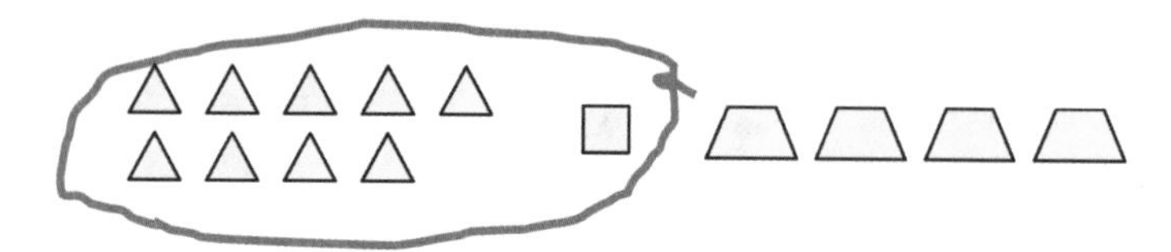

S: We have 14 blocks!

T: (Write 14 to complete the equation.)

T: Talk with your partner. Write the new number sentence explaining what we just did, starting with 10, on your personal white board.

S: (Discuss and write 10 + 4 = 14.)

T: Good! Now it's your turn.

Assign partners, and hand out blocks. The following is a suggested sequence of stories to tell as students work with a partner to represent each problem on their personal white boards. Students should put their boards next to one another to make a larger board. Together, they write the expression, circle 10, and solve for the unknown.

- At lunch, Marcus put 2 pepper slices, 8 carrots, and 6 banana pieces on his tray. When he reached the checkout, how many pieces of food did he have?
- Lena was playing basketball during recess. She made 4 jump shots, 7 layups, and 3 free throws. How many baskets did Lena make?
- We had 5 upper-grade buddies come and visit our classroom with 3 more buddies following them. Soon after that, 5 more buddies came to our classroom. How many total buddies came?

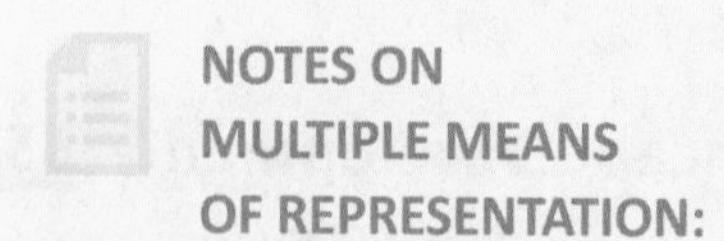

NOTES ON MULTIPLE MEANS OF REPRESENTATION:

Facilitate students' discovery of patterns and structure in math by allowing for a variety of responses to questions. For example, some students may use their pictorial representation and see 4 + 1 = 5 and then use the 5 triangles embedded in the 9 to make a ten.

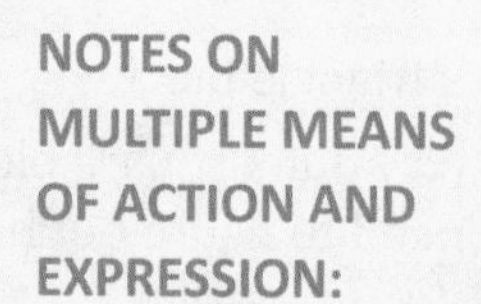

NOTES ON MULTIPLE MEANS OF ACTION AND EXPRESSION:

Having students act out number stories is a great way to provide math-they- can-see. This may help students who are hearing impaired. It also provides visual and kinesthetic learners an opportunity to engage in the lesson using their preferred style of learning.

Problem Set (10 minutes)

Students may work individually, in pairs, or in groups when completing Problem Sets. For Problem Sets that include word problems, it may be best to read problems aloud, particularly early in the year. Students should do their personal best to complete the Problem Set within the allotted 10 minutes. Some problems do not specify a method for solving. This is an intentional reduction of scaffolding that invokes MP.5, Use Appropriate Tools Strategically. Students should solve these problems using the RDW approach used for Application Problems.

For some classes, it may be appropriate to modify the assignment by specifying which problems students should work on first. With this option, let the careful sequencing of the Problem Set guide your selections so that problems continue to be scaffolded. Balance word problems with other problem types to ensure a range of practice. Consider assigning incomplete problems for homework or at another time during the day.

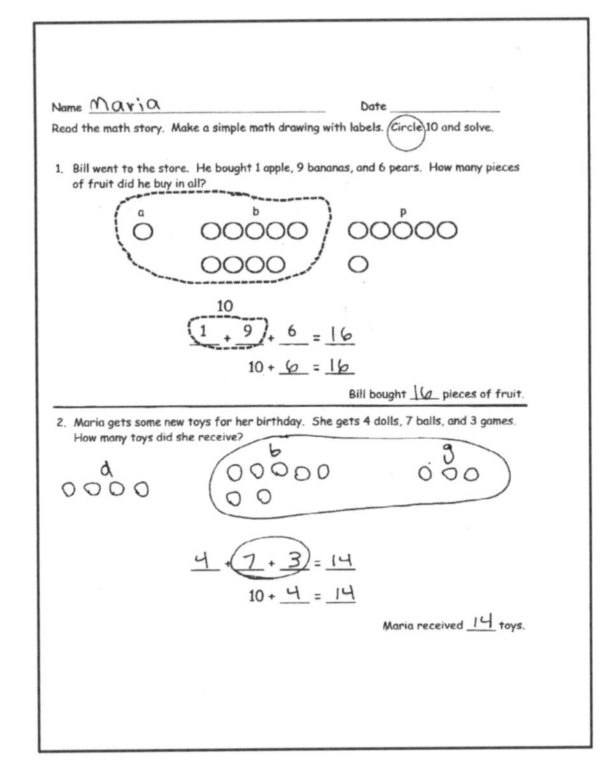
Name Maria Date

Read the math story. Make a simple math drawing with labels. Circle 10 and solve.

1. Bill went to the store. He bought 1 apple, 9 bananas, and 6 pears. How many pieces of fruit did he buy in all?

1 + 9 + 6 = 16

10 + 6 = 16

Bill bought 16 pieces of fruit.

2. Maria gets some new toys for her birthday. She gets 4 dolls, 7 balls, and 3 games. How many toys did she receive?

4 + 7 + 3 = 14

10 + 4 = 14

Maria received 14 toys.

Student Debrief (10 minutes)

Lesson Objective: Solve word problems with three addends, two of which make ten.

The Student Debrief is intended to invite reflection and active processing of the total lesson experience.

Invite students to review their solutions for the Problem Set. They should check work by comparing answers with a partner before going over answers as a class. Look for misconceptions or misunderstandings that can be addressed in the Debrief. Guide students in a conversation to debrief the Problem Set and process the lesson.

Any combination of the questions below may be used to lead the discussion.

- Earlier, we had 9 triangles, 1 square, and 4 trapezoid blocks on the floor. The teacher next door has 4 triangles and 10 squares in her bin of blocks. Does she have more, fewer, or the same number of blocks as we have? How do you know? (Re-create the configuration from the Concept Development if necessary.)

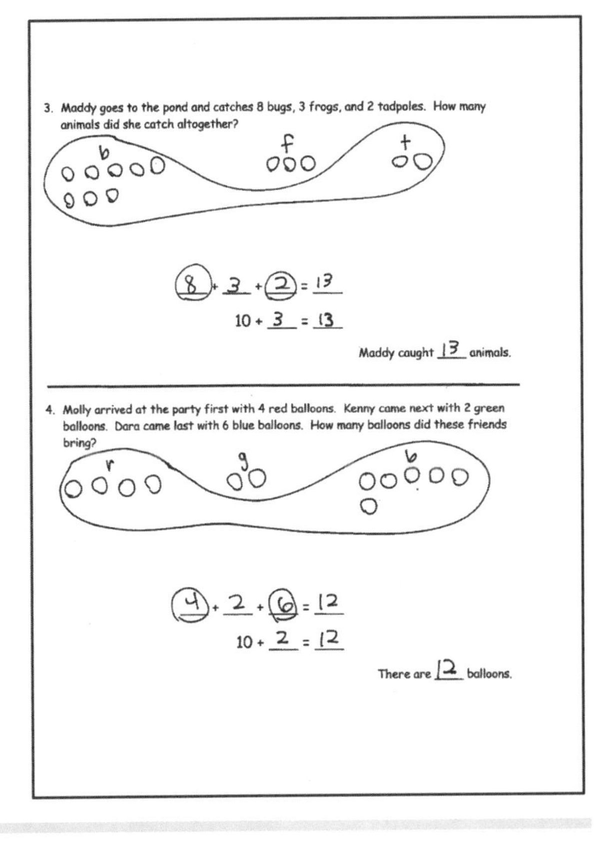
3. Maddy goes to the pond and catches 8 bugs, 3 frogs, and 2 tadpoles. How many animals did she catch altogether?

8 + 3 + 2 = 13

10 + 3 = 13

Maddy caught 13 animals.

4. Molly arrived at the party first with 4 red balloons. Kenny came next with 2 green balloons. Dara came last with 6 blue balloons. How many balloons did these friends bring?

4 + 2 + 6 = 12

10 + 2 = 12

There are 12 balloons.

- What similarities do you notice between Problem 3 and Problem 4?
- How did the Application Problem connect to today's lesson?
- What new way or strategy to add did we learn today? Talk with your partner. (Make ten.) Why is 10 such a friendly number?

Exit Ticket (3 minutes)

After the Student Debrief, instruct students to complete the Exit Ticket. A review of their work will help with assessing students' understanding of the concepts that were presented in today's lesson and planning more effectively for future lessons. The questions may be read aloud to the students.

Homework

Homework at the K–1 level is not a convention in all schools. In this curriculum, homework is an opportunity for additional practice of the content from the day's lesson. The teacher is encouraged, with the support of parents, administrators, and colleagues, to discern the appropriate use of homework for his or her students. Fluency exercises can also be considered as an alternative homework assignment.

Name ______________________ Date ____________

Read the math story. Make a simple math drawing with labels. Circle 10 and solve.

1. Bill went to the store. He bought 1 apple, 9 bananas, and 6 pears. How many pieces of fruit did he buy in all?

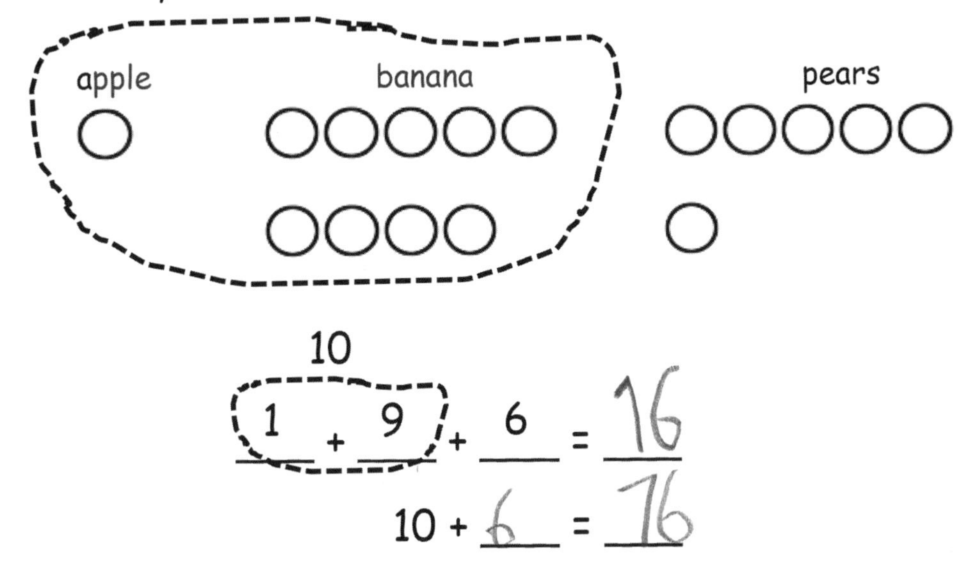

10

1 + 9 + 6 = 16

10 + 6 = 16

Bill bought ____ pieces of fruit.

---

2. Maria gets some new toys for her birthday. She gets 4 dolls, 7 balls, and 3 games. How many toys did she receive?

____ + ____ + ____ = ____

10 + ____ = ____

Maria received ____ toys.

3. Maddy goes to the pond and catches 8 bugs, 3 frogs, and 2 tadpoles. How many animals did she catch altogether?

____ + ____ + ____ = ____

10 + ____ = ____

Maddy caught _____ animals.

---

4. Molly arrived at the party first with 4 red balloons. Kenny came next with 2 green balloons. Dara came last with 6 blue balloons. How many balloons did these friends bring?

____ + ____ + ____ = ____

10 + ____ = ____

There are _____ balloons.

Name ______________________________ Date ______________

Read the math story. Make a simple math drawing with labels. (Circle) 10 and solve.

Toby has ice cream money. He has 2 dimes. He finds 4 more dimes in his jacket and 8 more on the table. How many dimes does Toby have?

____ + ____ + ____ = ____

10 + ____ = ____

Toby has ____ dimes.

Name ______________________________ Date ______________

Read the math story. Make a simple math drawing with labels. (Circle) 10 and solve.

1. Chris bought some treats. He bought 5 granola bars, 6 boxes of raisins, and 4 cookies. How many treats did Chris buy?

___ + ___ + ___ = ___

10 + ___ = ___

Chris bought ____ treats.

---

2. Cindy has 5 cats, 7 goldfish, and 5 dogs. How many pets does she have in all?

___ + ___ + ___ = ___

10 + ___ = ___

Cindy has ____ pets.

3. Mary gets stickers at school for good work. She got 7 puffy stickers, 6 smelly stickers, and 3 flat stickers. How many stickers did Mary get at school altogether?

____ + ____ + ____ = ____

10 + ____ = ____

Mary got ____ stickers at school.

---

4. Jim sat at a table with 4 teachers and 9 children. How many people were at the table after Jim sat down?

____ + ____ + ____ = ____

____ + ____ = ____

There were ____ people at the table after Jim sat down.

0	1	2	3
4	5	6	7
8	9	10	10
	10	5	5

5-group cards, first two pages double-sided, last page single-sided

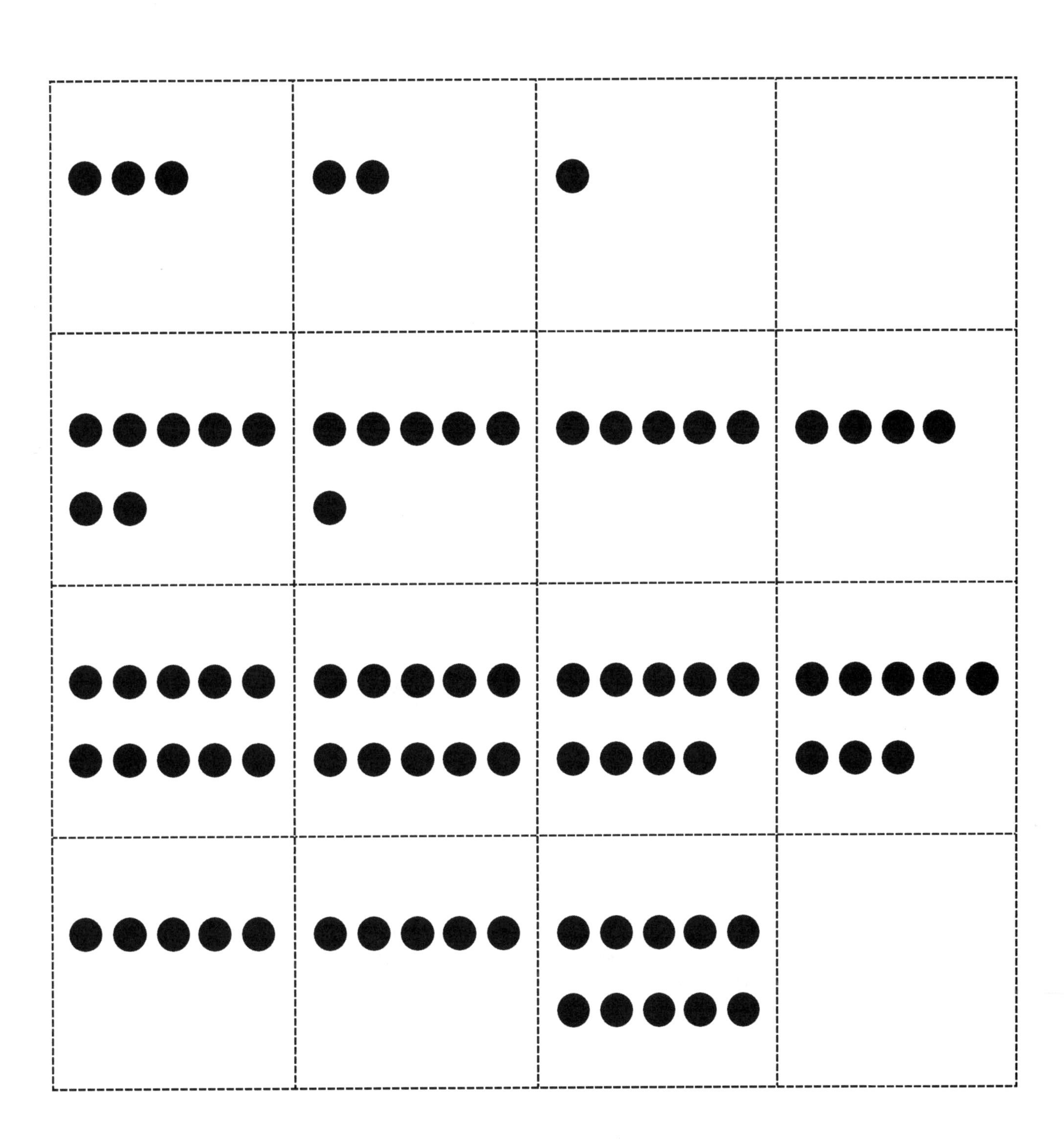

5-group cards, first two pages double-sided, last page single-sided

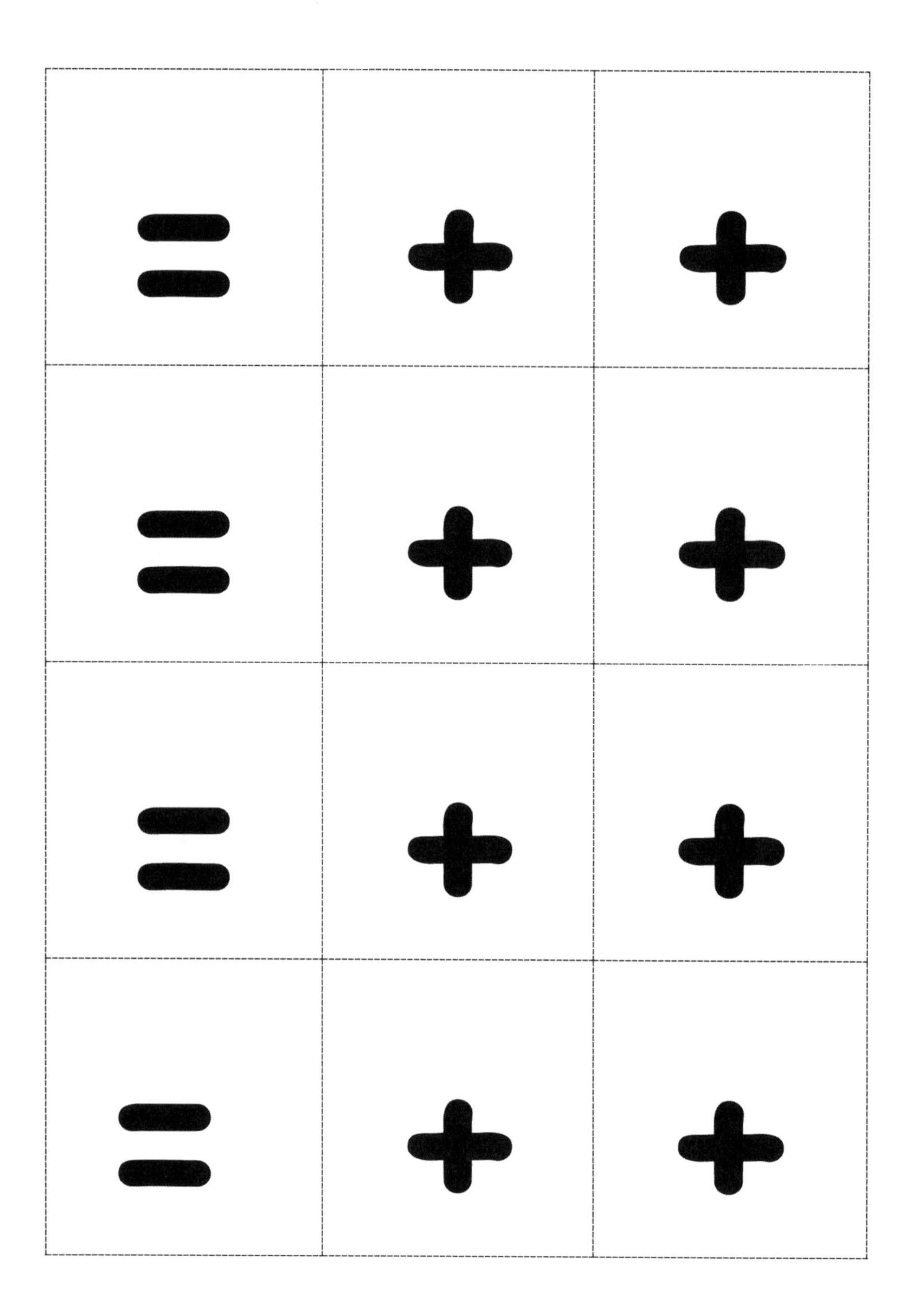

5-group cards, first two pages double-sided, last page single-sided

Lesson 2

Objective: Use the associative and commutative properties to make ten with three addends.

Suggested Lesson Structure

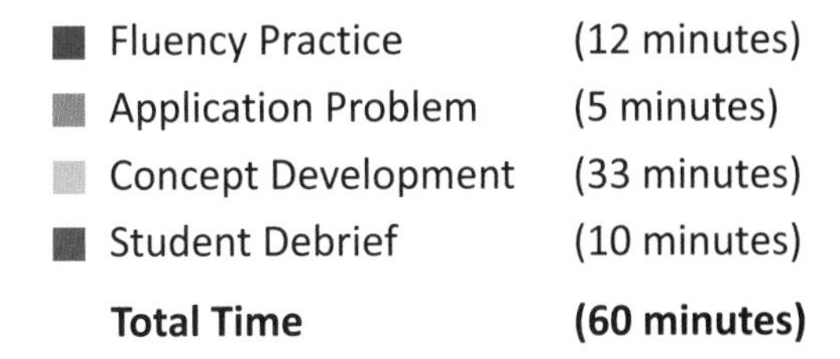

Fluency Practice	(12 minutes)
Application Problem	(5 minutes)
Concept Development	(33 minutes)
Student Debrief	(10 minutes)
Total Time	**(60 minutes)**

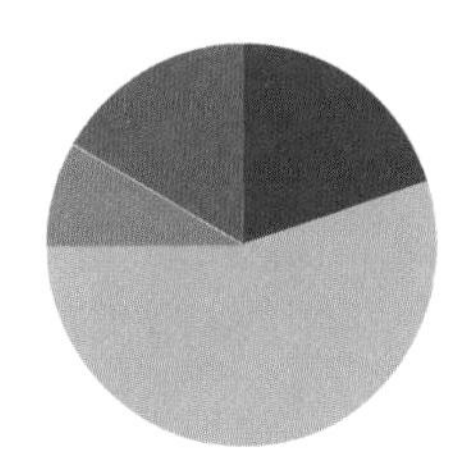

Fluency Practice (12 minutes)

- Take Out 1: Number Bonds **1.OA.6** (5 minutes)
- 5-Group Flash: Partners to Ten **1.OA.6** (5 minutes)
- Say Ten Co.nversion **1.NBT.2** (2 minutes)

Take Out 1: Number Bonds (5 minutes)

Materials: (S) Personal white board

Note: This is an anticipatory fluency activity for the make ten strategy with an addend of 9. Students take 1 from the other addend. The goal is for them to be able to do so quickly and accurately.

Say a number within 10. Students quickly write a number bond for the number said, using 1 as a part, and hold up their personal white boards when finished.

5-Group Flash: Partners to Ten (5 minutes)

Materials: (T) 5-group cards (Lesson 1 Fluency Template) (S) Personal white board

Note: This is a maintenance fluency activity with partners to ten to facilitate the make ten addition strategy.

Flash a card for 1 to 3 seconds (e.g., 9). Students write two expressions that make ten (e.g., 9 + 1 and 1 + 9).

Say Ten Conversion (2 minutes)

Note: This activity strengthens students' understanding of the place value system as it relates to counting.

Call out numbers between 10 and 20, alternating between saying the number the regular way or the Say Ten way. When the teacher uses the Say Ten way, students say the number the regular way. When the teacher uses the regular way, students say it the Say Ten way. Play for a minute, and then give students a chance to be the caller.

Application Problem (5 minutes)

Lisa was reading a book. She read 6 pages the first night, 5 pages the next night, and 4 pages the following night. How many pages did she read?

Make a drawing to show your thinking. Write a statement to go with your work.

Extension: If she read a total of 20 pages by the fifth night, how many pages could she have read on the fourth night and the fifth night?

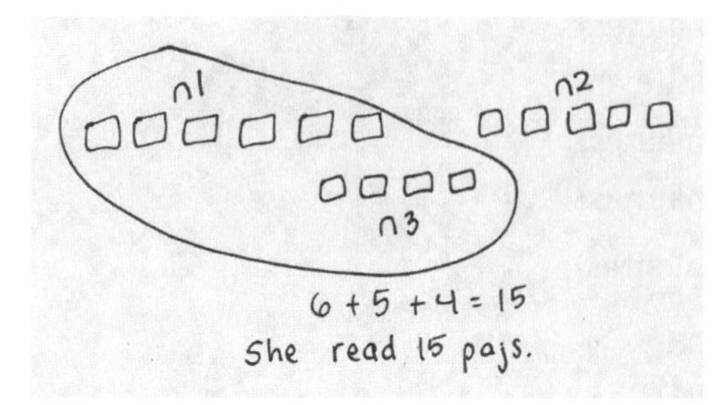

Note: This problem applies the Lesson 1 objective of adding three addends, two of which make ten. The two addends that make ten are separated within the story during the Student Debrief in connection with today's lesson.

Concept Development (33 minutes)

Materials: (S) Personal white board

Have students sit in a semicircle at the meeting area with their materials.

T: (Write 5 + 3 + 5 = ___ on the board.) Draw to solve for this unknown.

S: (Draw to solve as the teacher circulates and notices student strategies.)

T: Let's see how our friends solved this. (Select a student who added all in a row and a student who rearranged the addends to share their work.)

S: I added 5 + 3 and remembered that was 8. Then, I counted up 5 more from 8 and got 13. → I drew the groups of 5 together and added those first since I knew they made ten. Then I added. 10 and 3 is 13.

T: Talk with your partner. How were the strategies used by your classmates similar and different from one another? Which one was correct?

S: (Discuss as the teacher circulates and listens.) They were both correct! → Bob put the fives together and made ten, and Jo added them in order.

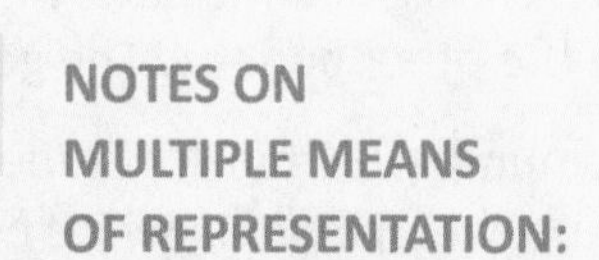

NOTES ON MULTIPLE MEANS OF REPRESENTATION:

During this lesson, it is important for students to articulate the way they chose to solve a problem so that other students can hear how they are thinking. This helps guide students toward the most efficient choice as they benefit from hearing strategies multiple times.

T: So, even though they added two different numbers together first, did they get the same total?

S: Yes!

T: Wow! Okay. Let's try this again. Let's use Bob's strategy of making ten from two of our addends. (Write 7 + 5 + 3 = ___.) Write the equation. Draw to show the three amounts.

S: (Draw to show the three quantities.)

T: What two numbers make ten?

S: 7 and 3.

MP.7

T: Good. Show that 7 and 3 make ten in your drawing by circling like we did yesterday with the string.

S: (Circle the 3 and the 7, making a group of 10.)

T: Here is a new number sentence that shows what numbers you added first. (Write 7 + 3 + 5 = ___.)

T: I'll make a number bond to show how you made ten from two numbers. (Bond the 7 and 3 to make ten.)

T: You just showed 10 and 5 more, which equals...?

S: 15.

T: Good. I'll show how we solved for the unknown. I'll write the new number sentence explaining what we just did, starting with 10.

S: (Solve 7 + 3 + 5 = ___ while the teacher writes 10 + 5 = 15.)

T: Jo showed us at the beginning of the lesson that she could solve from left to right, without moving the addends around, in order to get the same answer as Bob. Work and talk with your partner to see if this is true again!

NOTES ON MULTIPLE MEANS OF ENGAGEMENT:

Addends should be chosen so that students can easily identify the partners to ten, recognizing that they can add these two addends first, regardless of where they are positioned within the number sentence. If students are not fluent with 7 and 3, they may be replaced with 9 and 1, respectively.

Repeat this process using the following suggested sequence: 9 + 2 + 1, 2 + 4 + 8 (highlighting that students might begin with the 8 rather than the 2), 4 + 3 + 6, and 3 + 8 + 7. Students complete the number sentence while the teacher completes the drawing for the third example.

Problem Set (10 minutes)

Students should do their personal best to complete the Problem Set within the allotted 10 minutes. For some classes, it may be appropriate to modify the assignment by specifying which problems they work on first. Some problems do not specify a method for solving. Students should solve these problems using the RDW approach used for Application Problems.

Note: Look at the example for Problem 1 in the Problem Set. Discuss the importance of making a simple math drawing by drawing three different simple shapes to represent three different numbers in the equation, reminding students about their experience using different concrete materials during previous lessons. Model this drawing if necessary.

Student Debrief (10 minutes)

Lesson Objective: Use the associative and commutative properties to make ten with three addends.

The Student Debrief is intended to invite reflection and active processing of the total lesson experience.

Invite students to review their solutions for the Problem Set. They should check work by comparing answers with a partner before going over answers as a class. Look for misconceptions or misunderstandings that can be addressed in the Debrief. Guide students in a conversation to debrief the Problem Set and process the lesson.

Any combination of the questions below may be used to lead the discussion.

- Look at your Problem Set. We added amounts in different orders. When we did this, did we get the same amount? Is this always true?
- Talk with your partner. How did you organize your drawings to show the three different amounts? How did you show that you used the make ten strategy in your drawings?
- Look at Problem 1 and Problem 4. What similarities do you notice?
- Are there any problems in the Problem Set that you can solve using your knowledge of doubles?
- Look at Problem 9. How did you show the number bond for making ten? How is it different from some of your other bonds? (Students share strategies of the number bond above or below or rewrite the number sentence below to enable the addends that make ten to be adjacent.)

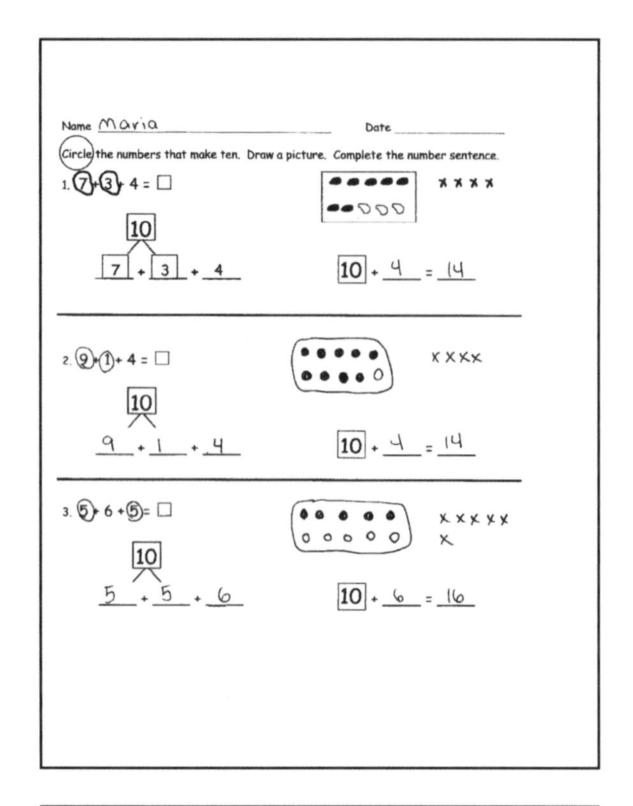
Name Maria Date

Circle the numbers that make ten. Draw a picture. Complete the number sentence.

1. 7 + 3 + 4 = □ 10; 7 + 3 + 4; 10 + 4 = 14

2. 9 + 1 + 4 = □ 10; 9 + 1 + 4; 10 + 4 = 14

3. 5 + 6 + 5 = □ 10; 5 + 5 + 6; 10 + 6 = 16

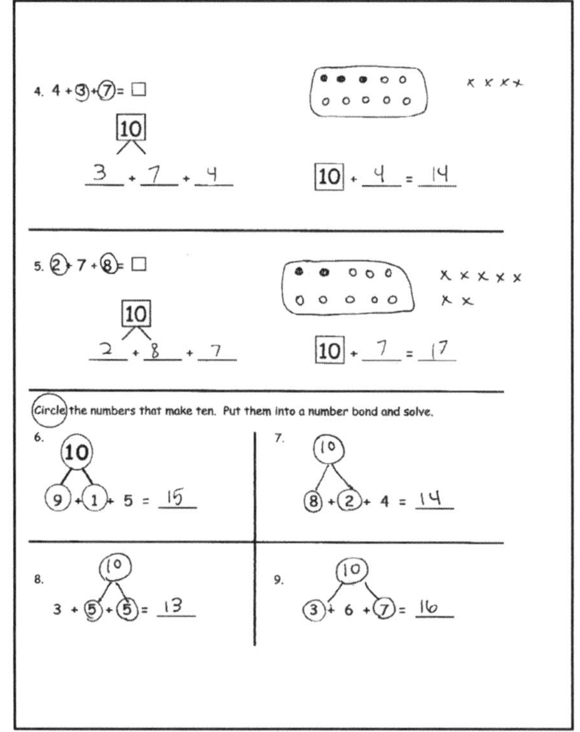
4. 4 + 3 + 7 = □ 10; 3 + 7 + 4; 10 + 4 = 14

5. 2 + 7 + 8 = □ 10; 2 + 8 + 7; 10 + 7 = 17

Circle the numbers that make ten. Put them into a number bond and solve.

6. 9 + 1 + 5 = 15

7. 8 + 2 + 4 = 14

8. 3 + 5 + 5 = 13

9. 3 + 6 + 7 = 16

Exit Ticket (3 minutes)

After the Student Debrief, instruct students to complete the Exit Ticket. A review of their work will help with assessing students' understanding of the concepts that were presented in today's lesson and planning more effectively for future lessons. The questions may be read aloud to the students.

Name ______________________ Date ____________

Circle the numbers that make ten. Draw a picture. Complete the number sentence.

1. 7 + 3 + 4 = □

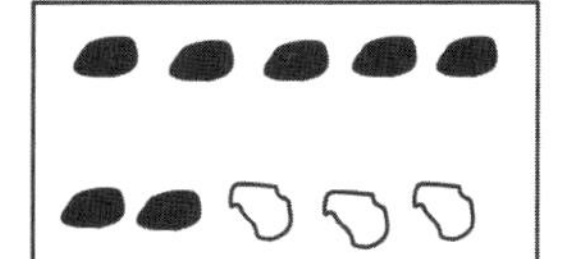

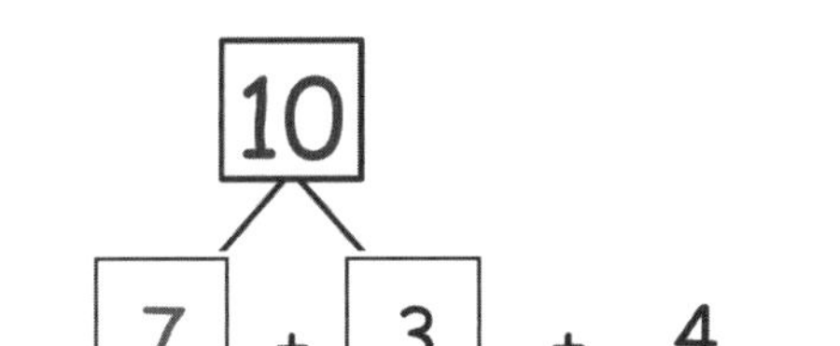

10

7 + 3 + 4

10 + ___ = ___

2. 9 + 1 + 4 = □

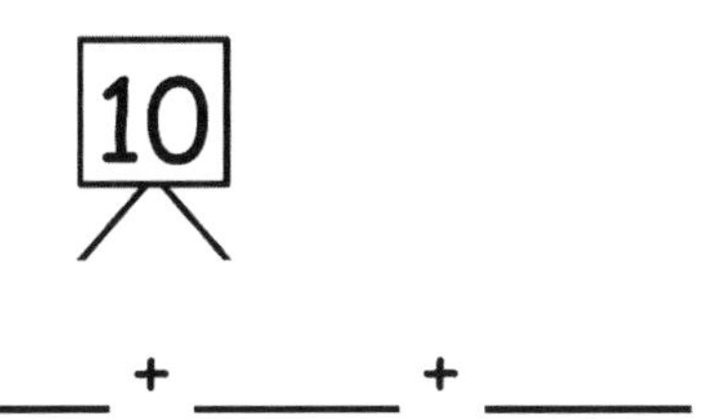

10

___ + ___ + ___

10 + ___ = ___

3. 5 + 6 + 5 = □

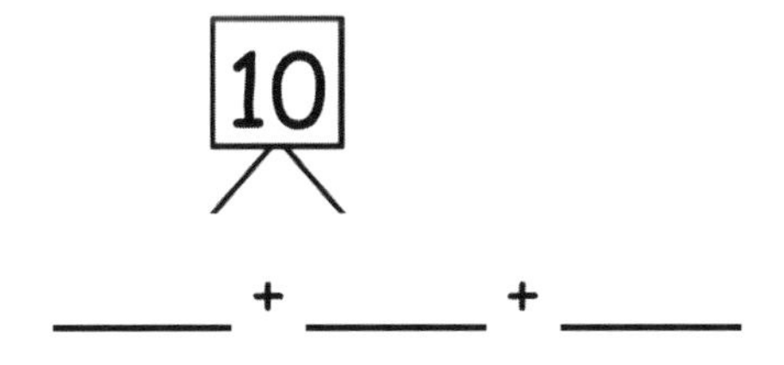

10

___ + ___ + ___

10 + ___ = ___

4. 4 + 3 + 7 = ☐

_____ + _____ + _____

10 + _____ = _____

5. 2 + 7 + 8 = ☐

10

_____ + _____ + _____

10 + _____ = _____

Circle the numbers that make ten. Put them into a number bond, and solve.

6.

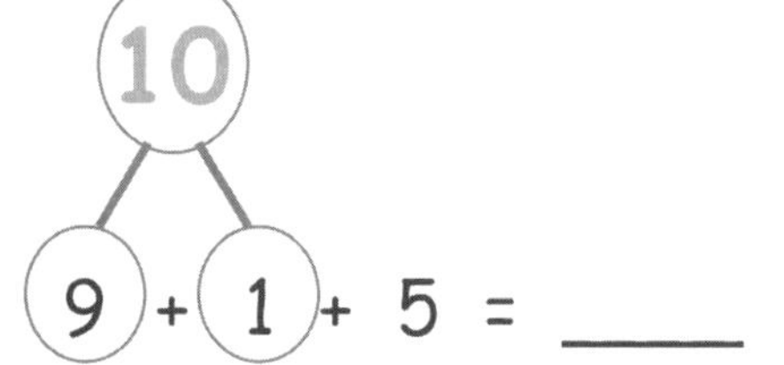

7. 8 + 2 + 4 = _____

8. 3 + 5 + 5 = _____

9. 3 + 6 + 7 = _____

Name ______________________________ Date ______________

Circle the numbers that make ten.

Draw a picture, and complete the number sentences to solve.

a. 8 + 2 + 3 = ____

____ + ____ = ____

10 + ____ = ____

b. 7 + 4 + 3 = ____

____ + ____ = ____

10 + ____ = ____

Name ______________________ Date ____________

Circle the numbers that make ten. Draw a picture. Complete the number sentence.

1. 6 + 2 + 4 = □

10

6 + ____ + 2

10 + ____ = ____

---

2. 5 + 3 + 5 = □

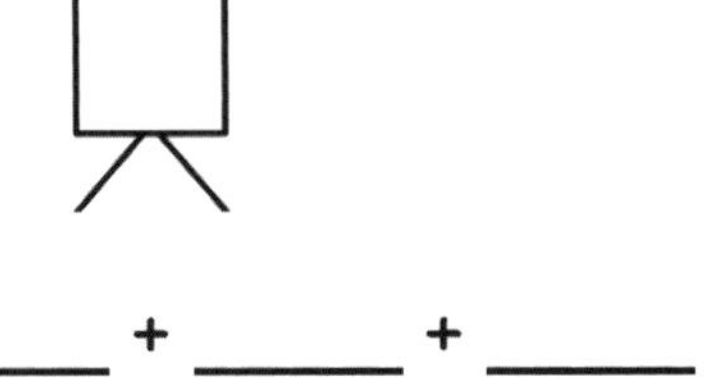

____ + ____ + ____

10 + ____ = ____

---

3. 5 + 2 + 8 = □

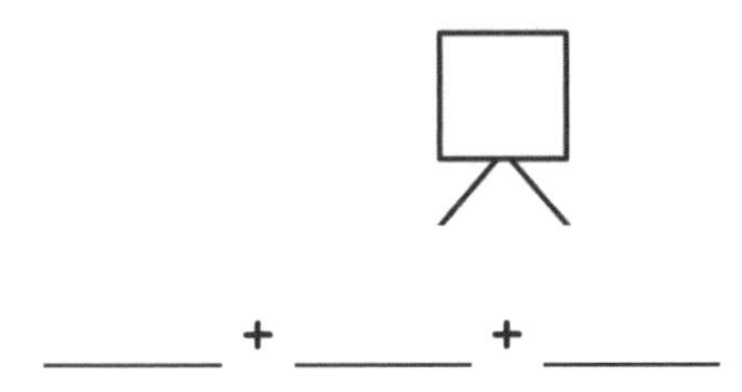

____ + ____ + ____

____ + 10 = ____

4. $2 + 7 + 3 = \square$

_____ + _____ + _____

_____ + 10 = _____

Circle the numbers that make ten, and put them into a number bond. Write a new number sentence.

5.

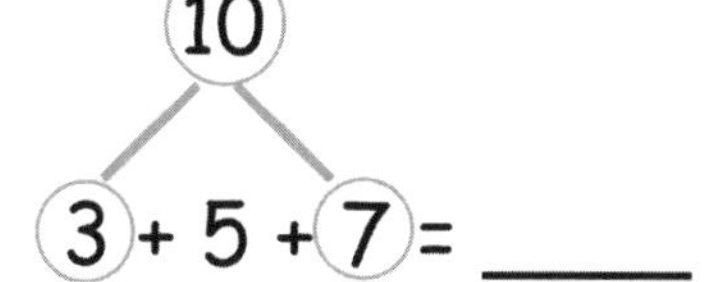

_____

_____ + _____ = _____

6.

$4 + 8 + 2 =$ _____

_____ + _____ = _____

Challenge: Circle the addends that make ten. Circle the true number sentences.

a. $5 + 5 + 3 = 10 + 3$

b. $4 + 6 + 6 = 10 + 6$

c. $3 + 8 + 7 = 10 + 6$

d. $8 + 9 + 2 = 9 + 10$

Lesson 3

Objective: Make ten when one addend is 9.

Suggested Lesson Structure

■ Fluency Practice	(10 minutes)
■ Application Problem	(5 minutes)
■ Concept Development	(35 minutes)
■ Student Debrief	(10 minutes)
Total Time	**(60 minutes)**

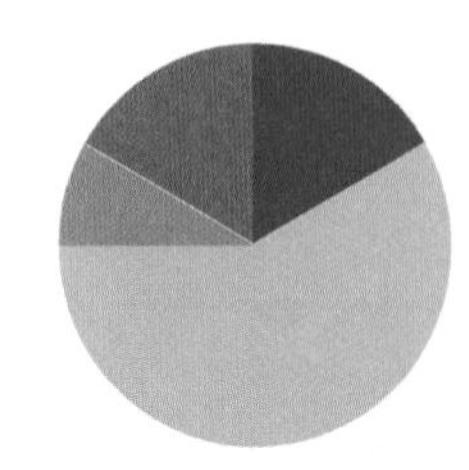

Fluency Practice (10 minutes)

- Take Out 1 **1.OA.6** (1 minute)
- Break Apart 10 **1.OA.6** (5 minutes)
- Add Partners of Ten First **1.OA.3** (4 minutes)

Take Out 1 (1 minute)

Materials: (S) Personal white board

Note: This is an anticipatory fluency activity for the make ten addition strategy, as students need to fluently take 1 out of the second addend when adding to 9.

Make the pace quicker now that students have done this for a few days. Celebrate their improvement.

Say a number between 1 and 9. Students say the number decomposed with one part as one.

Break Apart 10 (5 minutes)

Materials: (T) 5-group cards (Lesson 1 Fluency Template) (S) Personal white board

Students write the numeral 10 on their personal white boards. Flash a 5-group card. Students break apart 10 using the number flashed as a part, without making bubbles or boxes around the numerals.

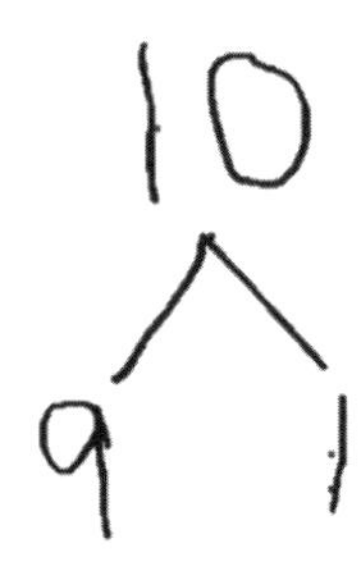

EUREKA MATH

Add Partners of Ten First (4 minutes)

Note: This activity reviews adding three numbers and prepares students for the make ten addition strategy when one addend is 9.

Build toward three addends. Begin with 9 + 1.

T: 9 + 1.
S: 10.
T: 10 + 5.
S: 15.
T: 9 + 1 (pause) + 5 is...?
S: 15.

Continue with the following suggested sequence: 9 + 1 + 6, 9 + 1 + 4, 9 + 1 + 3, 9 + 1 + 7, 8 + 2 + 7.

Application Problem (5 minutes)

Tom's mother gave him 4 pennies. His father gave him 9 pennies. His sister gave him enough pennies so that he now has a total of 14. How many pennies did his sister give him? Use a drawing, a number sentence, and a statement.

Extension: How many more would he need to have 19 pennies?

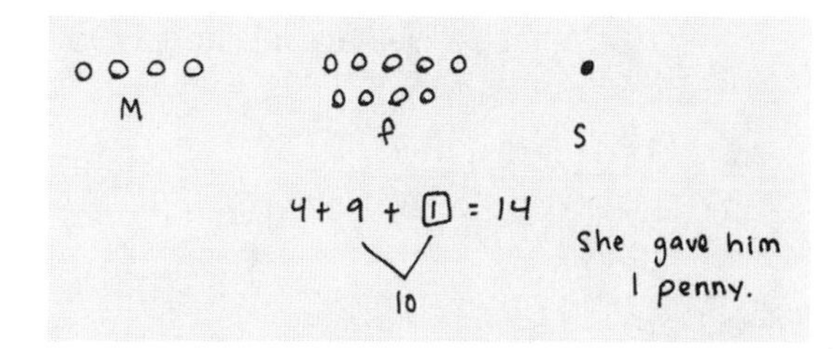

Note: This Application Problem challenges students to consider finding an unknown addend within a context with three addends. Students may add 4 and 9 together first, noticing that they need 1 more penny to make 14. Other students may recognize that 14 is made of 10 and 4 and realize that they are looking for the partner for 9 when making ten. During the Student Debrief, students explore how they could use making ten as a quick strategy to add the sets of pennies that Tom's parents gave him (9 pennies and 4 pennies).

Concept Development (35 minutes)

Materials: (T) 10 red and 10 green linking cubes (S) 10 red and 10 green linking cubes, personal white board

Have students sit at their seats with materials.

T: (Project and read aloud.) Maria has 9 snowballs, and Tony has 3. How many do they have altogether?
T: What is the expression to solve this problem?
S: 9 + 3.
T: Use your green linking cubes to show how many snowballs Maria has.
S: (Lay out 9 green linking cubes.)

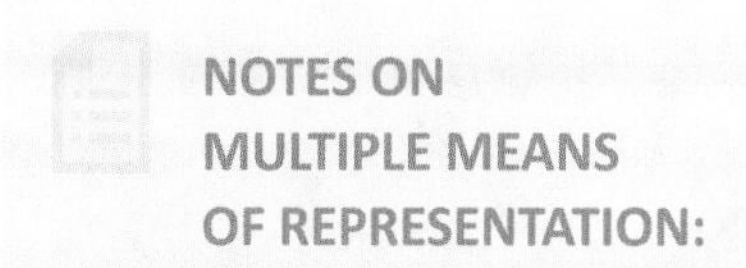

While some students are experts at solving **10+** number sentences, others may need pictorial support such as ten-frames (rather than numerals) to help develop mental calculations.

T: Using the red cubes, show how many snowballs Tony has. Put them in a separate pile.

S: (Lay out 3 linking cubes.)

T: How would you solve this problem?

S: Count on!

T/S: Niiiine, 10, 11, 12.

T: (Complete the equation 9 + 3 = 12.)

T: Is there a way to make ten with the amounts we have in front of us? Turn and talk to your partner.

S: (Discuss while the teacher circulates.)

T: (Choose a student who used the strategy below.)

S: I made ten by moving 1 red cube to the green pile. I had 9 cubes in that pile, but now I have 10.

T: You made ten! Everyone, make ten.

S: (Move 1 red cube to the green pile.)

T: Now we have 10 here. (Gesture to the pile of 10.) What do we have left here? (Point to the other pile.)

S: 2.

T: Look at your new piles. What is our new number sentence?

S: 10 + 2 = 12.

T: (Write 10 + 2 = 12 on the board.) Did we change the *amount* of linking cubes we have?

S: No.

T: So, 9 + 3 is the same as what addition expression?

S: 10 + 2.

T: (Write 9 + 3 = 10 + 2.)

T: What is 10 + 2?

S: 12.

T: What is 9 + 3? Say the number sentence.

S: 9 + 3 = 12.

T: How many snowballs do Maria and Tony have?

S: 12 snowballs.

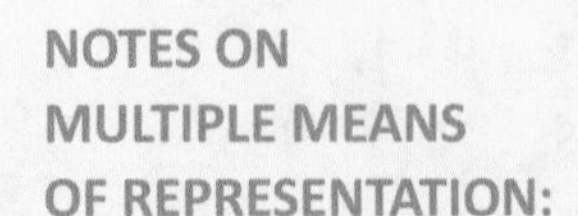

NOTES ON MULTIPLE MEANS OF REPRESENTATION:

For those students who can fluently solve math facts within 20, cultivate excitement by connecting on-level math to higher math. Continue on the same line as numbers to 100 and add (e.g., 29 + 3 or 39 + 13).

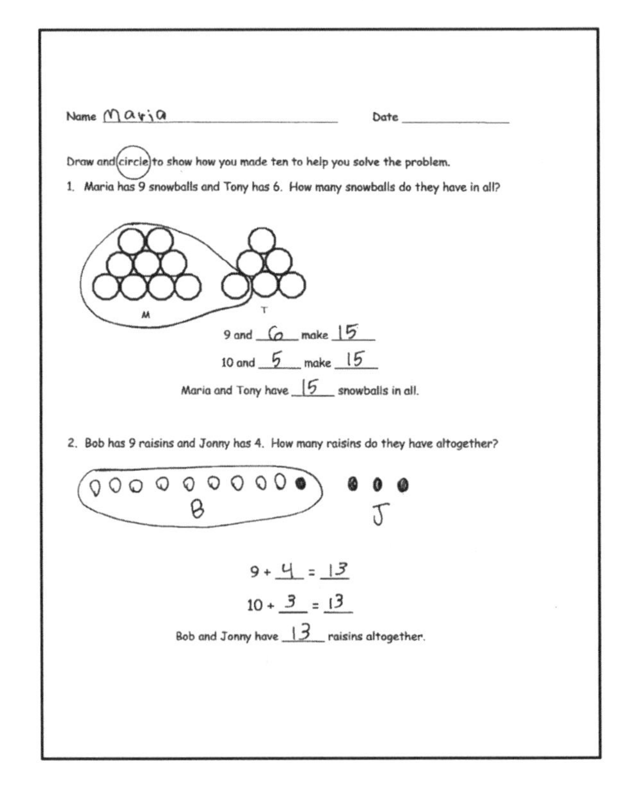

Name Maria Date ______

Draw and circle to show how you made ten to help you solve the problem.

1. Maria has 9 snowballs and Tony has 6. How many snowballs do they have in all?

M T

9 and 6 make 15

10 and 5 make 15

Maria and Tony have 15 snowballs in all.

2. Bob has 9 raisins and Jonny has 4. How many raisins do they have altogether?

B J

9 + 4 = 13

10 + 3 = 13

Bob and Jonny have 13 raisins altogether.

Repeat the process with snowball situations for 9 + 2 and 9 + 4. Then, change to 5-group drawings instead of cubes. Continue to repeat the process with the following suggested sequence: 9 + 5, 9 + 8, and 9 + 7. Create different story situations for 9 + 6, 8 + 9, and 9 + 9. Be sure to have students label their pictures, circle 10, and write three number sentences (e.g., 9 + 6 = 15, 10 + 5 = 15, 9 + 6 = 10 + 5).

Problem Set (10 minutes)

Students should do their personal best to complete the Problem Set within the allotted 10 minutes. For some classes, it may be appropriate to modify the assignment by specifying which problems they work on first. Some problems do not specify a method for solving. Students should solve these problems using the RDW approach used for Application Problems.

Note: Students should save the Problem Sets from this lesson through Lesson 6. They provide comparisons for students when they begin making ten when one addend is 8. Setting up a portfolio of past Problem Sets and strategies helps students access these readily.

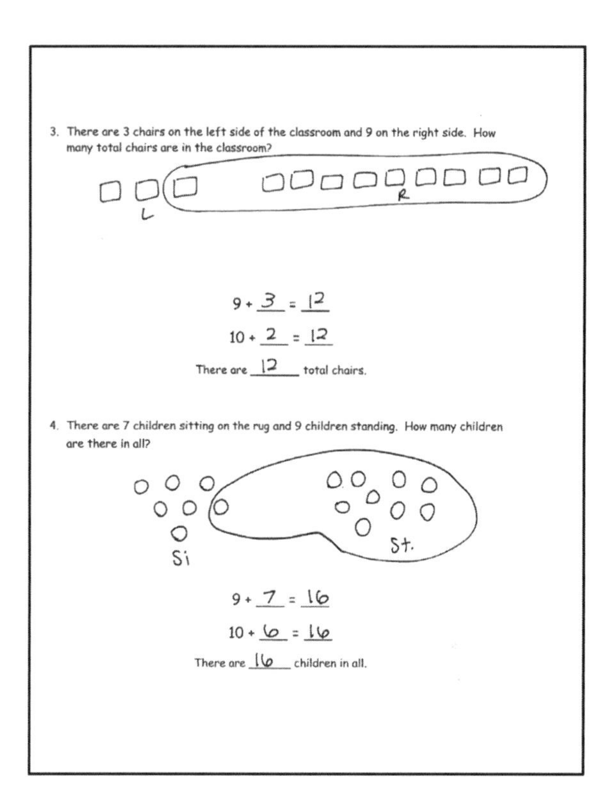

3. There are 3 chairs on the left side of the classroom and 9 on the right side. How many total chairs are in the classroom?

L R

9 + 3 = 12

10 + 2 = 12

There are 12 total chairs.

4. There are 7 children sitting on the rug and 9 children standing. How many children are there in all?

Si St.

9 + 7 = 16

10 + 6 = 16

There are 16 children in all.

Student Debrief (10 minutes)

Lesson Objective: Make ten when one addend is 9.

The Student Debrief is intended to invite reflection and active processing of the total lesson experience.

Invite students to review their solutions for the Problem Set. They should check work by comparing answers with a partner before going over answers as a class. Look for misconceptions or misunderstandings that can be addressed in the Debrief. Guide students in a conversation to debrief the Problem Set and process the lesson.

Any combination of the questions below may be used to lead the discussion.

- Look at Problem 1. What are the two number sentences that show your work?
- Look at Problem 1 and Problem 3 with a partner. How was setting up the problem to complete Problem 1 different from setting up Problem 3? What did you need to be sure to do? Why?
- How can solving Problem 1 help you solve Problem 4?
- After you made ten, what did you notice about the addend you broke apart? (The other addend is left with 1 less!)
- What new strategy did we use today to solve math problems? How is it more efficient than counting on to add?
- Look at your Application Problem. How could you use the make ten strategy to solve the problem?

Exit Ticket (3 minutes)

After the Student Debrief, instruct students to complete the Exit Ticket. A review of their work will help with assessing students' understanding of the concepts that were presented in today's lesson and planning more effectively for future lessons. The questions may be read aloud to the students.

Name ______________________________ Date ______________

Draw and circle to show how you made ten to help you solve the problem.

1. Maria has 9 snowballs, and Tony has 6. How many snowballs do they have in all?

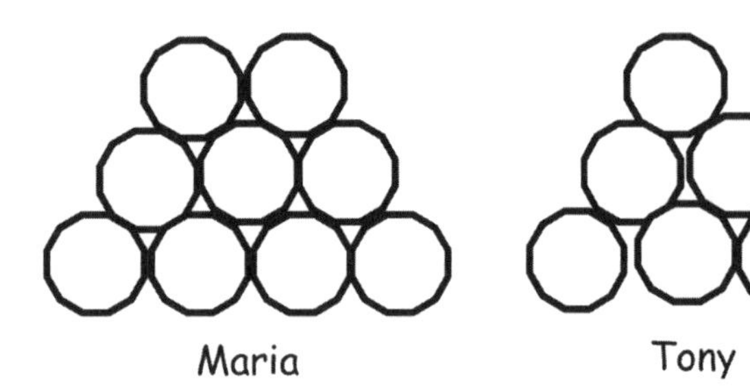

9 and ______ make ______.

10 and ______ make ______.

Maria and Tony have ______ snowballs in all.

2. Bob has 9 raisins, and Jonny has 4. How many raisins do they have altogether?

9 + ____ = ____

10 + ____ = ____

Bob and Jonny have ______ raisins altogether.

3. There are 3 chairs on the left side of the classroom and 9 on the right side. How many total chairs are in the classroom?

9 + ____ = ____

10 + ____ = ____

There are ______ total chairs.

4. There are 7 children sitting on the rug and 9 children standing. How many children are there in all?

9 + ____ = ____

10 + ____ = ____

There are ______ children in all.

Name ______________________________ Date ______________

Draw and circle to show how to make ten to solve. Complete the number sentences.

Tammy has 4 books, and John has 9 books. How many books do Tammy and John have altogether?

____ + ____ = ____

____ + ____ = ____

Tammy and John have ____ books.

Name ______________________________ Date ______________

Draw, label, and circle to show how you made ten to help you solve.
Complete the number sentences.

1. Ron has 9 marbles, and Sue has 4 marbles.
   How many marbles do they have in all?

9 and ______ make ______.

10 and ______ make ______.

Ron and Sue have _____ marbles.

---

2. Jim has 5 cars, and Tina has 9. How many cars do they have altogether?

9 and ______ make ______.

10 and ______ make ______.

Jim and Tina have ___ cars.

3. Stan has 6 fish, and Meg has 9. How many fish do they have in all?

9 + ____ = ____

10 + ____ = ____

Stan and Meg have ____ fish.

---

4. Rick made 7 cookies, and Mom made 9. How many cookies did Rick and Mom make?

9 + ____ = ____

10 + ____ = ____

Rick and Mom made _____ cookies.

---

5. Dad has 8 pens, and Tony has 9. How many pens do Dad and Tony have in all?

9 + ____ = ____

10 + ____ = ____

Dad and Tony have ____ pens.

Lesson 4

Objective: Make ten when one addend is 9.

Suggested Lesson Structure

■ Fluency Practice	(12 minutes)
■ Application Problem	(5 minutes)
■ Concept Development	(33 minutes)
■ Student Debrief	(10 minutes)
Total Time	**(60 minutes)**

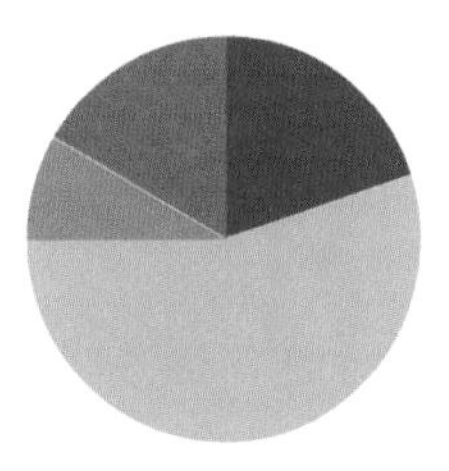

Fluency Practice (12 minutes)

- Happy Counting the Say Ten Way **1.NBT.2** (2 minutes)
- Sprint: Add Three Numbers **1.OA.3** (10 minutes)

Happy Counting the Say Ten Way (2 minutes)

Note: Say Ten counting strengthens student understanding of place value.

Tell students to look at your thumb and count up and down between 10 and 120 the Say Ten way. When your thumb points and motions up, students count up. When your thumb is to the side, students stop. When your thumb points and motions down, students count down (see example below).

T: 

T/S: 4 ten 4 ten 1 4 ten 2 (pause) 4 ten 1 4 ten (pause) 4 ten 1 4 ten 2 4 ten 3

Choose numbers based on student skill level. If students are very proficient up to 40, start at 40, and quickly go up to 80. If they are proficient between 40 and 80, Happy Count between 80 and 120. Alternate at times between regular and Say Ten counting, too.

Sprint: Add Three Numbers (10 minutes)

Note: This Sprint provides practice with adding three numbers by making ten first.

Materials: (S) Add Three Numbers Sprint

Application Problem (5 minutes)

Michael plants 9 flowers in the morning. He then plants 4 flowers in the afternoon. How many flowers did he plant by the end of the day? Make a drawing, a number bond, and a statement.

Note: Students can apply the make ten strategy from Lesson 3 as they solve this problem. During the Student Debrief, the teacher discusses how using rows to show the plants can create a clear and quick visual for identifying the compositions and decompositions needed to apply the make ten strategy.

Concept Development (33 minutes)

Materials: (T) 10 green and 10 red linking cubes, a ten-frame border (S) 10 green and 10 red linking cubes, personal white board

Have students come to the meeting area with linking cubes and personal white boards.

T: (Project and read aloud.) Maria has 9 green cubes. Tony has 3 red cubes. How many cubes do Maria and Tony have?

T: What is the expression to solve this story problem?

S: 9 + 3.

T: (Show two piles: 9 scattered green cubes and 3 scattered red cubes.)

T: How can you check that I have the correct number of cubes representing Maria's cubes?

S: We can count, one at a time.

T: Okay, but that's not very efficient. Is there a way to organize my green cubes so we can tell there are 9 cubes faster?

S: Put them in a 5-group!

T: Great idea. When we arrange or draw things in a 5-group, we are all going to follow these steps. Just like reading, we'll start with the top row and from the left. (Place 5 green cubes in a row.)

T: We start in the next line with 6 and try to match it up to the top as closely as we can. (Place 4 in the bottom row.)

T: Now can you see we have 9 cubes right away?

S: Yes!

T: (Arrange the 3 red cubes in a 5-group on the other side.) The red cubes are also organized.

T: What do we do to solve 9 + 3?

S: Make ten.

It is important to make the connection between concrete math and math models. This helps English language learners and struggling learners understand the math without getting bogged down with language acquisition.

T: (Circle the 9 green cubes and 1 red cube with a finger.)

T: Here's another way to show ten. (Move 1 red cube to add to 9 green cubes.)

T: (Place a red cube in the tenth slot.) We made ten!

MP.4

T: I'm going to put a frame around it. (Place the frame around ten.) We are going to call this a **ten-frame**. It looks just like our 5-group drawings, but now that we are making ten, we can call it a ten-frame. Whenever we make ten, we make or draw a frame around it. That way, we can see ten right away.

Ten-Frame

T: Look at the new piles. What new expression do you see?

S: 10 + 2.

T: So, 9 + 3 is the same as...?

S: 10 + 2.

T: (Write 9 + 3 = 10 + 2.)

T: What is 10 + 2?

S: 12.

T: What is 9 + 3?

S: 12.

T: How many cubes do Maria and Tony have?

S: 12 cubes.

T: Where are the 9 green cubes? Point to them.

S: (Point to 9.)

T: Where are the 3 red cubes? Point to them.

S: (Point to 1 and 2.)

T: You are pointing to two different places. Why?

S: We broke 3 apart into 1 and 2.

T: Let's use a number bond to show how we broke apart 3.

T: Just like we framed the ten in our picture, we'll frame the numbers that make ten. (Circle 9 and 1.)

T: 9 and 1 make...?

S: 10.

T: 10 and 2 make...?

S: 12.

T: So, 9 plus 3 equals...?

S: 12.

NOTES ON MULTIPLE MEANS OF REPRESENTATION:

Be aware of the different learning needs in the class, and adjust the lesson as necessary. As some students may need to work at the concrete level for a longer period of time, allow students access to manipulatives.

Repeat the process by having students work with cubes. Be sure to guide students when organizing their cubes into a ten-frame. The following is a suggested sequence: 9 + 2 (pictured to the right), 4 + 9, and 5 + 9. Note that the smaller addend sometimes appears first. Guide students to realize that they can still compose ten from the 9 for efficiency during the last two problems.

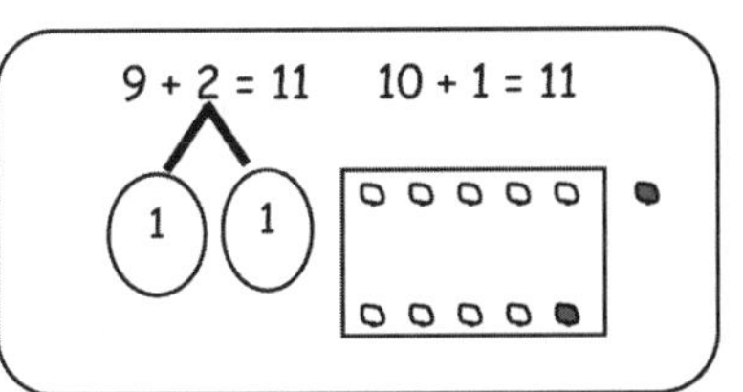

Next, repeat the process by having students use math drawings to solve the following in this suggested sequence: 9 + 6, 3 + 9, and 7 + 9. The 9 should be drawn with open circles. The other addend should be drawn with filled-in circles. Before students add dark circles to their math drawings, ask them, "How many does 9 need to make ten?" and "How many do you have when you take away 1 from [the other addend]?" to guide how they decompose the addend. Additionally, encourage students to place the 1 closer to the 9 as they write the number bond below the other addend, making it easier to make ten with 9.

Name Maria Date

Change the picture to make ten. Write the easier number sentence and solve.

1. Tom has 9 red pencils and 5 yellow. How many pencils does Tom have in all?

9 + 5 = 14

10 pencils + 4 pencils = 14 pencils

Circle 10 and solve.

2. 9 + 3

10 + 2 = 12

3. 4 + 9

10 + 3 = 13

Problem Set (10 minutes)

Students should do their personal best to complete the Problem Set within the allotted 10 minutes. For some classes, it may be appropriate to modify the assignment by specifying which problems they work on first. Some problems do not specify a method for solving. Students should solve these problems using the RDW approach used for Application Problems.

Student Debrief (10 minutes)

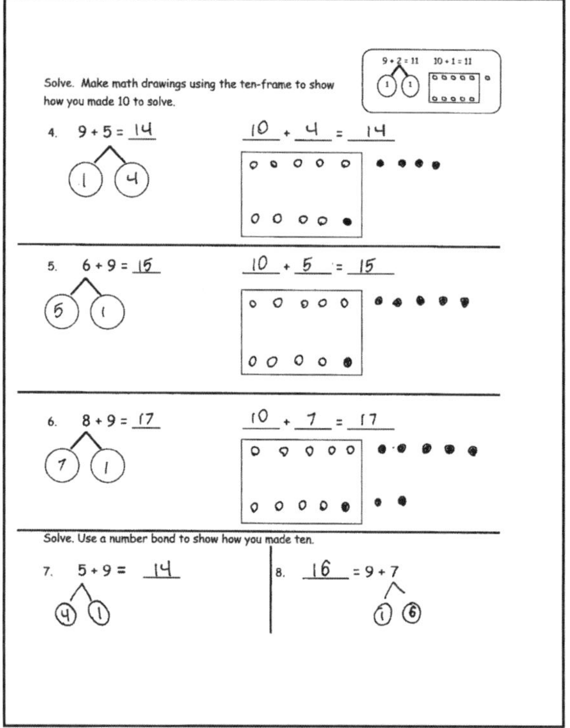
Solve. Make math drawings using the ten-frame to show how you made 10 to solve.

4. 9 + 5 = 14 10 + 4 = 14

5. 6 + 9 = 15 10 + 5 = 15

6. 8 + 9 = 17 10 + 7 = 17

Solve. Use a number bond to show how you made ten.

7. 5 + 9 = 14

8. 16 = 9 + 7

Lesson Objective: Make ten when one addend is 9.

The Student Debrief is intended to invite reflection and active processing of the total lesson experience.

Invite students to review their solutions for the Problem Set. They should check work by comparing answers with a partner before going over answers as a class. Look for misconceptions or misunderstandings that can be addressed in the Debrief. Guide students in a conversation to debrief the Problem Set and process the lesson.

Any combination of the questions below may be used to lead the discussion.

- How did solving Problem 4 help you solve Problem 5?
- What new (or significant) math vocabulary did we use today to make our pictures precise?
- What were some strategies we learned today to solve addition problems efficiently? (Organizing materials and drawings in **ten-frame**, making ten, starting with the 9 to add.)

- Look at your Problem Set. What pattern did you notice when adding 9 to a number? Why is it always a ten and the number that is 1 less than the other addend?
- Look at the Application Problem. Share your drawing with a partner. How could you use the ten-frame to show your work? How does the ten-frame help you see your total amount?

Exit Ticket (3 minutes)

After the Student Debrief, instruct students to complete the Exit Ticket. A review of their work will help with assessing students' understanding of the concepts that were presented in today's lesson and planning more effectively for future lessons. The questions may be read aloud to the students.

A

Number Correct:

Name ____________________ Date ____________

*Make a ten to add.

1.	$9 + 1 + 3 = \square$		16.	$6 + 4 + 5 = \square$	
2.	$9 + 1 + 5 = \square$		17.	$6 + 4 + 6 = \square$	
3.	$1 + 9 + 5 = \square$		18.	$4 + 6 + 6 = \square$	
4.	$1 + 9 + 1 = \square$		19.	$4 + 6 + 5 = \square$	
5.	$5 + 5 + 4 = \square$		20.	$4 + 5 + 6 = \square$	
6.	$5 + 5 + 6 = \square$		21.	$5 + 3 + 5 = \square$	
7.	$5 + 5 + 5 = \square$		22.	$6 + 5 + 5 = \square$	
8.	$8 + 2 + 1 = \square$		23.	$1 + 4 + 9 = \square$	
9.	$8 + 2 + 3 = \square$		24.	$9 + 1 + \square = 14$	
10.	$8 + 2 + 7 = \square$		25.	$8 + 2 + \square = 11$	
11.	$2 + 8 + 7 = \square$		26.	$\square + 3 + 4 = 13$	
12.	$7 + 3 + 3 = \square$		27.	$2 + \square + 6 = 16$	
13.	$7 + 3 + 6 = \square$		28.	$1 + 1 + \square = 11$	
14.	$7 + 3 + 7 = \square$		29.	$19 = 5 + \square + 9$	
15.	$3 + 7 + 7 = \square$		30.	$18 = 2 + \square + 6$	

B

Number Correct:

Name ____________________ Date __________

*Make a ten to add.

1.	5 + 5 + 4 = □		16.	6 + 4 + 2 = □	
2.	5 + 5 + 6 = □		17.	6 + 4 + 3 = □	
3.	5 + 5 + 5 = □		18.	4 + 6 + 3 = □	
4.	9 + 1 + 1 = □		19.	4 + 6 + 6 = □	
5.	9 + 1 + 2 = □		20.	4 + 7 + 6 = □	
6.	9 + 1 + 5 = □		21.	5 + 4 + 5 = □	
7.	1 + 9 + 5 = □		22.	8 + 5 + 5 = □	
8.	1 + 9 + 6 = □		23.	1 + 7 + 9 = □	
9.	8 + 2 + 4 = □		24.	9 + 1 + □ = 11	
10.	8 + 2 + 7 = □		25.	8 + 2 + □ = 12	
11.	2 + 8 + 7 = □		26.	□ + 3 + 4 = 14	
12.	7 + 3 + 7 = □		27.	3 + □ + 7 = 20	
13.	7 + 3 + 8 = □		28.	7 + 8 + □ = 17	
14.	7 + 3 + 9 = □		29.	16 = 3 + □ + 6	
15.	3 + 7 + 9 = □		30.	19 = 2 + □ + 7	

Name ______________________ Date ____________

Change the picture to make ten. Write the easier number sentence and solve.

1. Tom has 9 red pencils and 5 yellow. How many pencils does Tom have in all?

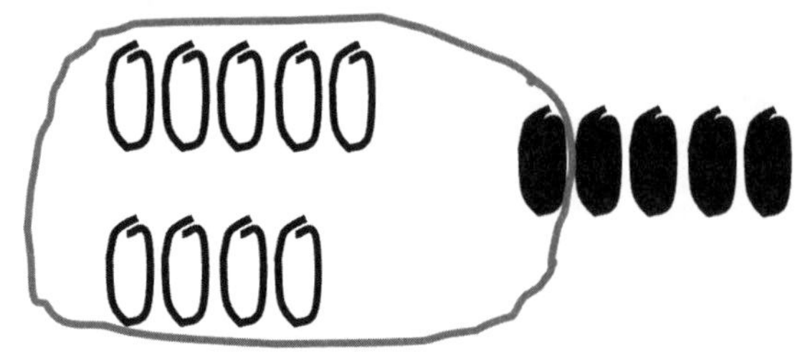

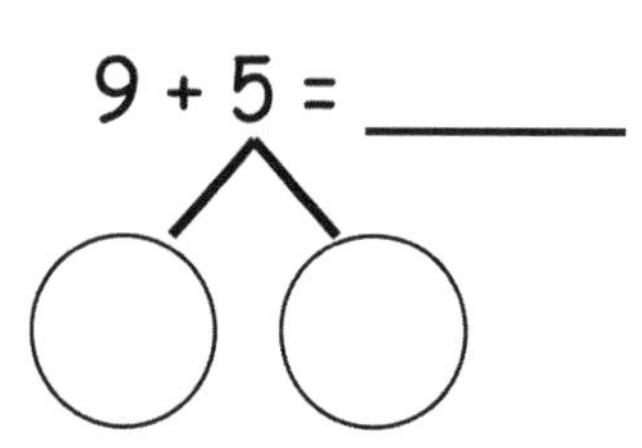

$9 + 5 =$ ______

10 pencils + ____ pencils = _____ pencils

Circle 10 and solve.

2. $9 + 3$

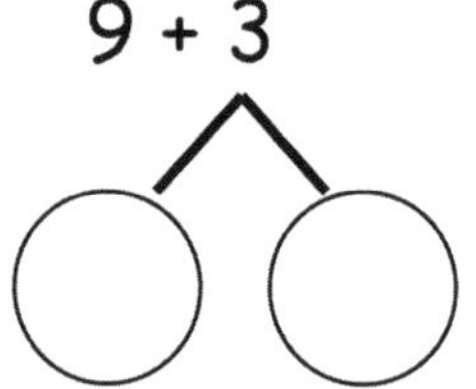

$10 +$ ____ $=$ _____

3. $4 + 9$

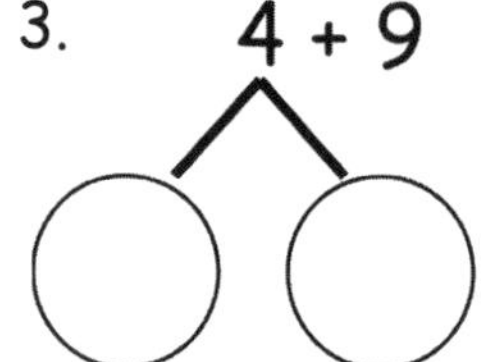

$10 +$ ____ $=$ _____

Solve. Make math drawings using the ten-frame to show how you made 10 to solve.

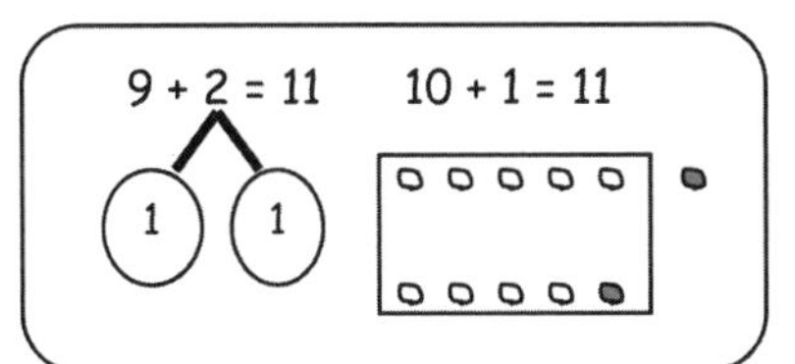

4. 9 + 5 = ____

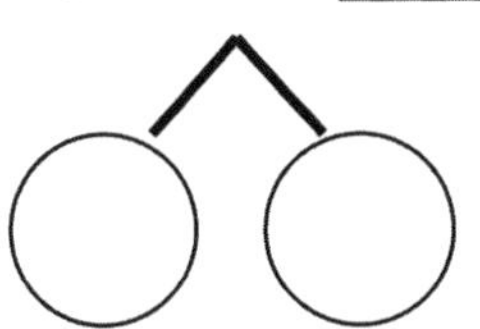

____ + ____ = ____

5. 6 + 9 = ____

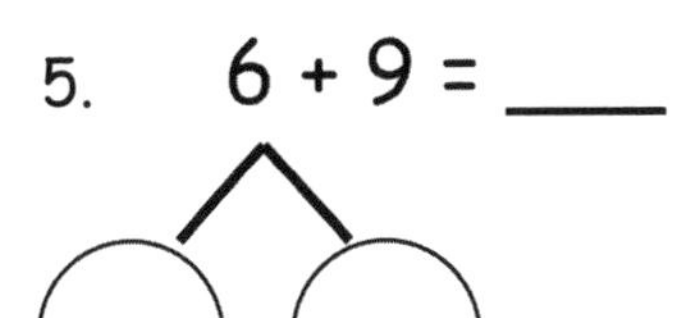

____ + ____ = ____

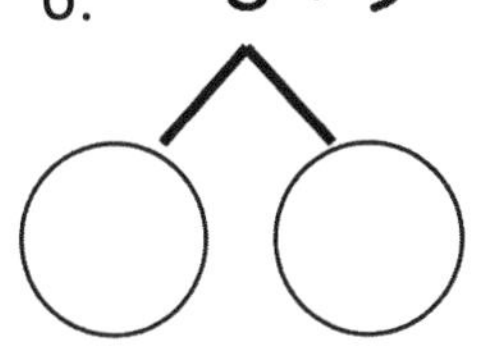

6. 8 + 9 = ____

____ + ____ = ____

Solve. Use a number bond to show how you made ten.

7. 5 + 9 = ____

8. ____ = 9 + 7

Name ______________________________ Date ______________

Solve.

Make math drawings using the ten-frame to show how you made 10 to solve.

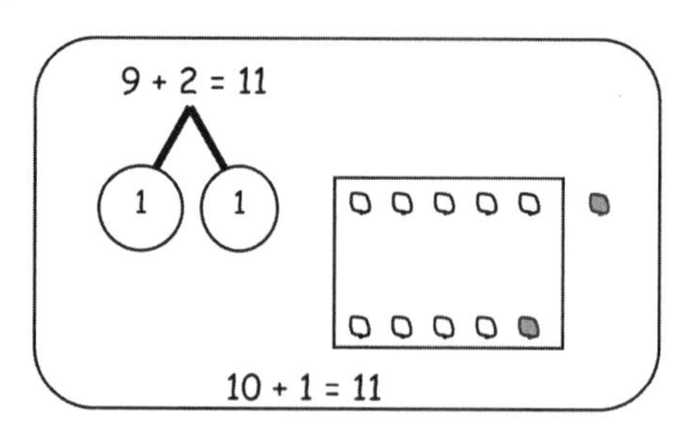

1. 6 + 9 = ___

10 + ___ = ___

2. ___ = 4 + 9

___ + ___ = ___

Name ______________________ Date ____________

Solve. Make math drawings using the ten-frame to show how you made 10 to solve.

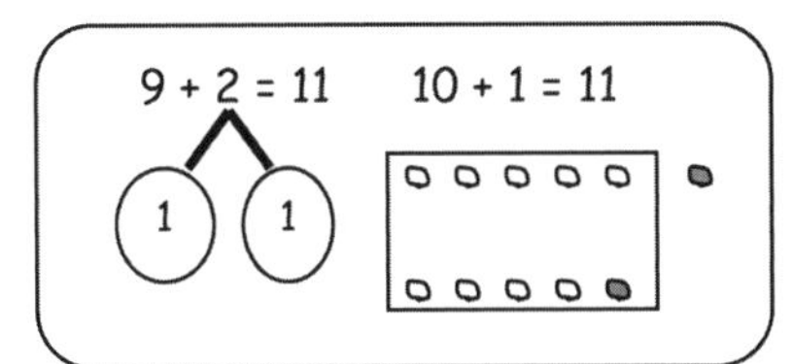

1. 9 + 3 = ____ ____ + ____ = ____

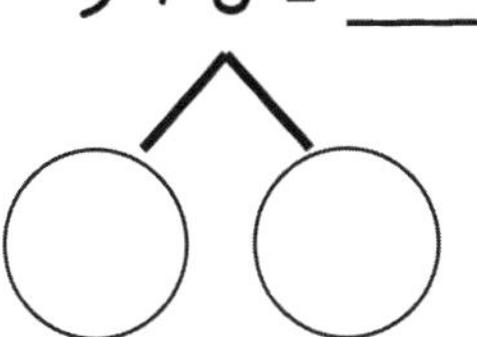

2. 9 + 6 = ____ ____ + ____ = ____

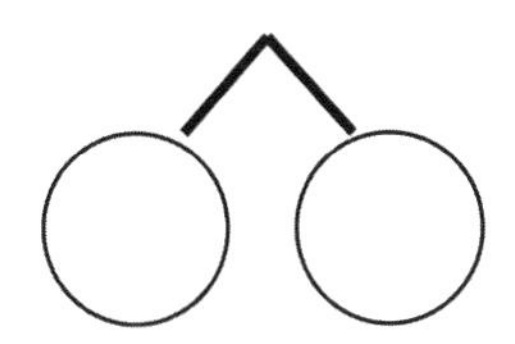

3. 7 + 9 = ____ ____ + ____ = ____

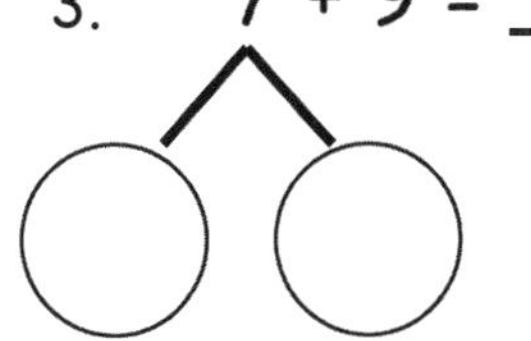

4. Match the number sentences to the bonds you used to help you make ten.

a. 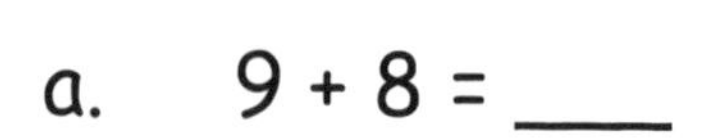
9 + 8 = ___

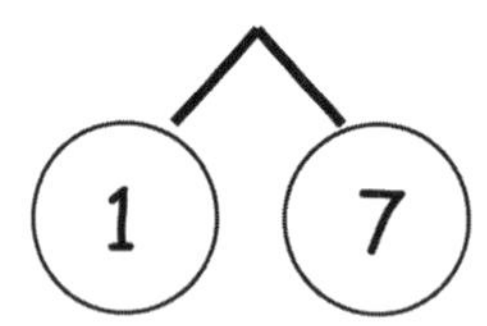

b. ___ = 9 + 6

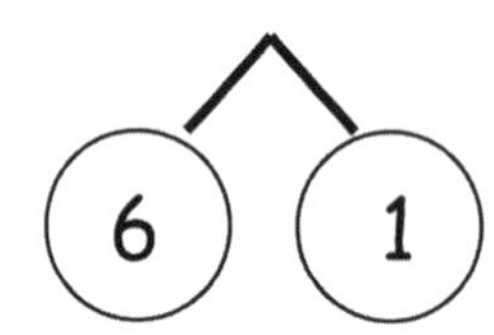

c. 7 + 9 =___

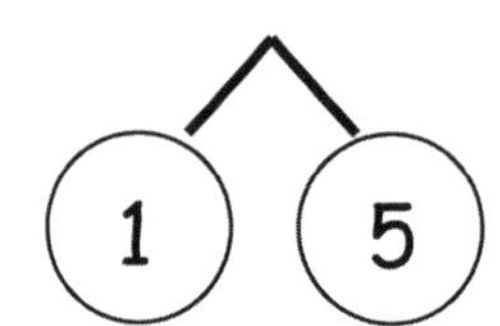

---

5. Show how the expressions are equal.

Use numbers bonds to make ten in the 9+ *fact* expression within the true number sentence. Draw to show the total.

a. 9 + 2 = 10 + 1

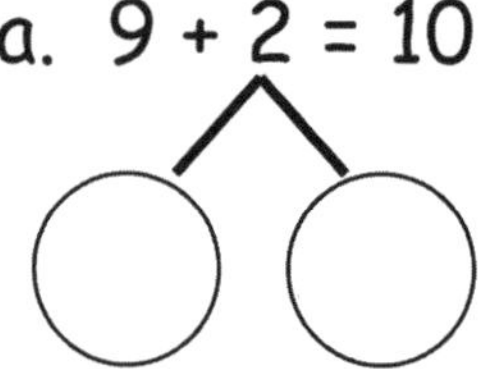

● ● ● ● ● 0
● ● ● ● 0

b. 10 + 3 = 9 + 4

c. 5 + 10 = 6 + 9

EUREKA MATH

Lesson 5

Objective: Compare efficiency of counting on and making ten when one addend is 9.

Suggested Lesson Structure

- Fluency Practice (13 minutes)
- Application Problem (5 minutes)
- Concept Development (32 minutes)
- Student Debrief (10 minutes)

Total Time (60 minutes)

Fluency Practice (13 minutes)

- Partners to Ten **1.OA.6** (5 minutes)
- Add Partners of Ten First **1.OA.6** (4 minutes)
- Take Out 2 **1.OA.6** (4 minutes)

Partners to Ten (5 minutes)

Materials: (S) Numeral cards (Lesson 1 Fluency Template 5-group cards with numeral-side only copied), personal white board

Note: This fluency activity provides maintenance with partners to ten while applying the commutative property.

Students put 5-group cards facedown and write 10 on their boards. Each partner takes a 5-group card and then draws a number bond without bubbles using the selected card as one part. Students write two addition sentences for the number bond and check each other's work.

Add Partners of Ten First (4 minutes)

Note: This activity reviews adding three numbers and prepares students for the make ten addition strategy when one addend is 9.

Conduct the activity as outlined in Lesson 3.

Take Out 2 (4 minutes)

Note: This is an anticipatory fluency activity for making ten when one addend is 8 since 8 needs 2 to make ten.

T: Take out 2 on my signal. For example, if I say "5," you say "2 and 3."
T: 3.
S: 2 and 1.
T: 10.
S: 2 and 8.

Continue with all numbers within 10 for about a minute, and then give students about 30 seconds of practice with a partner. Repeat the set as a whole class, and celebrate improvement.

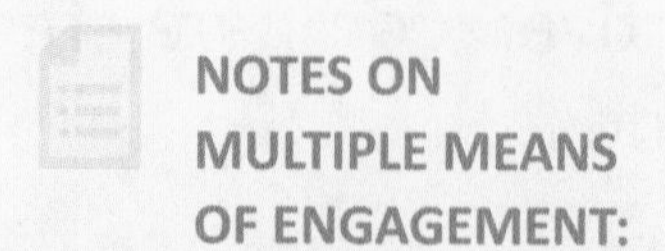

NOTES ON MULTIPLE MEANS OF ENGAGEMENT:

Fluency games and activities provide most students the opportunity to gain math confidence by experiencing daily math success. Be sure to highlight students' math successes frequently in order to facilitate continued effort and persistence.

Application Problem (5 minutes)

There are 9 red birds and 6 blue birds in a tree. How many birds are in the tree? Use a ten-frame drawing and a number sentence. Write a number bond to match the story and a number bond to show the matching 10+ fact. Write a statement.

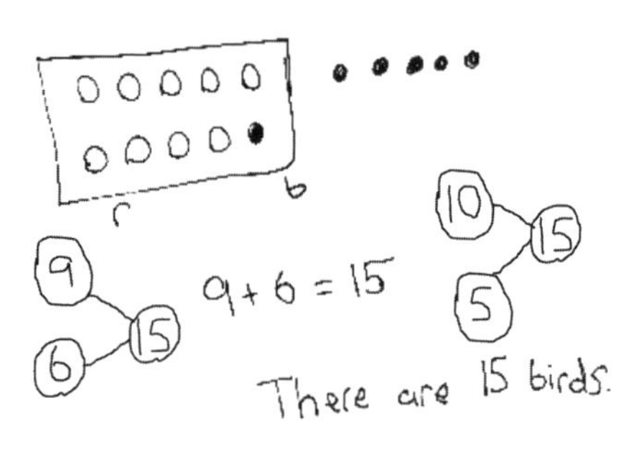

Note: This problem continues to provide contextual practice of solving addition situations where one addend is 9. By drawing a number bond to match the story and drawing a number bond to match the ten-frame drawing, students continue to relate the addition facts of 9 with the addition facts of 10. Students consider the problem's relationship to today's lesson during the Student Debrief.

Concept Development (32 minutes)

Materials: (S) Personal white board

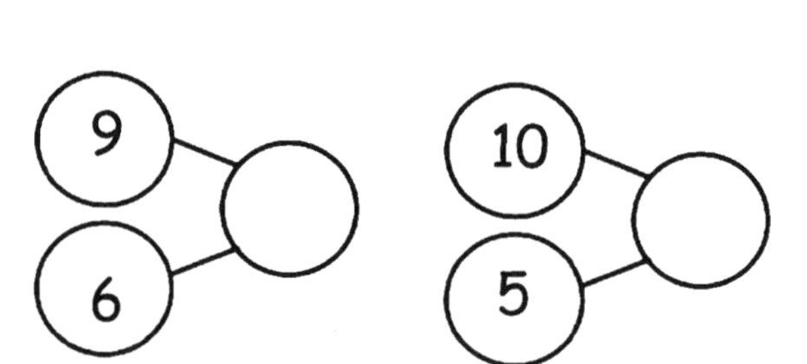

Have students sit at their desks or the meeting area.

T: (Project or write the two number bonds shown here.) Which number bond is easier to solve?
S: 10 and 5.
T: (Write 10 + 5 = ____.) 10 + 5 = ...?
S: 15.
T: (Record the solution.) How did you know that so quickly?
S: Because we know our 10+ facts. → Because 10 is a friendly number.

T: (Write 9 + 6 = ____.) Now let's count on to solve 9 + 6.

T/S: Niiiine, 10, 11, 12, 13, 14, 15. 15.

T: (Record the solution.) Wait. 9 + 6 is equal to 10 + 5?

S: Yes!

T: Both number bonds have the same total, but when one part is 10, our solution came to us automatically.

T: (Read aloud.) Sergio and Lila were getting ready to go to recess. They both had to solve 9 + 8. The first one to solve it got to go to recess first! Sergio decided he was going to count on to solve it. (Pause.) Was there another way to solve 9 + 8 that Sergio could have used? (Circulate and listen.)

S: (Discuss.) Make ten! → Take 1 out from 8, and give it to the 9, in order to make ten.

T: Some of you said that you would make ten. Well, that is just what Lila decided to do. (Assign partners.) Partner A, use your personal white board to show how Sergio solved 9 + 8 by counting on. Partner B, show how Lila solved 9 + 8 by making ten.

S: (Solve 9 + 8 using counting on or making ten.)

T: Talk to your partner about the strategy you used to solve 9 + 8.

S: (Discuss and share as the teacher circulates.)

T: Help me make a number bond to show what Sergio did. What were the parts that Sergio used?

17
9 8

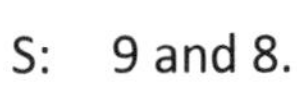

S: 9 and 8.

T: (Write the bond.) What was the total?

S: 17.

T: (Complete the bond.) Help me make a number bond to show what Lila did. What were the parts that Lila used?

S: 10 and 7.

T: (Write the bond.) What was the total?

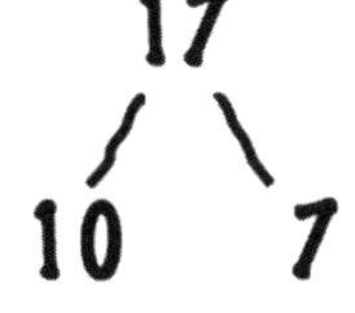

S: 17.

T: (Complete the bond.) Which number bond will help you solve more efficiently or quickly?

S: 10 and 7.

T: So, based on these number bonds and the work you and your partner just did, who do you think got to go to recess first?

S: Lila!

T: You're right! By using the make ten strategy, she was able to solve for the unknown quickly or efficiently.

Continue with partners solving each problem, showing how to solve using counting on and making ten: 9 + 6, 9 + 5, 9 + 2 (counting on may actually be more efficient here), and 9 + 9.

NOTES ON MULTIPLE MEANS OF ENGAGEMENT:

It is important to partner important vocabulary with captions or pictorial representations for all students. It is especially beneficial to English language learners and students with hearing impairments. Have students model or demonstrate their understanding of more difficult vocabulary such as *efficient*.

Problem Set (10 minutes)

Students should do their personal best to complete the Problem Set within the allotted 10 minutes. For some classes, it may be appropriate to modify the assignment by specifying which problems they work on first. Some problems do not specify a method for solving. Students should solve these problems using the RDW approach used for Application Problems.

Note: Students should save the Problem Set to provide the opportunity to compare making ten when adding 8.

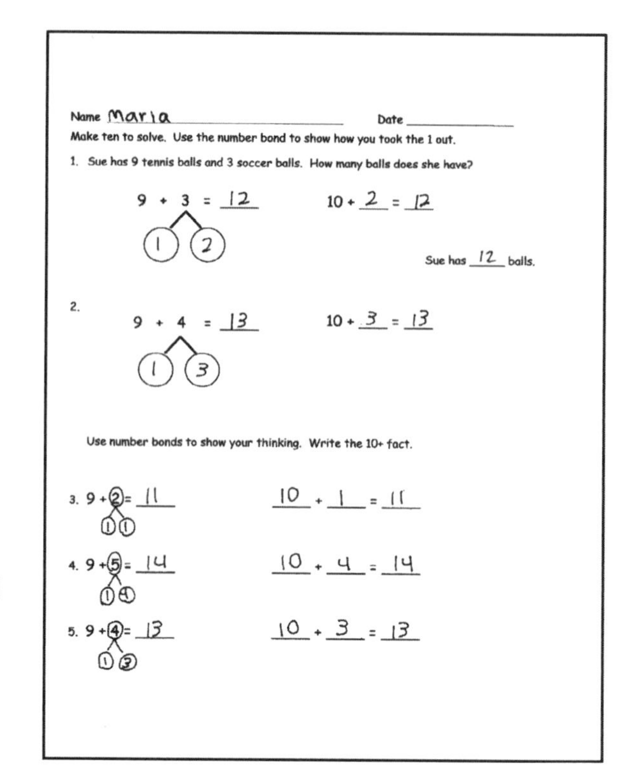

Name Maria Date ____

Make ten to solve. Use the number bond to show how you took the 1 out.

1. Sue has 9 tennis balls and 3 soccer balls. How many balls does she have?

9 + 3 = 12   10 + 2 = 12

Sue has 12 balls.

2. 9 + 4 = 13   10 + 3 = 13

Use number bonds to show your thinking. Write the 10+ fact.

3. 9 + 2 = 11   10 + 1 = 11

4. 9 + 5 = 14   10 + 4 = 14

5. 9 + 4 = 13   10 + 3 = 13

Student Debrief (10 minutes)

Lesson Objective: Compare efficiency of counting on and making ten when one addend is 9.

The Student Debrief is intended to invite reflection and active processing of the total lesson experience.

Invite students to review their solutions for the Problem Set. They should check work by comparing answers with a partner before going over answers as a class. Look for misconceptions or misunderstandings that can be addressed in the Debrief. Guide students in a conversation to debrief the Problem Set and process the lesson.

Any combination of the questions below may be used to lead the discussion.

- Which problems could you solve more efficiently by making ten?
- Why was that a more efficient way to solve?
- Were there any problems that you think could have been solved more efficiently using counting on? Why?
- Look at Problems 8–10. What do you notice about the number bonds? How does knowing your 10+ facts help you with your 9+ facts?
- Look at your Application Problem. What is the related 10+ fact for this problem? How does your drawing show both the 9+ fact and the related 10+ fact?
- Look at Problems 3–6. Think about these statements: 9 and ____ make ____, and 10 and ____ make____. (For example, 9 and 2 make 11, and 10 and 1 make 11.) What pattern do you notice?

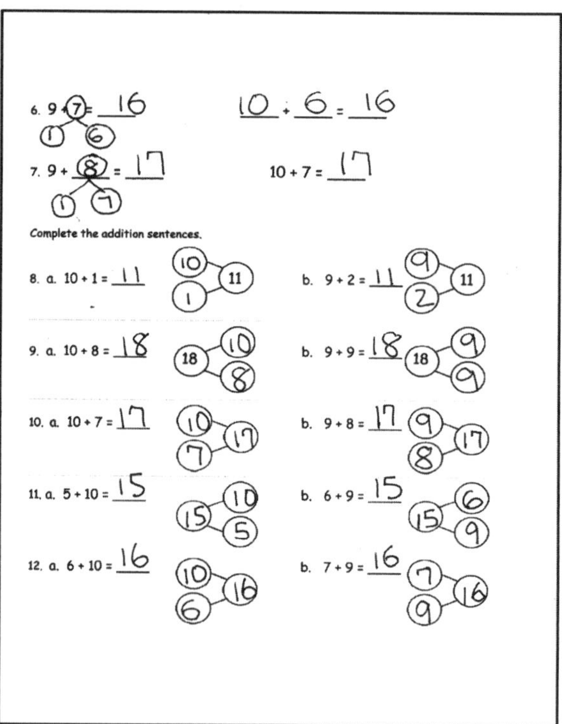

6. 9 + 7 = 16   10 + 6 = 16

7. 9 + 8 = 17   10 + 7 = 17

Complete the addition sentences.

8. a. 10 + 1 = 11   b. 9 + 2 = 11

9. a. 10 + 8 = 18   b. 9 + 9 = 18

10. a. 10 + 7 = 17   b. 9 + 8 = 17

11. a. 5 + 10 = 15   b. 6 + 9 = 15

12. a. 6 + 10 = 16   b. 7 + 9 = 16

Exit Ticket (3 minutes)

After the Student Debrief, instruct students to complete the Exit Ticket. A review of their work will help with assessing students' understanding of the concepts that were presented in today's lesson and planning more effectively for future lessons. The questions may be read aloud to the students.

Name ________________________ Date ____________

Make ten to solve. Use the number bond to show how you took the 1 out.

1. Sue has 9 tennis balls and 3 soccer balls. How many balls does she have?

9 + 3 = ____ 10 + ____ = ____

Sue has _____ balls.

2. 9 + 4 = ____ 10 + ____ = ____

Use number bonds to show your thinking. Write the 10+ fact.

3. 9 + 2 = _____ _____ + _____ = _____

4. 9 + 5 = _____ _____ + _____ = _____

5. 9 + 4 = _____ _____ + _____ = _____

6. 9 + 7 = ______ ______ + ______ = ______

7. 9 + ______ = ______ 10 + 7 = ______

Complete the addition sentences

8. a. 10 + 1 = _____ (11) b. 9 + 2 = _____ (11)

9. a. 10 + 8 = _____

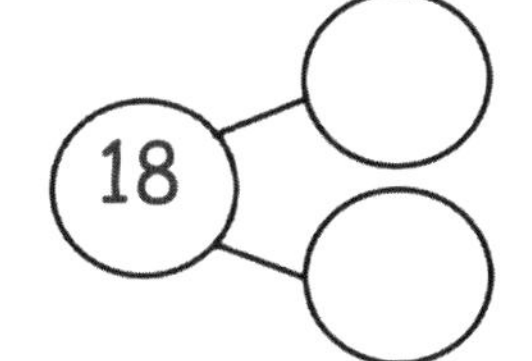

b. 9 + 9 = _____

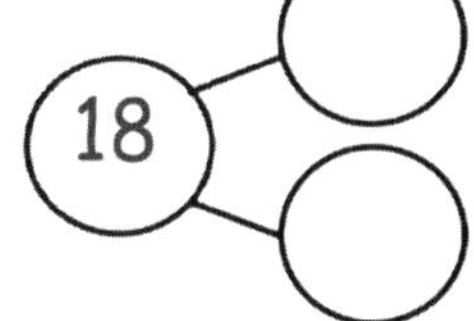

10. a. 10 + 7 = _____

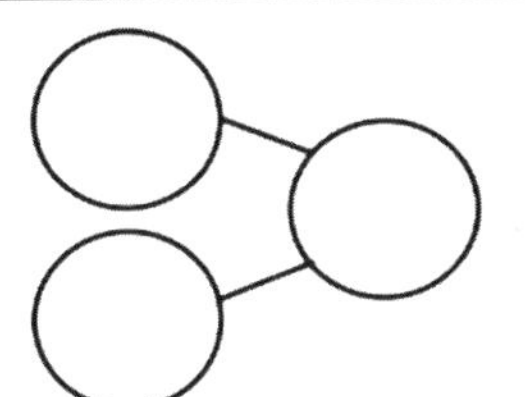

b. 9 + 8 = _____

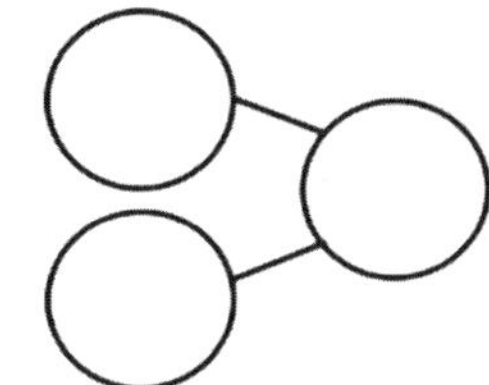

11. a. 5 + 10 = _____

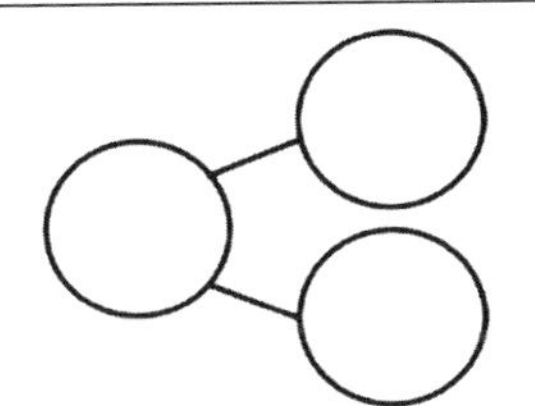

b. 6 + 9 = _____

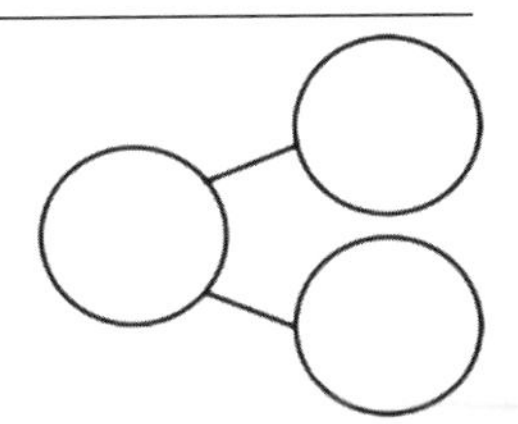

12. a. 6 + 10 = _____

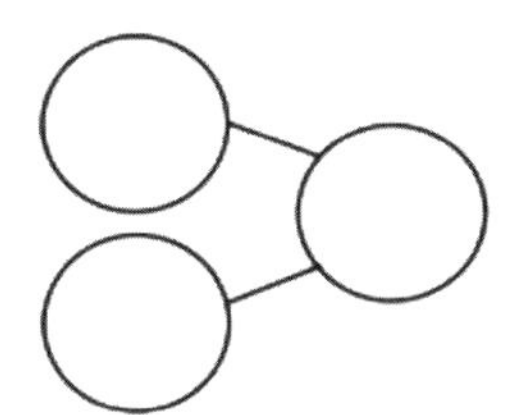

b. 7 + 9 = _____

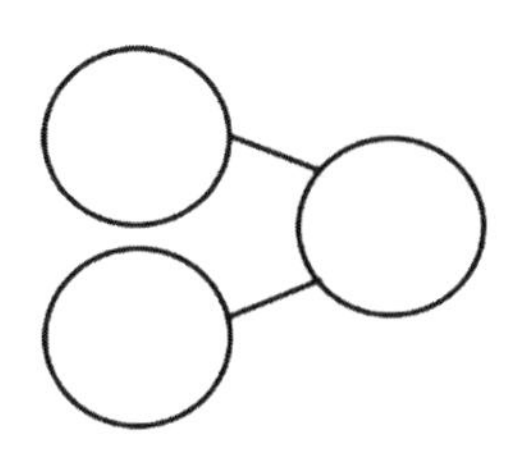

Name ____________________ Date ____________

Complete the number sentence.
Use an efficient strategy to solve the number sentences.

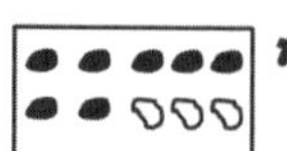

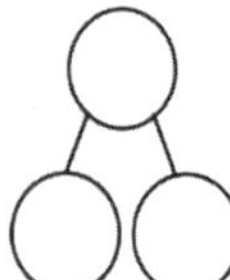

1. $9 + 2 =$ ____

2. $7 + 9 =$ ____

3. ____ $= 9 + 5$

Name ______________________________ Date ______________

Solve the number sentences. Use number bonds to show your thinking. Write the 10+ fact and new number bond.

1. 9 + 6 = _____ 10 + _____ = _____

2. 9 + 8 = _____ _____ + _____ = _____

3. 5 + 9 = _____ _____ + _____ = _____

4. 7 + 9 = _____ _____ + _____ = _____

5. Solve. Match the number sentence to the 10+ number bond.

a. 9 + 5 = _____ b. 9 + 6 = _____ c. 9 + 8 = ____

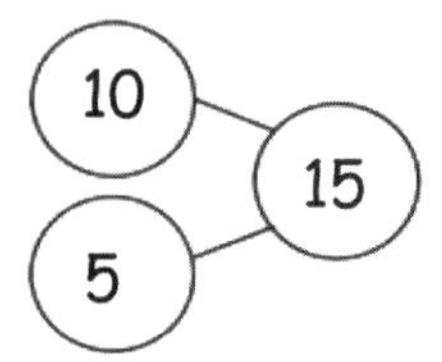

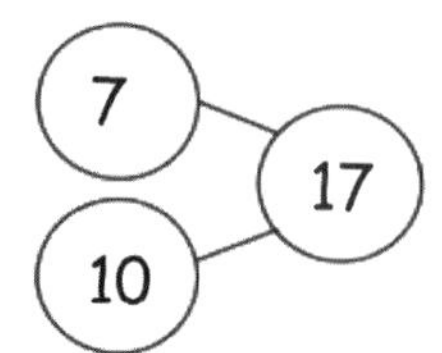

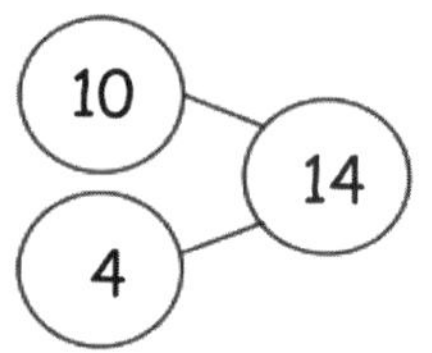

Use an efficient strategy to solve the number sentences.

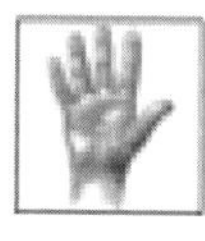
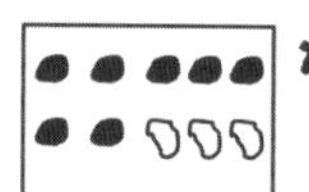

6. 9 + 7 = _____ 7. 9 + 2 = _____ 8. 9 + 1 = _____

9. 8 + 9 = _____ 10. 4 + 9 = _____ 11. 9 + 9 = _____

Lesson 6

Objective: Use the commutative property to make ten.

Suggested Lesson Structure

■ Fluency Practice	(10 minutes)
■ Application Problem	(5 minutes)
■ Concept Development	(35 minutes)
■ Student Debrief	(10 minutes)
Total Time	**(60 minutes)**

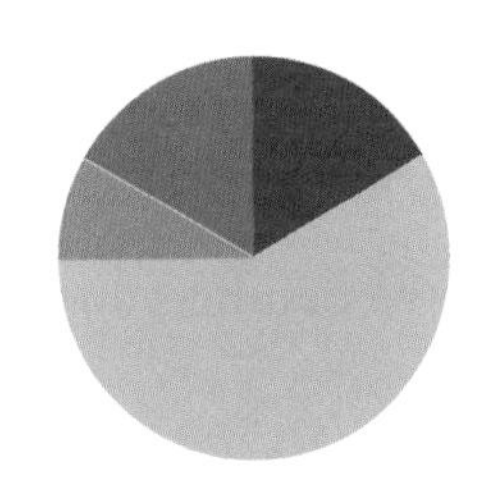

Fluency Practice (10 minutes)

- Happy Counting by Twos **1.OA.5** (2 minutes)
- Take Out 2: Number Bonds **1.OA.6** (4 minutes)
- Decompose Addition Sentences into Three Parts **1.OA.6** (4 minutes)

Happy Counting by Twos (2 minutes)

Materials: (T) Rekenrek, if available

Note: Reviewing counting on allows students to maintain fluency with adding and subtracting 2.

Repeat the Happy Counting activity from Module 1, Lesson 3, counting by twos from 0 to 20 and back.

Note: As it relates to addition and subtraction, counting forward and backward by twos affords students review with this strategy. This fluency activity may be challenging for students at first. A Rekenrek helps students visualize numbers and makes it easier for students to change direction as they count. Rekenreks can be made simply and inexpensively with cardboard, elastic, and beads. If these are not available, there are also interactive Rekenreks owned and provided by The Math Learning Center: https://apps.mathlearningcenter.org/number-rack/.

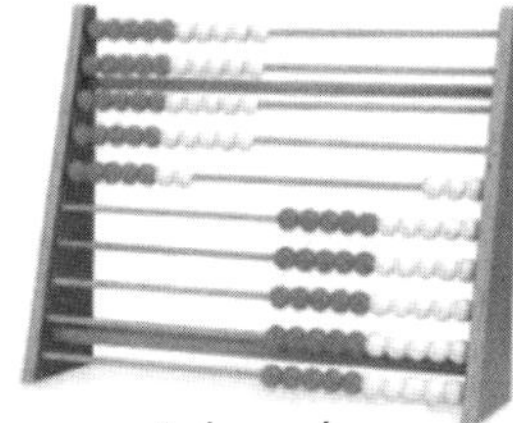

Rekenrek

Move the beads on the Rekenrek to model counting forward and backward by twos within twenty. Students count along with the beads (e.g., 2, 4, 6, 8, 10, 8, 6, 4).

When students are ready, put the Rekenrek away, and tell them to look at your thumb to count forward and backward by twos. When your thumb points and motions up, students count up. When your thumb is to the side, students stop. When your thumb points and motions down, students count down (see the illustration below).

T: 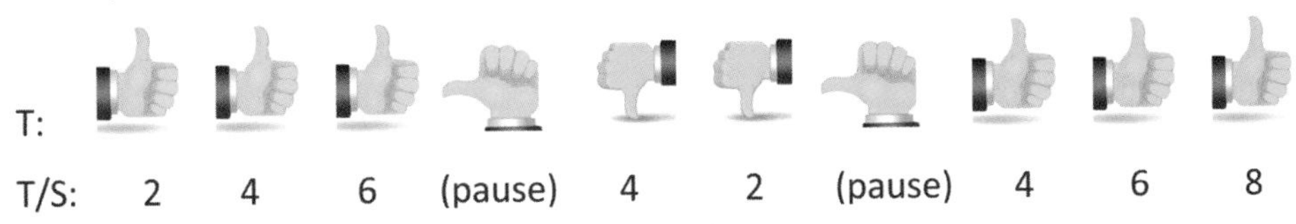

T/S: 2 4 6 (pause) 4 2 (pause) 4 6 8

Take Out 2: Number Bonds (4 minutes)

Materials: (S) Personal white board

Note: This is an anticipatory fluency activity for the make ten addition strategy when one addend is 8.

Say a number within 10. Students quickly write a number bond for the number said, using 2 as a part, and hold up their boards when finished.

Decompose Addition Sentences into Three Parts (4 minutes)

Note: This fluency activity reviews adding three numbers and making ten when one addend is 9.

Decompose addition sentences into three steps.

T: (Write 9 + 3.) Say 3 as an addition sentence starting with 1.

S: 1 + 2.

T: (Write 1 + 2 below 3.) Say 9 + 3 as a three-part addition sentence.

S: 9 + 1 + 2 = 12.

Write out the equation for students to see if necessary. Repeat the process for other problems.

Application Problem (5 minutes)

There are 6 children on the swings and 9 children playing tag. How many children are playing on the playground? Make ten to solve. Create a drawing, a number bond, and a number sentence along with your statement.

6 + 9 = 15

10 + 5 = 15

There are 15 children on the playground.

Note: This problem gives students the chance to apply learning from Lessons 3, 4, and 5 as they solve problems with 9 as an addend. During the Student Debrief, discuss how the commutative property is applied to solve the problem efficiently.

Concept Development (35 minutes)

Materials: (S) Personal white board

Students sit in partnerships at tables or in the meeting area.

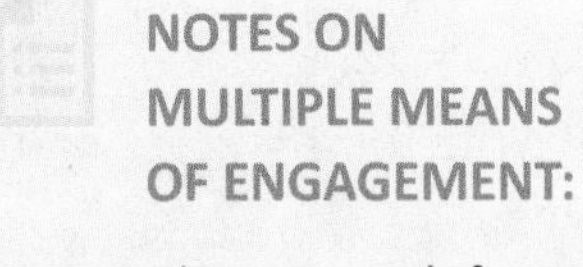

NOTES ON MULTIPLE MEANS OF ENGAGEMENT:

Some students are ready for more challenging numbers. Adjust the lesson structure as appropriate by providing just right numbers, such as 13 and 9, where students can continue to apply the making ten strategy in a more complex way.

T: (Write 5 + 9 = _____ on the board.) Turn and talk to your partner. What strategy should we use to solve efficiently?

S: Make ten.

T: Should we make ten with 5 or with 9? Let's have each partner try it a different way. Partner A, solve this by making ten with 5. Partner B, solve this by making ten with 9.

S: (Solve on personal white boards as the teacher circulates.)

T: Share your solution with your partner. Did you get the same total or a different total? Discuss how you solved it.

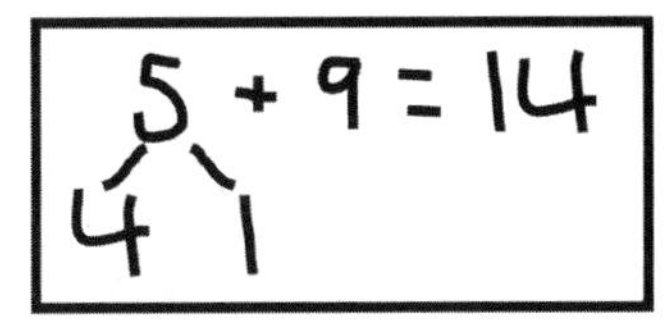

S: (Share solutions and how they broke apart the numbers.)

T: How much is 5 + 9?

S: 14.

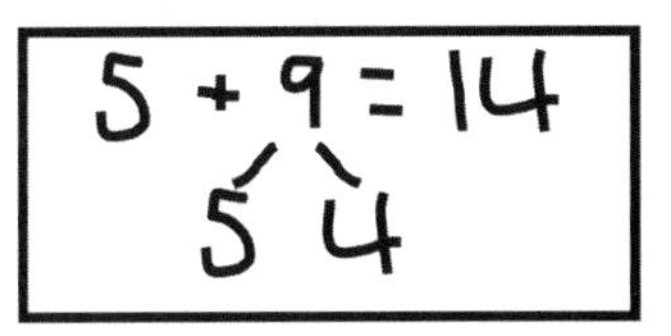

T: Did you solve for the total using the same way? How did you and your partner solve this?

S: We used different ways. I broke apart the 9 into 5 and 4 so I could make ten with 5 + 5, and my partner broke apart the 5 into 4 and 1 so she could make ten with 9 + 1.

T: (Write students' solutions on the board, including bonds.) So, Partner A added 5 + 9 using 5 + 5 + 4. (Point to the number bond.) You're saying that this is the same as Partner B's work where she added 5 + 9 using 9 + 1 + 4. (Point to the number bond.) So, 5 + 5 + 4 is the same as 9 + 1 + 4? (Point to the number bonds.)

S: Yes!

T: Which way did you prefer? Why?

S: I know 9 is made from 5 and 4, so taking apart 9 was fast for me. → Making 10 with 9 was fast and easy for me. It's just 1 away from 10. It's easy to take away 1 from a number.

T: Do we always have to start with the first addend when we are adding?

S: No. We can add in any order, as long as we add all of the parts.

T: (Project 3 + 9.) Which number should we start with?

S: 9, because all we have to do is take the 1 out of 3 to make ten.

T: On your personal white board, find the total, and show your bonds.

S: (Write 3 + 9 = 12, showing bonds of 2 and 1 under 3.)

T: What is the related 10+ fact to help you solve 3 + 9?

S: 10 + 2 = 12.

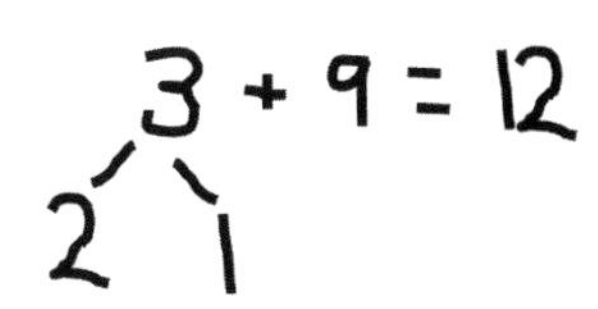

T: So, what is 3 + 9? Say the number sentence.

S: 3 + 9 = 12.

T: (Write 9 + 4 = _____ on the board.) Which number should we make ten with?

S: 9.

T: Which number should we break apart?

S: 4.

T/S: (Repeat the process to find the sum.)

9 + 4 = 13
1 3

Repeat the process using the suggested sequence: 9 + 6, 8 + 9, and 7 + 9. For each problem, have students make ten to solve and alternate writing the related 10+ fact as a number bond and a number sentence.

Problem Set (10 minutes)

Students should do their personal best to complete the Problem Set within the allotted 10 minutes. For some classes, it may be appropriate to modify the assignment by specifying which problems they work on first. Some problems do not specify a method for solving. Students should solve these problems using the RDW approach used for Application Problems.

Note: Students should save the Problem Set so it is available as a comparison during Student Debriefs focusing on making ten when one addend is 8.

Student Debrief (10 minutes)

Lesson Objective: Use the commutative property to make ten.

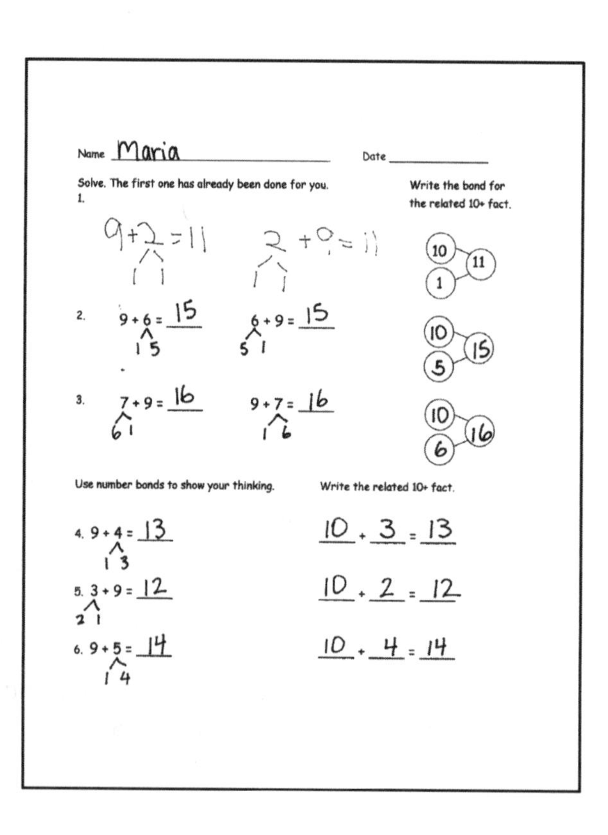
Name Maria Date ______

Solve. The first one has already been done for you.

Write the bond for the related 10+ fact.

1. 9 + 2 = 11 2 + 9 = 11 (10, 1, 11)
2. 9 + 6 = 15 6 + 9 = 15 (10, 5, 15)
3. 7 + 9 = 16 9 + 7 = 16 (10, 6, 16)

Use number bonds to show your thinking.

Write the related 10+ fact.

4. 9 + 4 = 13 10 + 3 = 13
5. 3 + 9 = 12 10 + 2 = 12
6. 9 + 5 = 14 10 + 4 = 14

The Student Debrief is intended to invite reflection and active processing of the total lesson experience.

Invite students to review their solutions for the Problem Set. They should check work by comparing answers with a partner before going over answers as a class. Look for misconceptions or misunderstandings that can be addressed in the Debrief. Guide students in a conversation to debrief the Problem Set and process the lesson.

Any combination of the questions below may be used to lead the discussion.

- Look at Problem 8. Find as many related equal equations as you can.
- Look at Problem 8. In which problem can you use your doubles + 1 fact to help you solve?
- How did we apply the make ten strategy today to solve addition problems efficiently?
- To solve 3 + 9, which addend should we make ten with? Why?

- Look at your Application Problem. Turn and talk to your partner about which addend we should break apart to solve the problem more efficiently.

Exit Ticket (3 minutes)

After the Student Debrief, instruct students to complete the Exit Ticket. A review of their work will help with assessing students' understanding of the concepts that were presented in today's lesson and planning more effectively for future lessons. The questions may be read aloud to the students.

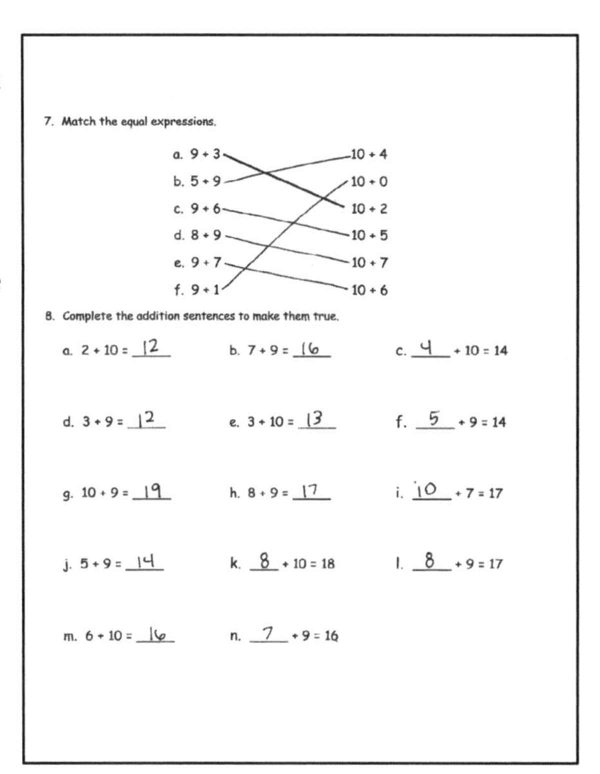

7. Match the equal expressions.

a. 9 + 3 — 10 + 4
b. 5 + 9 — 10 + 0
c. 9 + 6 — 10 + 2
d. 8 + 9 — 10 + 5
e. 9 + 7 — 10 + 7
f. 9 + 1 — 10 + 6

8. Complete the addition sentences to make them true.

a. 2 + 10 = 12
b. 7 + 9 = 16
c. 4 + 10 = 14
d. 3 + 9 = 12
e. 3 + 10 = 13
f. 5 + 9 = 14
g. 10 + 9 = 19
h. 8 + 9 = 17
i. 10 + 7 = 17
j. 5 + 9 = 14
k. 8 + 10 = 18
l. 8 + 9 = 17
m. 6 + 10 = 16
n. 7 + 9 = 16

Name ______________________ Date ____________

Solve. The first one has already been done for you.

Write the bond for the related 10+ fact.

1.

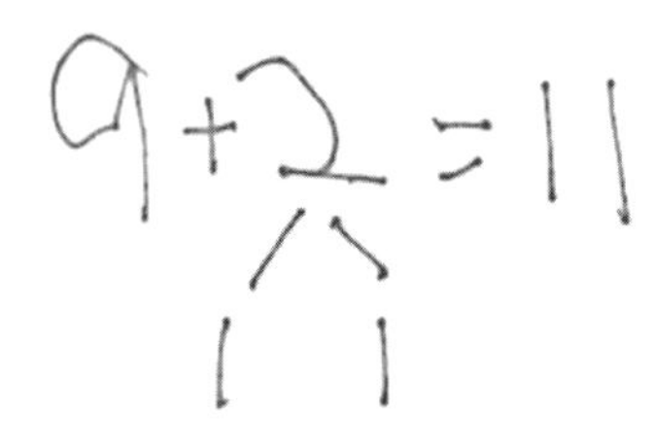

2 + 9 = 11

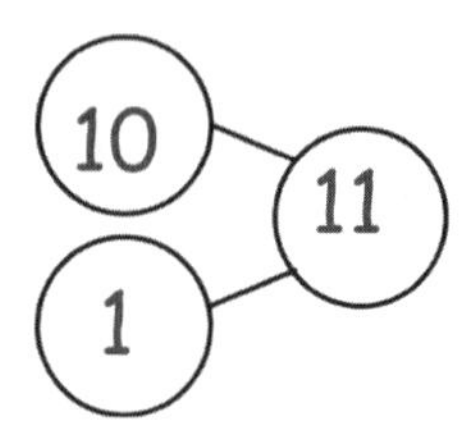

2. 9 + 6 = _____ 6 + 9 = _____

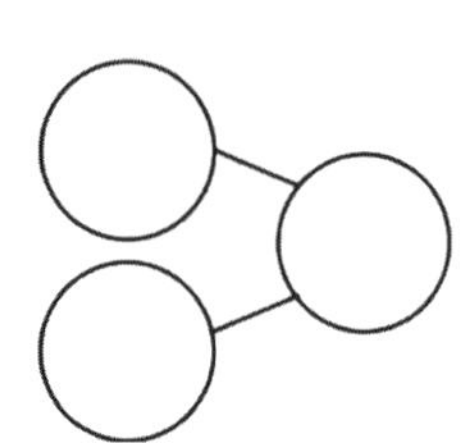

3. 7 + 9 = _____ 9 + 7 = _____

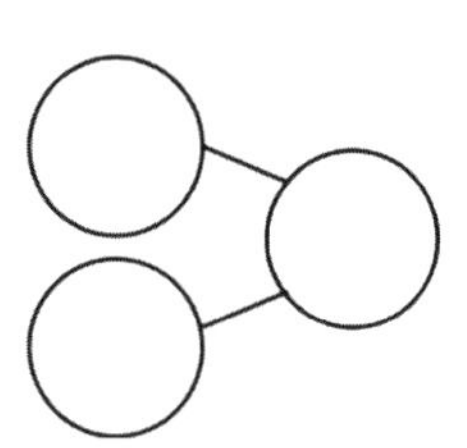

Use number bonds to show your thinking.

Write the related 10+ fact.

4. 9 + 4 = _____ _____ + _____ = _____

5. 3 + 9 = _____ _____ + _____ = _____

6. 9 + 5 = _____ _____ + _____ = _____

7. Match the equal expressions.

a. 9 + 3	10 + 4
b. 5 + 9	10 + 0
c. 9 + 6	10 + 2
d. 8 + 9	10 + 5
e. 9 + 7	10 + 7
f. 9 + 1	10 + 6

8. Complete the addition sentences to make them true.

a. 2 + 10 = ______ b. 7 + 9 = _____ c. _____ + 10 = 14

d. 3 + 9 = ______ e. 3 + 10 = _____ f. _____ + 9 = 14

g. 10 + 9 = ______ h. 8 + 9 = _____ i. _____ + 7 = 17

j. 5 + 9 = ______ k. _____ + 10 = 18 l. _____ + 9 = 17

m. 6 + 10 = ______ n. ______ + 9 = 16

Name ____________________ Date __________

1. Solve. Use number bonds to show your thinking. Write the bond for the related 10+ fact.

$9 + 5 =$ ____ $5 + 9 =$ ____

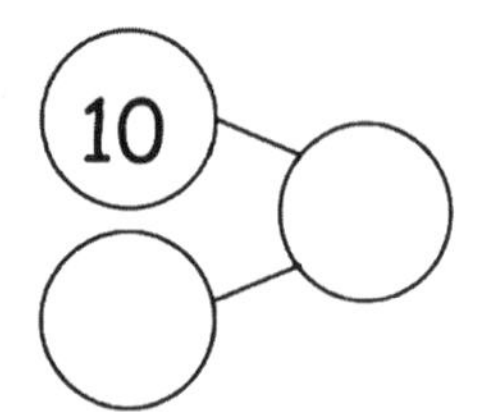

2. Solve. Draw a line to match the related facts and write the related 10+ fact.

a.	$9 + 7 =$ ____	____ $= 9 + 8$	____
b.	____ $= 6 + 9$	$7 + 9 =$ ____	$10 + 6 = 16$
c.	$8 + 9 =$ ____	$9 + 6 =$ ____	____

Name ______________________________ Date ______________

1. Solve. Use your number bonds. Draw a line to match the related facts. Write the related 10+ fact.

a.	9 + 6 = ____	____ = 9 + 8	________
b.	____ = 3 + 9	____ = 7 + 9	________
c.	____ = 9 + 5	6 + 9 = ____	10 + 5 = 15
d.	8 + 9 = ____	9 + 3 = ____	________
e.	9 + 7 = ____	5 + 9 = ____	________

2. Complete the addition sentences to make them true.

a. 3 + 10 = ______

b. 4 + 9 = ______

c. 10 + 5 = ______

d. 9 + 6 = ______

e. 7 + 10 = ______

f. ______ = 7 + 9

g. 10 + ______ = 18

h. 9 + 8 = ______

i. ______ + 9 = 19

j. 5 + 9 = ______

3. Find and color the expression that is equal to the expression on the snowman's hat. Write the true number sentence below.

a.

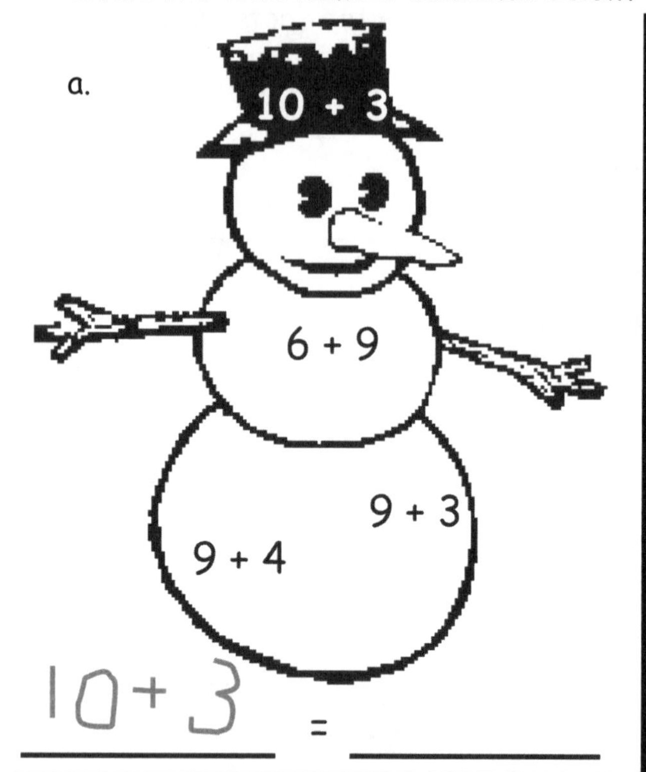

10 + 3 = ______

b.

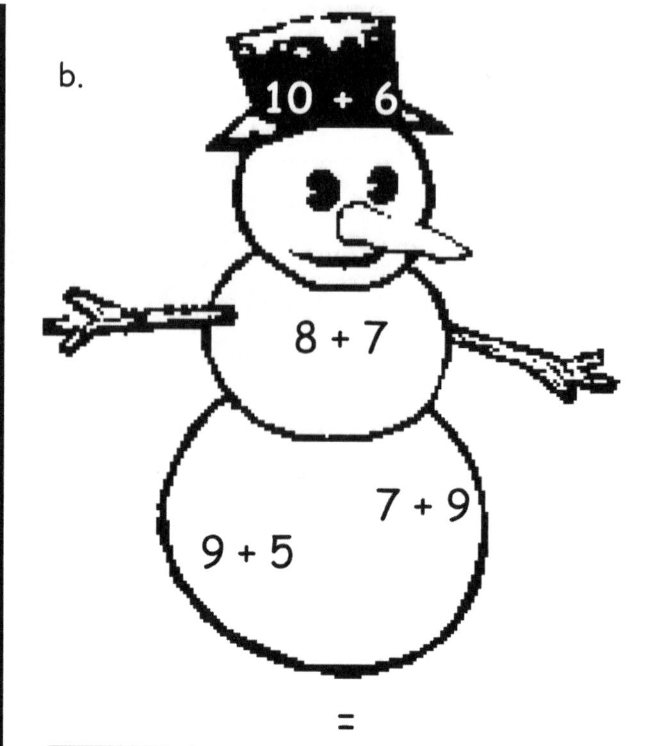

______ = ______

c.

______ = ______

d.

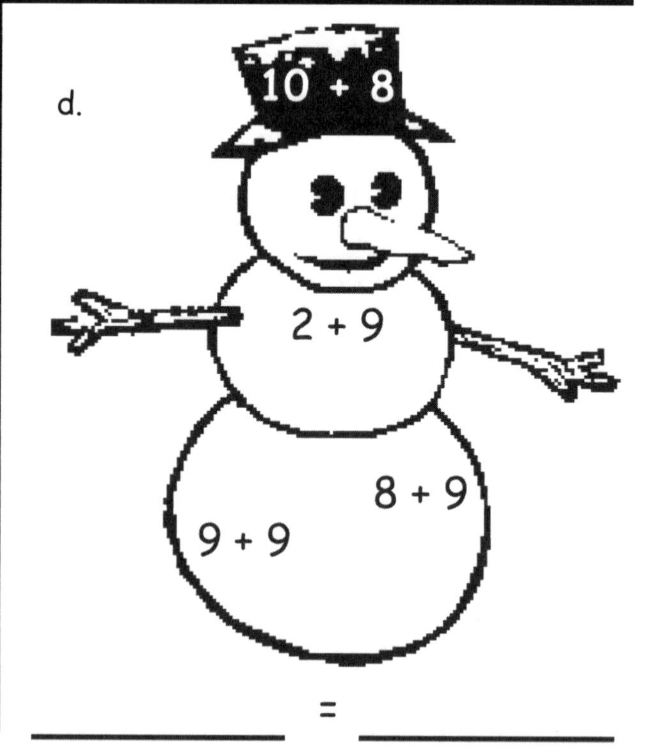

______ = ______

Lesson 7

Objective: Make ten when one addend is 8.

Suggested Lesson Structure

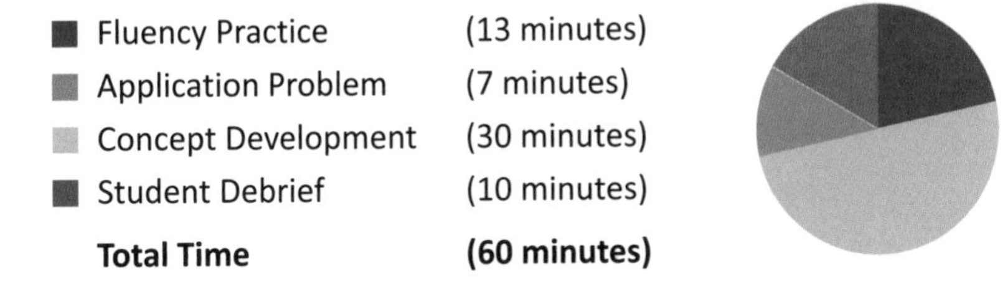

Fluency Practice	(13 minutes)
Application Problem	(7 minutes)
Concept Development	(30 minutes)
Student Debrief	(10 minutes)
Total Time	**(60 minutes)**

Fluency Practice (13 minutes)

- Add to 9 **1.OA.6** (5 minutes)
- Friendly Fact Go Around: Make It Equal **1.OA.6** (5 minutes)
- Take Out 2: Addition Sentences **1.OA.6** (3 minutes)

Add to 9 (5 minutes)

Materials: (T) 9 + *n* addition cards (Fluency Template 1) (S) Personal white board

Note: This activity supports the make ten addition strategy as students need to fluently decompose an addend in order to make ten.

Show an addition flash card (e.g., 9 + 3). Students write the three-addend equation (9 + 1 + 2 = 12).

Friendly Fact Go Around: Make It Equal (5 minutes)

Materials: (T) Friendly fact go around: make it equal (Fluency Template 2)

Note: This activity reinforces the make ten adding strategy and promotes an understanding of equality.

Project the Friendly Fact Go Around: Make It Equal (or make and display a poster). Point to a problem, and call on a student: 9 + 6 = 10 + □. The student answers "five." The class says the number sentence aloud with the answer: "9 + 6 = 10 + 5." If a student gives an incorrect answer, he then repeats the correct equation that the class has given. The teacher can adapt the problem to individual students, pointing to easier problems for students who are less fluent.

Take Out 2: Addition Sentences (3 minutes)

Note: This activity supports the make ten addition strategy when one addend is 8 since 8 needs 2 to make ten.

Say a number between 2 and 10 (e.g., 3). Students say an addition sentence beginning with 2 (e.g., 2 + 1 = 3).

Application Problem (7 minutes)

Stacy made 6 drawings. Matthew made 2 drawings. Tim made 4 drawings. How many drawings did they make altogether? Use a drawing, a number sentence, and a statement to match the story.

Note: Some students may actually create detailed drawings. Continue discussing how simple shapes, such as squares or circles, can be used to efficiently represent the story's drawings rather than spending time and thought on elaborate pictures.

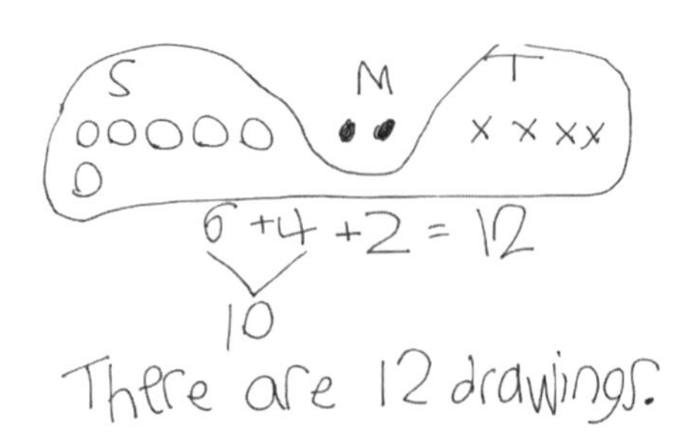

Concept Development (30 minutes)

NOTES ON MULTIPLE MEANS OF REPRESENTATION:

When using colors in lessons, be sensitive to those students who have difficulty seeing certain colors. Use primary colors and typically sharp contrast, like green (or red) and yellow, that can be distinguished by these students. Be sure to adjust the color names to align when implementing the Concept Development.

Materials: (T) 10 blue and 10 yellow linking cubes, a ten-frame border (S) 10 blue and 10 yellow linking cubes, personal white board

Have students sit at their seats with the materials.

T: (Project and read aloud.) Peter has 8 books, and Willie has 5. How many books do they have altogether?

T: What is the expression to solve this problem?

S: 8 + 5.

T: On your personal white board, use your blue linking cubes in 5-groups to show how many books Peter has.

S: (Organize 8 blue linking cubes.)

T: Use your yellow cubes to show how many books Willie has. Put them in a line of five next to your board.

S: (Organize 5 yellow linking cubes.)

MP.7

T: What are the different ways we can solve 8 + 5?

S: Count on! → Make ten with 5. → Make ten with 8.

T: (Call on students to demonstrate each of these strategies, saving making 10 with 8 for the end. As a student volunteer makes ten, use the ten-frame border to physically group the ten.)

T: Let's use the last strategy to solve 8 + 5. Everyone, make ten with 8.

S: (Move 2 yellow cubes to the blue pile.)

T: With your marker, draw a frame around your 10 cubes.

S: (Frame 10 cubes.)

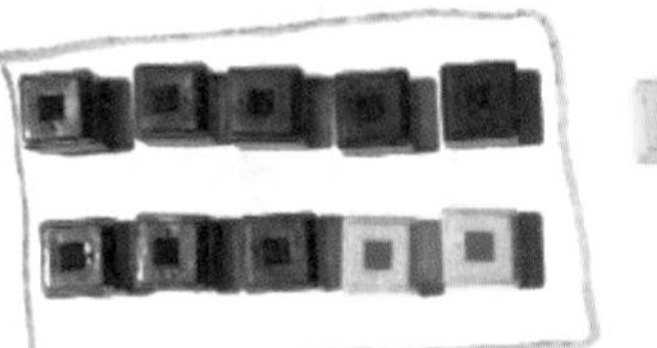

T: We have 10 here. (Gesture to the 10.) What do we have left here? (Point to the other pile.)

S: 3.

T: Look at your new groups. What is our new number sentence?

S: 10 + 3 = 13.

MP.7

T: (Write 10 + 3 = 13 on the board.) Did we change the *number* of linking cubes we have?
S: No.

T: So, 8 + 5 is the same as what addition expression?
S: 10 + 3.
T: (Write 8 + 5 = 10 + 3.)
T: What is 10 + 3?
S: 13.
T: What is 8 + 5? Say the number sentence.
S: 8 + 5 = 13.
T: How many books do Peter and Willie have?
S: 13 books.

Repeat the process with the following suggested sequence: 8 + 3, 8 + 6, 4 + 8, 8 + 7, 8 + 8. Be sure to have students make ten with 8, reinforcing the concept of commutativity for efficient problem solving. Write both number sentences (8 + 6 = 14, 10 + 4 = 14) and a number sentence equating the equivalent expressions (8 + 6 = 10 + 4).

Problem Set (10 minutes)

Students should do their personal best to complete the Problem Set within the allotted 10 minutes. For some classes, it may be appropriate to modify the assignment by specifying which problems they work on first. Some problems do not specify a method for solving. Students should solve these problems using the RDW approach used for Application Problems.

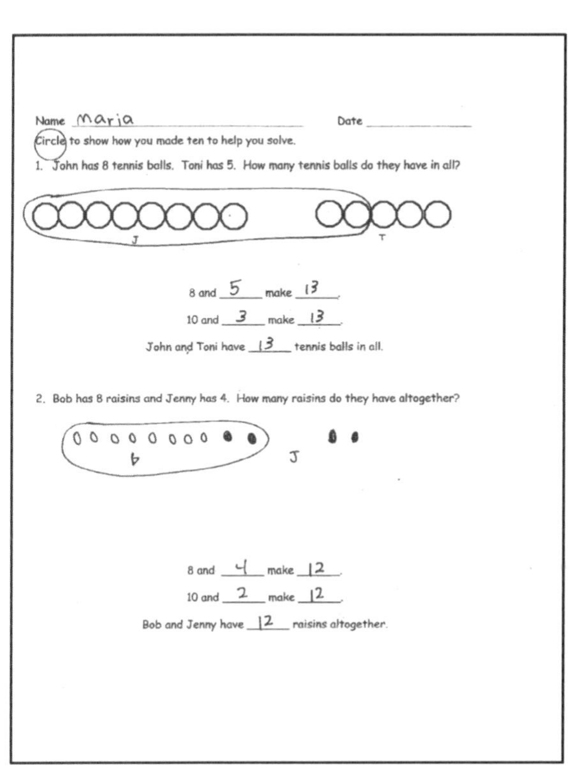
Name maria Date

Circle to show how you made ten to help you solve.

1. John has 8 tennis balls. Toni has 5. How many tennis balls do they have in all?

8 and 5 make 13.

10 and 3 make 13.

John and Toni have 13 tennis balls in all.

2. Bob has 8 raisins and Jenny has 4. How many raisins do they have altogether?

8 and 4 make 12.

10 and 2 make 12.

Bob and Jenny have 12 raisins altogether.

Student Debrief (10 minutes)

Lesson Objective: Make ten when one addend is 8.

Note: Distribute the student Problem Set from Lesson 3 or Lesson 4 for comparing with today's Problem Set.

The Student Debrief is intended to invite reflection and active processing of the total lesson experience.

Invite students to review their solutions for the Problem Set. They should check work by comparing answers with a partner before going over answers as a class. Look for misconceptions or misunderstandings that can be addressed in the Debrief. Guide students in a conversation to debrief the Problem Set and process the lesson.

Any combination of the questions below may be used to lead the discussion.

- Look at Problem 1. What are the two number sentences that match the statements? (Repeat with other problems as necessary.)
- (Write 8 + 5 = 13 and 10 + 3 = 13 on the board.) How can you make one true number sentence from the two number sentences on the board? (8 + 5 = 10 + 3)
- When you had 8 as an addend, how many objects did you circle from the other addend?
- Look at your Problem Set from Lesson 3 or Lesson 4. How are these problems similar to today's Problem Set? How are they different? What do you notice about the answers when you have 9 as an addend compared to 8 as an addend? Why do you think this is?
- Look at the Application Problem. What did you add first? Why? (Some students may have added 6 + 4 because it is an efficient way to make ten. Some students may still be adding the numbers in order. If students added 6 + 2 first, ask them to use today's lesson to show making ten to solve.)

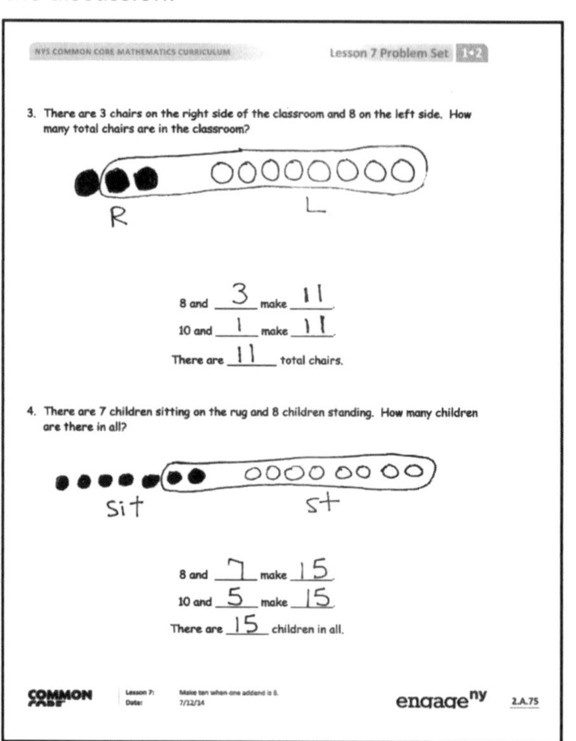
NYS COMMON CORE MATHEMATICS CURRICULUM Lesson 7 Problem Set 1•2

3. There are 3 chairs on the right side of the classroom and 8 on the left side. How many total chairs are in the classroom?

R L

8 and 3 make 11.
10 and 1 make 11.
There are 11 total chairs.

4. There are 7 children sitting on the rug and 8 children standing. How many children are there in all?

sit st

8 and 7 make 15.
10 and 5 make 15.
There are 15 children in all.

COMMON CORE engage^ny 2.A.75

Exit Ticket (3 minutes)

After the Student Debrief, instruct students to complete the Exit Ticket. A review of their work will help with assessing students' understanding of the concepts that were presented in today's lesson and planning more effectively for future lessons. The questions may be read aloud to the students.

Name ______________________________ Date ______________

Circle to show how you made ten to help you solve.

1. John has 8 tennis balls. Toni has 5. How many tennis balls do they have in all?

John

Toni

8 and ______ make ______.

10 and ______ make ______.

John and Toni have ______ tennis balls in all.

2. Bob has 8 raisins, and Jenny has 4. How many raisins do they have altogether?

8 and ______ make ______.

10 and ______ make ______.

Bob and Jenny have ______ raisins altogether.

3. There are 3 chairs on the right side of the classroom and 8 on the left side. How many total chairs are in the classroom?

8 and ______ make ______.

10 and ______ make ______.

There are ______ total chairs.

4. There are 7 children sitting on the rug and 8 children standing. How many children are there in all?

8 and ______ make ______.

10 and ______ make ______.

There are ______ children in all.

Name ______________________________ Date ______________

Draw, label, and (Circle) to show how you made ten to help you solve.

Write the number sentences you used to solve.

Nick picks some peppers. He picks 5 green peppers and 8 red peppers. How many peppers does he pick in all?

8 and ______ make ______.

10 and ______ make ______.

Nick picks ______ peppers.

Name ______________________________ Date ______________

Draw, label, and (Circle) to show how you made ten to help you solve.

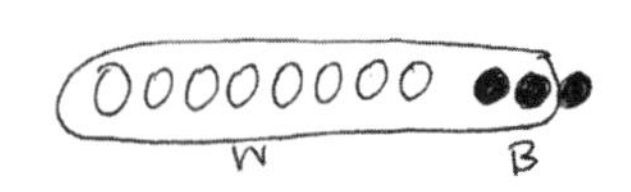

Write the number sentences you used to solve.

8 + 3 = 11
10 + 1 = 11

1. Meg gets 8 toy animals and 4 toy cars at a party.
   How many toys does Meg get in all?

$8 + 4 =$ _____

$10 +$ _____ $=$ _____

Meg gets _____ toys.

2. John makes 6 baskets in his first basketball game and 8 baskets in his second.
   How many baskets does he make altogether?

_____ + _____ = _____

_____ + _____ = _____

John makes _____ baskets.

3. May has a party. She invites 7 girls and 8 boys. How many friends does she invite in all?

_____ + _____ = _____

_____ + _____ = _____ May invites _____ friends.

4. Alec collects baseball hats. He has 9 Mets hats and 8 Yankees hats. How many hats are in his collection?

_____ + _____ = _____

_____ + _____ = _____ Alec has _____ hats.

9 + 2 =	3 + 9 =
9 + 4 =	5 + 9 =
9 + 6 =	7 + 9 =
9 + 8 =	9 + 9 =

9 + n addition cards, print on cardstock and cut

Friendly Fact Go Around: Make It Equal

9 + 1 = 10 + □	9 + 3 = 10 + □	9 + 5 = 10 + □
9 + 4 = 10 + □	9 + 7 = 10 + □	9 + 6 = 10 + □

---

3 + 9 = 10 + □	2 + 9 = 10 + □	8 + 9 = 10 + □
5 + 9 = 10 + □	4 + 9 = 10 + □	9 + 9 = 10 + □

---

9 + 4 = □ + 10	9 + 6 = □ + 10	9 + 5 = □ + 10
9 + 2 = □ + 10	9 + 7 = □ + 10	9 + 9 = □ + 10
9 + □ = 10 + 5	9 + □ = 10 + 7	9 + □ = 10 + 8
9 + □ = 10 + 3	9 + □ = 10 + 4	9 + □ = 10 + 6

friendly fact go around: make it equal

Lesson 8

Objective: Make ten when one addend is 8.

Suggested Lesson Structure

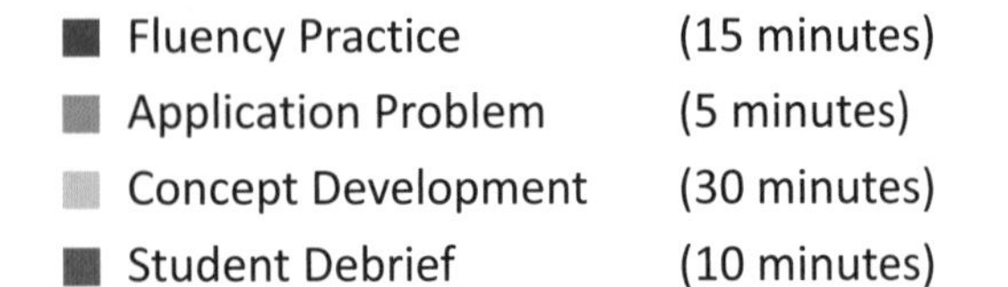

- Fluency Practice (15 minutes)
- Application Problem (5 minutes)
- Concept Development (30 minutes)
- Student Debrief (10 minutes)

Total Time (60 minutes)

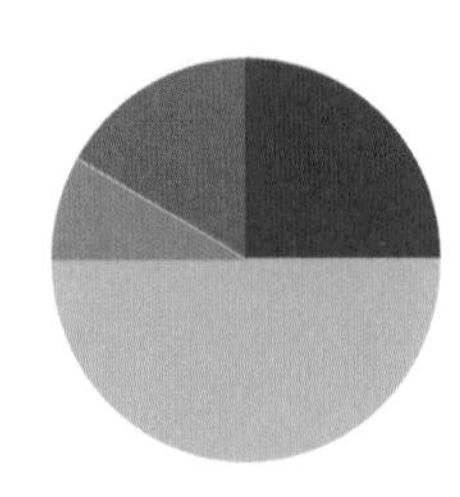

Fluency Practice (15 minutes)

- Sprint: 9 + *n* Using Make Ten **1.OA.6** (10 minutes)
- Happy Counting by Twos **1.OA.5** (2 minutes)
- Take Out 2: Addition Sentences **1.OA.6** (3 minutes)

Sprint: 9 + *n* Using Make Ten (10 minutes)

Materials: (S) 9 + *n* Using Make Ten Sprint

Note: This Sprint provides practice with the make ten addition strategy, when one addend is 9.

Happy Counting by Twos (2 minutes)

Note: This reviewing of counting on allows students to maintain fluency with adding and subtracting 2.

Repeat the Happy Counting activity from Lesson 4, counting by twos from 0 to 20 and back (this range may be adjusted to meet the needs of students). As students strengthen their skills, start with other numbers such as 1, 7, 11, or 8.

Take Out 2: Addition Sentences (3 minutes)

Note: This activity supports the make ten addition strategy when one addend is 8 since 8 needs 2 to make ten.

Say a number between 2 and 10 (e.g., 3). Students say an addition sentence beginning with 2 (e.g., 2 + 1 = 3).

Application Problem (5 minutes)

A tree lost 8 leaves one day and 4 leaves the next. How many leaves did the tree lose at the end of the two days? Use a number bond, a number sentence, and a statement to match the story.

Extension: On the third day, the tree lost 6 leaves. How many leaves did it lose by the end of the third day?

Note: This problem revisits the idea of making ten when one addend is 8. It also challenges students to use addition, although the leaves are being lost.

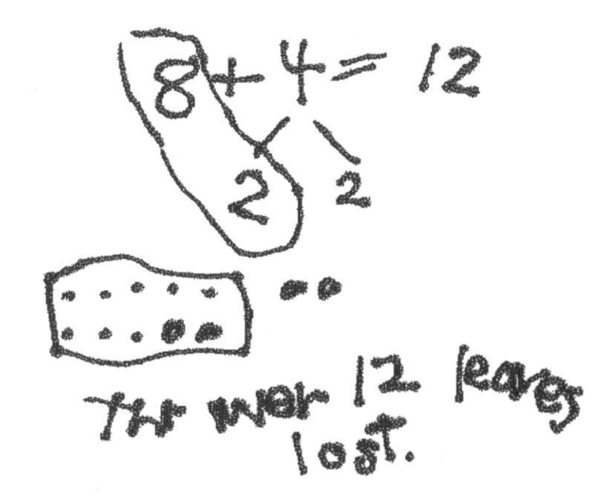

Concept Development (30 minutes)

Materials: (T) 10 blue and 10 yellow linking cubes, ten-frame border (S) Personal white board

Have students come to the meeting area with their personal white boards.

T: (Project and read aloud.) Amy wrote 8 letters to her friends. Peter wrote 3 to his friends. How many letters did they write? (Pause.) What is the expression to solve this story?

S: 8 + 3.

T: How many blue cubes do I need to represent the number of letters Amy wrote? How should I arrange it?

S: 8 cubes. Put them in a 5-group.

T: Why should I organize them in 5-group?

S: It's easy for everyone to see that there are 8 instead of counting the cubes.

T: With your partner, figure out how many letters Amy and Peter wrote. Use your personal white board to record your work.

S: (Discuss and solve the problem while the teacher circulates and listens.)

T: How many letters did Amy and Peter write?

S: 11 letters!

T: How did you solve the problem?

S: I counted on from 8. Eiiight, 9, 10, 11. → I put 2 cubes with the 8 blue ones and had 1 cube left. That made 11. → I broke apart the 3 into 2 and 1 to make 10 and 1.

T: Let's all try using this last strategy of making ten to solve this problem.

T: (Lay out 8 blue cubes.) How many yellow cubes do I need to represent the number of letters Peter wrote?

S: 3 cubes.

T: (Lay out 3 yellow cubes as a separate pile.) What should we do to add 8 and 3 efficiently?

S: Make ten!

T: How many does 8 need to make ten?

S: 2.

T: (Place 2 yellow cubes into a 5-group arrangement.)

T: Now that we have 10 here, we can put a frame around it. (Frame it.) Look at the new piles. What expression is 8 + 3 equal to?

S: 10 + 1.

T: Let's write a true number sentence using these expressions. (Write 8 + 3 = 10 + 1.)

T: What's 10 + 1?

S: 11.

T: (Write 10 + 1 = 11.) So, what is 8 + 3? Say the number sentence.

S: 8 + 3 = 11.

T: (Write 8 + 3 = 11.) How many letters did Amy and Peter write?

S: 11 letters.

T: Show me on your board how we solved 8 + 3. Remember, it's easy to show how we are solving 8 + 3 if we organize our math drawings just like the way we organized the cubes. Use empty circles to represent 8 and dark circles to represent 3. Don't forget to put a frame around the 10 cubes!

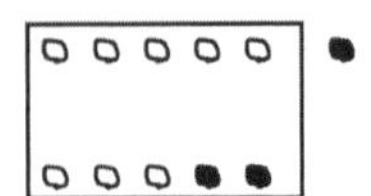

> 
>
> **NOTES ON MULTIPLE MEANS OF ENGAGEMENT:**
>
> Adjust lesson structure to suit specific learning needs remembering that some students may need to keep practicing with their linking cubes as they complete problems.

S: (Draw.)

T: Where is the 3 in your picture?

S: (Point to 2 and 1.)

T: You are pointing to two different places. Why?

S: We broke 3 apart into 2 and 1.

T: Let's use a number bond to show how we broke apart 3.

T: Just like we framed the ten in our picture, we'll frame the numbers that make ten. (Ring 8 and 2.)

T: 8 and 2 make...?

S: 10.

T: 10 and 1 make...?

S: 11.

T: So, 8 plus 3 equals...?

S: 11.

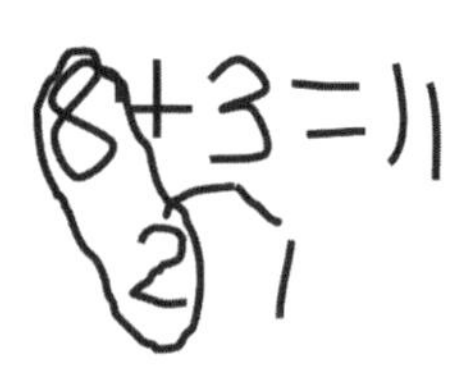

> 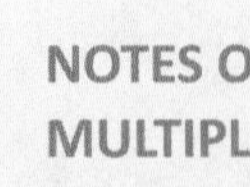
>
> **NOTES ON MULTIPLE MEANS OF ENGAGEMENT:**
>
> Offer opportunities for student leadership as "teacher." Have students demonstrate for the class how they are breaking apart and joining their linking cubes. Listen for accurate use of math vocabulary in their descriptions. Math language in this lesson includes *expression, organize, join, break apart,* and *frame*.

Repeat the process adding the numbers 4–9 in order as time allows, alternating 8 as the first and the second addend. For the first example, use linking cubes to illustrate what the math drawings should look like. For the remainder of the examples, move toward having students draw without the visual aid. Before students add dark circles to their math drawings, ask them, "How many does 8 need to make ten?" and "How many do you have when you take away 2 from [the other addend]?" to guide how they can decompose the addend when drawing.

Be sure to have students make ten with 8, reinforcing the concept of commutativity for efficient problem solving. Be sure that they also write two number sentences (8 + 6 = 14, 10 + 4 = 14) and the equivalent expression (8 + 6 = 10 + 4).

Problem Set (10 minutes)

Students should do their personal best to complete the Problem Set within the allotted 10 minutes. For some classes, it may be appropriate to modify the assignment by specifying which problems they work on first. Some problems do not specify a method for solving. Students should solve these problems using the RDW approach used for Application Problems.

Student Debrief (10 minutes)

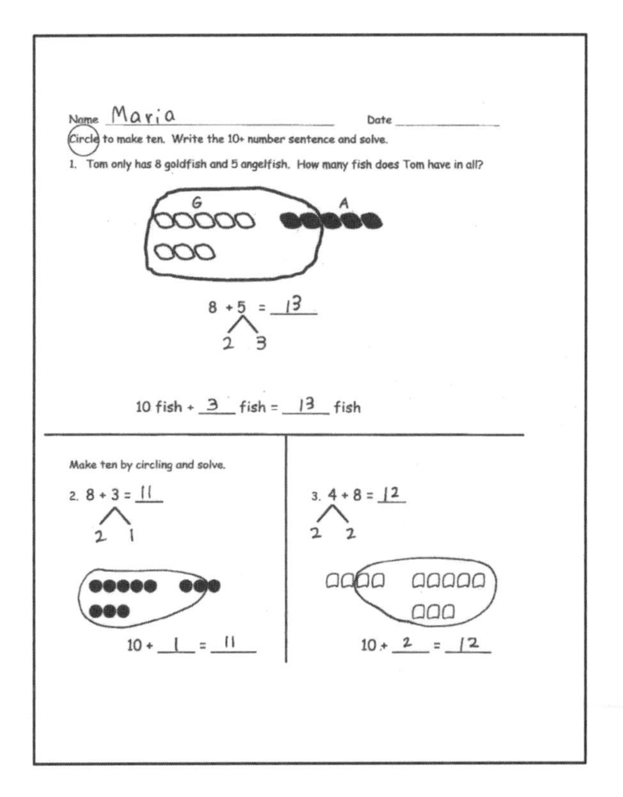

Lesson Objective: Make ten when one addend is 8.

Note: Distribute the student Problem Set from Lesson 4 for comparing with today's Problem Set.

The Student Debrief is intended to invite reflection and active processing of the total lesson experience.

Invite students to review their solutions for the Problem Set. They should check work by comparing answers with a partner before going over answers as a class. Look for misconceptions or misunderstandings that can be addressed in the Debrief. Guide students in a conversation to debrief the Problem Set and process the lesson.

Any combination of the questions below may be used to lead the discussion.

- Look at Problem 1 and Problem 6. How are your drawings different? Which drawing shows how you solved 8 + 5 more easily?
- What did you notice about having 8 as an addend? What happens to the other addend when it gets broken apart?
- How did Problem 6 help you solve Problem 7?
- Look at your Problem Set from a few days ago. What do you notice about the answers when you have 9 as an addend compared to 8 as an addend? Why do you think this is?

- How would you solve 8 + 9? Turn and talk to your partner. Explain your strategy.
- Why is it important to make our math drawings in an organized way?
- Look at your Application Problem. Draw an organized picture to show how you can solve this problem.

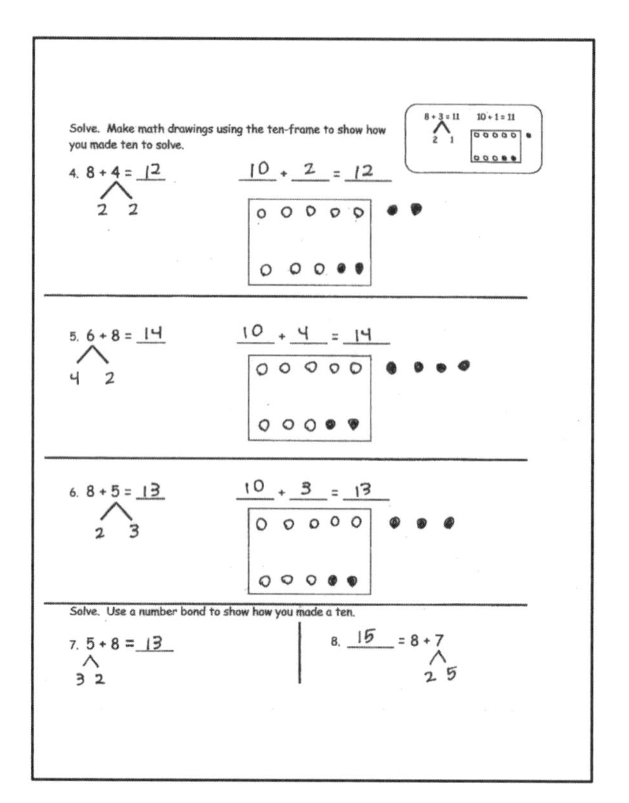
Solve. Make math drawings using the ten-frame to show how you made ten to solve.

8 + 3 = 11   10 + 1 = 11

4. 8 + 4 = 12   10 + 2 = 12

5. 6 + 8 = 14   10 + 4 = 14

6. 8 + 5 = 13   10 + 3 = 13

Solve. Use a number bond to show how you made a ten.

7. 5 + 8 = 13

8. 15 = 8 + 7

Exit Ticket (3 minutes)

After the Student Debrief, instruct students to complete the Exit Ticket. A review of their work will help with assessing students' understanding of the concepts that were presented in today's lesson and planning more effectively for future lessons. The questions may be read aloud to the students.

A

Number Correct:

Name ____________________ Date __________

*Write the missing number.

1.	9 + 1 = □		16.	9 + 5 = □	
2.	10 + 1 = □		17.	9 + 6 = □	
3.	9 + 2 = □		18.	6 + 9 = □	
4.	9 + 1 = □		19.	9 + 4 = □	
5.	10 + 2 = □		20.	4 + 9 = □	
6.	9 + 3 = □		21.	9 + 8 = □	
7.	9 + 1 = □		22.	9 + 9 = □	
8.	10 + 4 = □		23.	9 + □ = 18	
9.	9 + 5 = □		24.	□ + 6 = 15	
10.	9 + 1 = □		25.	□ + 6 = 16	
11.	10 + 6 = □		26.	13 = 9 + □	
12.	9 + 7 = □		27.	17 = 8 + □	
13.	9 + 1 = □		28.	10 + 2 = 9 + □	
14.	10 + 8 = □		29.	9 + 5 = 10 + □	
15.	9 + 9 = □		30.	□ + 7 = 8 + 9	

B

Number Correct:

Name ______________________ Date ____________

*Write the missing number.

1.	$9 + 1 = \square$		16.	$5 + 9 = \square$	
2.	$10 + 2 = \square$		17.	$6 + 9 = \square$	
3.	$9 + 3 = \square$		18.	$9 + 6 = \square$	
4.	$9 + 1 = \square$		19.	$9 + 7 = \square$	
5.	$10 + 1 = \square$		20.	$7 + 9 = \square$	
6.	$9 + 2 = \square$		21.	$9 + 8 = \square$	
7.	$9 + 1 = \square$		22.	$9 + 9 = \square$	
8.	$10 + 3 = \square$		23.	$9 + \square = 17$	
9.	$9 + 4 = \square$		24.	$\square + 5 = 14$	
10.	$9 + 1 = \square$		25.	$\square + 4 = 14$	
11.	$10 + 5 = \square$		26.	$15 = 9 + \square$	
12.	$9 + 6 = \square$		27.	$16 = 7 + \square$	
13.	$9 + 1 = \square$		28.	$10 + 4 = 9 + \square$	
14.	$10 + 4 = \square$		29.	$9 + 6 = 10 + \square$	
15.	$9 + 5 = \square$		30.	$\square + 6 = 7 + 9$	

Name ______________________ Date ____________

Circle to make ten. Write the 10+ number sentence and solve.

1. Tom only has 8 goldfish and 5 angelfish. How many fish does Tom have in all?

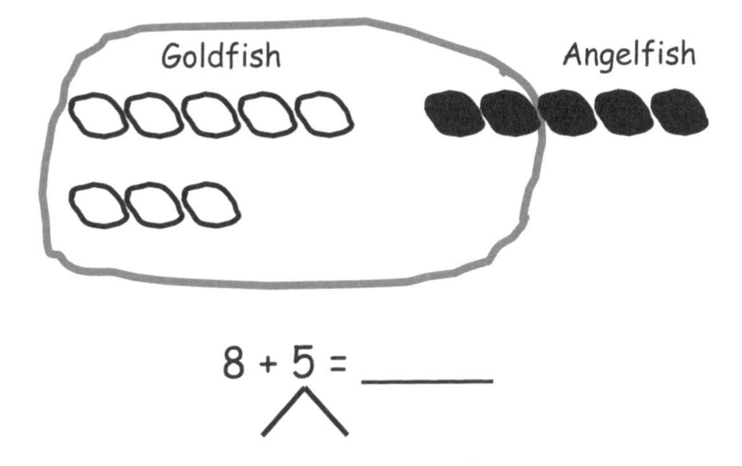

8 + 5 = _____

10 fish + _____ fish = _____ fish

Make ten by circling and solve.

2. 8 + 3 = ___

10 + _____ = _____

3. 4 + 8 = ___

10 + _____ = _____

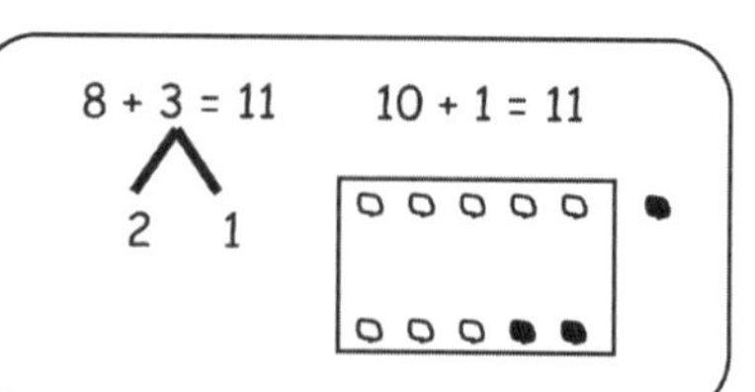

Solve. Make math drawings using the ten-frame to show how you made ten to solve.

4. 8 + 4 = ____ ____ + ____ = ____

5. 6 + 8 = ____ ____ + ____ = ____

6. 8 + 5 = ____ ____ + ____ = ____

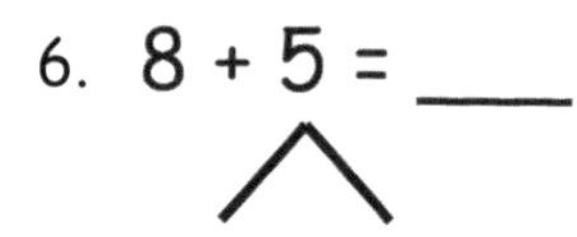

Solve. Use a number bond to show how you made a ten.

7. 5 + 8 = ____

8. ____ = 8 + 7

Name ______________________________ Date ______________

Make math drawings using the ten-frame to solve. Rewrite as a 10+ number sentence.

1. 6 + 8 = ____

10 + ____ = ____

2. ____ = 4 + 8

____ + ____ = ____

Name ______________________________ Date ______________

Solve. Make math drawings using the ten-frame to show how you made ten to solve.

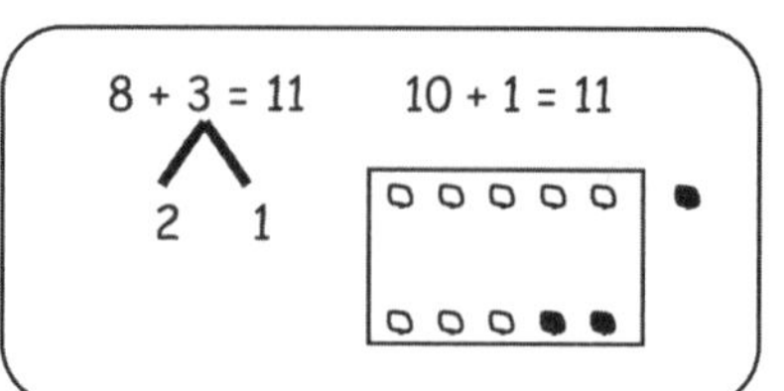

1. 8 + 4 = ___ ___ + ___ = ___

2. 8 + 6 = ___ ___ + ___ = ___

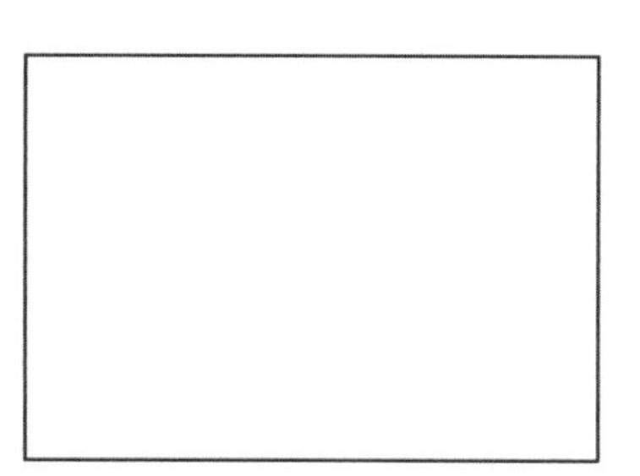

3. 7 + 8 = ___ ___ + ___ = ___

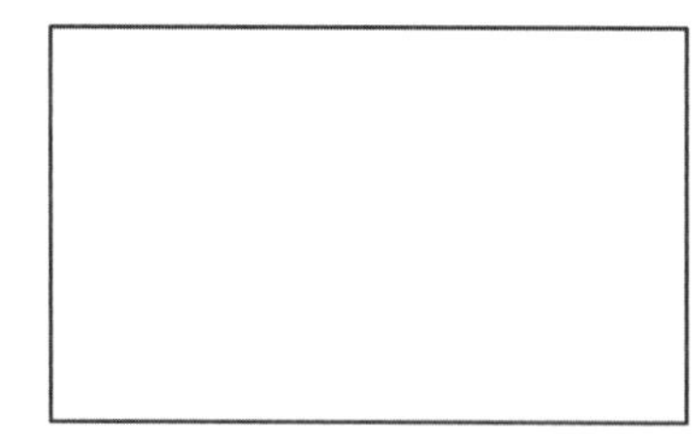

4. Make math drawings using ten-frames to solve. (Circle) the true number sentences.

Write an X to show number sentences that are not true.

a. 8 + 4 = 10 + 2

b. 10 + 6 = 8 + 8

c. 7 + 8 = 10 + 6

d. 5 + 10 = 5 + 8

e. 2 + 10 = 8 + 3

f. 8 + 9 = 10 + 7

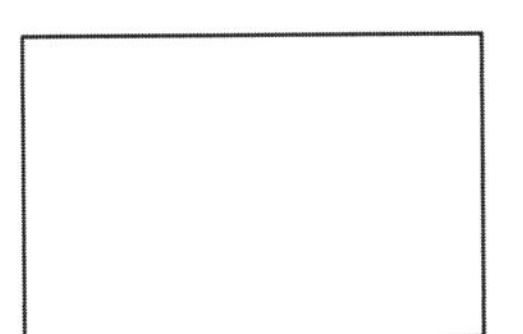

Lesson 9

Objective: Compare efficiency of counting on and making ten when one addend is 8.

Suggested Lesson Structure

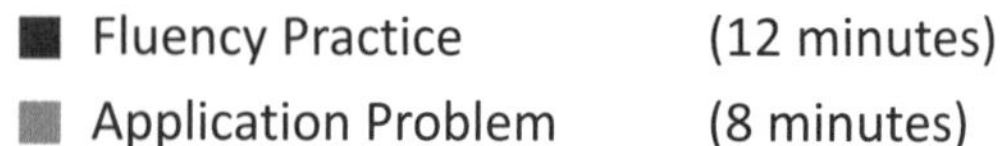

- Fluency Practice (12 minutes)
- Application Problem (8 minutes)
- Concept Development (30 minutes)
- Student Debrief (10 minutes)

Total Time (60 minutes)

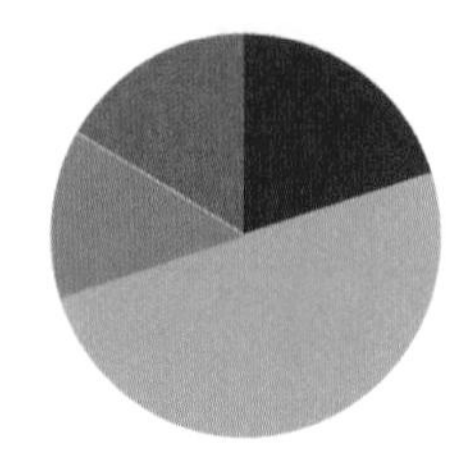

Fluency Practice (12 minutes)

- Decompose Addition Sentences into Three Parts **1.OA.6** (5 minutes)
- Cold Call: Break Apart Numbers **1.OA.6** (2 minutes)
- Make It Equal **1.OA.6** (5 minutes)

Decompose Addition Sentences into Three Parts (5 minutes)

Note: This fluency activity reviews adding three numbers and making ten when one addend is 8.

Decompose addition sentences into three addends that are more efficient to add.

T: (Write 8 + 3.) How many do we need from 3 to make ten?
S: 2.
T: Say 3 as an addition expression, starting with 2.
S: 2 + 1.
T: (Write 2 + 1 below the 3, showing the decomposition of 3.) Say 8 + 3 as a three-part addition sentence.
S: 8 + 2 + 1 = 11.

Repeat the process for other problems.

Cold Call: Break Apart Numbers (2 minutes)

Note: This is an anticipatory fluency activity for making ten when one addend is 7 since 7 needs 3 to make ten.

Say a number between 3 and 10. Tell students they are going to be cold called to say the number bond with 3 as a part. Alternate between calling on individual students, the whole class, and groups of students (e.g., only boys, only girls). Use the example below as a reference.

T: 4. (Pause to provide thinking time.) Everybody.
S: 3 and 1.
T: 6. (Pause.) Boys.
S: (Only boys.) 3 and 3.

Repeat with numbers 3 through 10.

Make It Equal (5 minutes)

Materials: (S) 5-group cards, one "=" card, and two "+" cards (Lesson 1 Fluency Template) per set of partners

Note: This activity reinforces the make ten addition strategy as students relate 10 + *n* addition sentences to an equivalent sentence with an addend of 8 or 9. Students ready to use the numeral side of the 5-group cards should be encouraged to do so.

Assign students partners of equal ability. Students arrange 5-group cards from 0 to 10, including the extra 5, and place the "=" card between them.

Write four numbers on the board (e.g., 10, 9, 1, and 2). Partners take the 5-group cards that match the numbers written to make two equivalent expressions (e.g., 10 + 1 = 9 + 2).

Suggested sequence: 10, 9, 1, 2; 10, 3, 9, 2; 10, 4, 5, 9; 10, 8, 1, 3; 10, 8, 4, 2.

Application Problem (8 minutes)

A squirrel found 8 nuts in the morning, 5 nuts in the afternoon, and 2 nuts in the evening. How many nuts did the squirrel find in all?

Extension: The next day, the squirrel found 3 more nuts in the morning, 1 more in the afternoon, and 1 more in the evening. How many did he collect over the two days?

Note: This problem uses three addends, revisiting the associative and commutative properties from earlier in this topic. During the Student Debrief, students who used making ten as a strategy to solve share their work, supporting students' development toward independent use of the strategy.

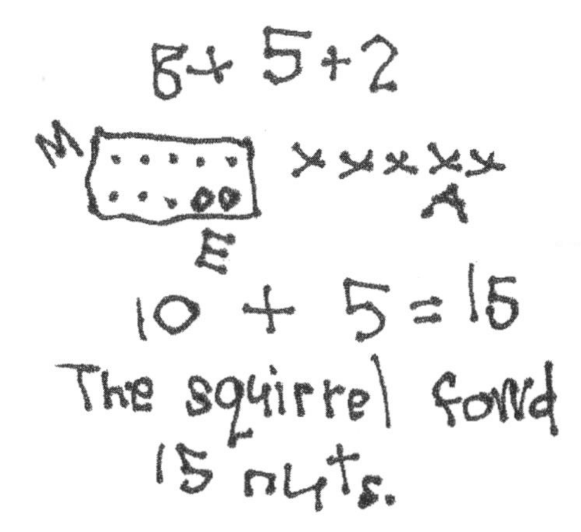

Concept Development (30 minutes)

Materials: (S) Personal white board

Have students sit at their desks or the meeting area with their personal white boards.

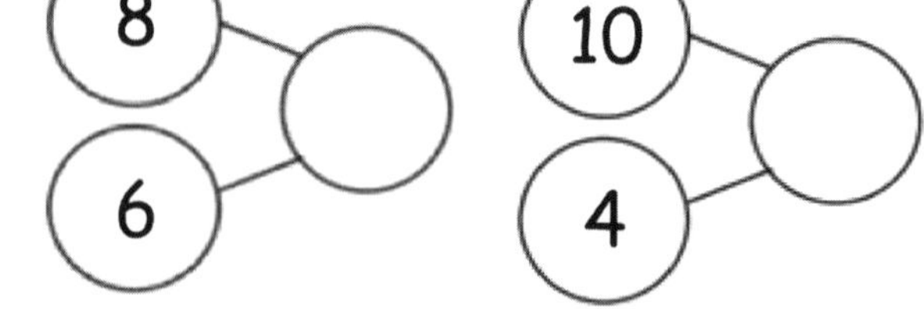

T: (Project or write the two number bonds shown here.) Which number bond are you able to solve faster?

S: 10 and 4.

T: (Write 10 + 4 = ____.) 10 + 4 is...?

S: 14.

T: (Record the solution.) How did you know that so quickly?

S: Because we know our 10+ facts. → Because 10 is a friendly number.

T: (Write 8 + 6 = ____.) Let's count on to solve 8 + 6.

T/S: Eeeiiiight, 9, 10, 11, 12, 13, 14. 14.

T: (Record the solution.)

T: (On another line, write 8 + 6 = ____.) What expression is equal to 8 + 6?

S: 10 + 4.

T: (Record this to make the true number sentence 8 + 6 = 10 + 4.) Use your personal white board to show how you can solve 8 + 6 by making ten to be sure that this is a true number sentence.

S: (Solve by making ten with 8, taking apart 6 into 2 and 4, etc.)

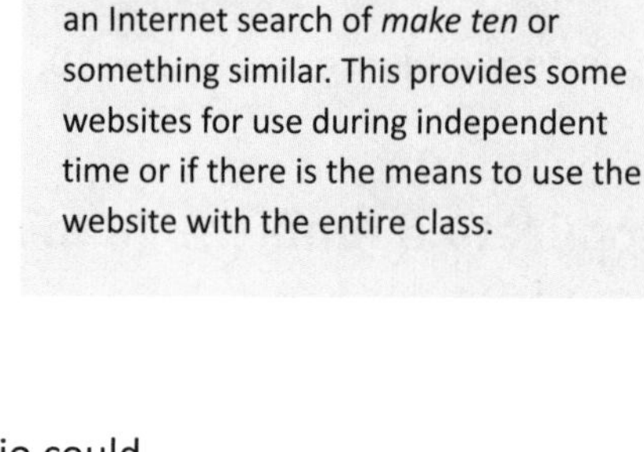

Students enjoy the use of interactive technology in the classroom. Do an Internet search of *make ten* or something similar. This provides some websites for use during independent time or if there is the means to use the website with the entire class.

T: (Read aloud.) Our friends Sergio and Lila are back again! They were getting ready to go to P.E. They both had to solve 8 + 7. The first one to solve it got to go to P.E. first! Sergio decided he was going to count on to solve it again. (Pause.) Was there another way to solve 8 + 7 that Sergio could have used?

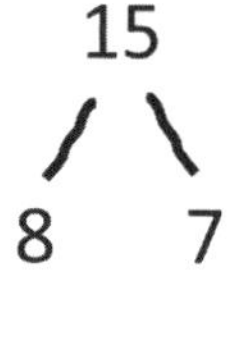

S: (Discuss as the teacher circulates and listens.) Make ten! → Take 2 from 7 to make ten with 8.

T: Some of you said that you would make ten. Well, that is just what Lila decided to do again. (Assign partners.) Partner A, explain to your partner how Sergio solved 8 + 7 by counting on. Partner B, explain to your partner how Lila solved 8 + 7 by making ten. Use your personal white board if it helps you share your thoughts.

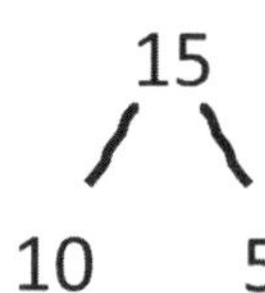

S: (Discuss as the teacher circulates and listens.)

T: Help me make a number bond to show what Sergio did. What were the parts that Sergio used?

S: 8 and 7. (Write the bond.)

T: What was the whole?

S: 15. (Complete the bond.)

T: Help me make a number bond to show what Lila did. What were the parts that Lila used?

S: 10 and 5. (Write the bond.)

T: What was the whole?

S: 15. (Complete the bond.)

T: Which number bond will help you solve more efficiently?

S: 10 and 5.

T: So, based on these number bonds and the work you and your partner just did, who do you think got to go to P.E. first?

S: Lila!

T: Again, you're right! Since Lila really knows how to use the make ten strategy, she was able to solve for the unknown very quickly or efficiently. Sometimes it takes practice before we can use a strategy quickly. When a strategy is new to us, it can take longer for us to use it until we get better at it. Let's keep practicing.

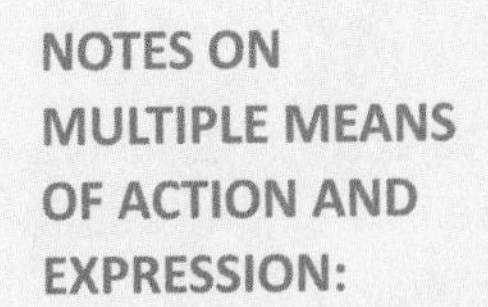

NOTES ON MULTIPLE MEANS OF ACTION AND EXPRESSION:

While encouraging students to use the most efficient strategy when solving number sentences, some may be able to use different number combinations as efficiently. For example, some might see $7 + 8$ as $7 + 7 + 1$ or $8 + 8 - 1$, and some may already know $7 + 8$ but benefit from the strategy discussion. Use this opportunity to show students how everyone thinks differently. Have students communicate their mathematical thinking to the class.

Continue with partners solving each problem, showing how to solve using counting on and making ten. The following is a suggested sequence of problems: $8 + 5$, $8 + 4$, $8 + 8$, $8 + 3$ (counting on may actually be more efficient here), and $8 + 9$.

Problem Set (10 minutes)

Students should do their personal best to complete the Problem Set within the allotted 10 minutes. For some classes, it may be appropriate to modify the assignment by specifying which problems they work on first. Some problems do not specify a method for solving. Students should solve these problems using the RDW approach used for Application Problems.

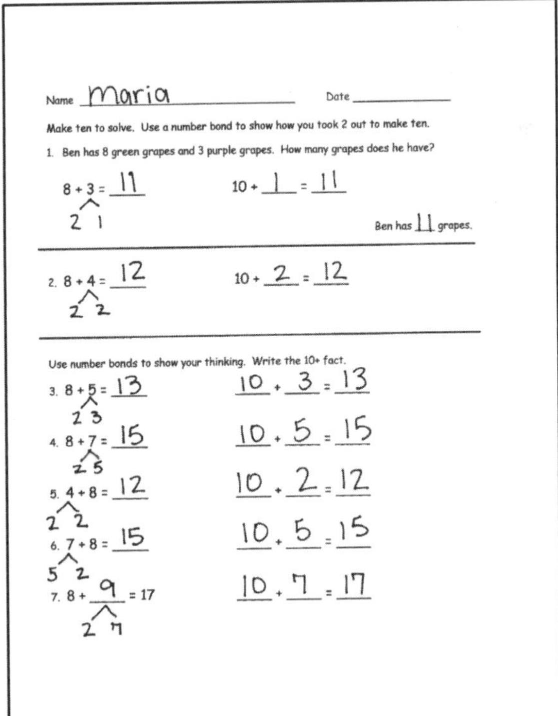

Name Maria Date

Make ten to solve. Use a number bond to show how you took 2 out to make ten.

1. Ben has 8 green grapes and 3 purple grapes. How many grapes does he have?

8 + 3 = 11 (2 1) 10 + 1 = 11

Ben has 11 grapes.

2. 8 + 4 = 12 (2 2) 10 + 2 = 12

Use number bonds to show your thinking. Write the 10+ fact.

3. 8 + 5 = 13 (2 3) 10 + 3 = 13

4. 8 + 7 = 15 (2 5) 10 + 5 = 15

5. 4 + 8 = 12 (2 2) 10 + 2 = 12

6. 7 + 8 = 15 (5 2) 10 + 5 = 15

7. 8 + 9 = 17 (2 7) 10 + 7 = 17

Student Debrief (10 minutes)

Lesson Objective: Compare efficiency of counting on and making ten when one addend is 8.

Note: Distribute the student Problem Set from Lesson 5 for comparing with today's Problem Set.

The Student Debrief is intended to invite reflection and active processing of the total lesson experience.

Invite students to review their solutions for the Problem Set. They should check work by comparing answers with a partner before going over answers as a class. Look for misconceptions or misunderstandings that can be addressed in the Debrief. Guide students in a conversation to debrief the Problem Set and process the lesson.

Any combination of the questions below may be used to lead the discussion.

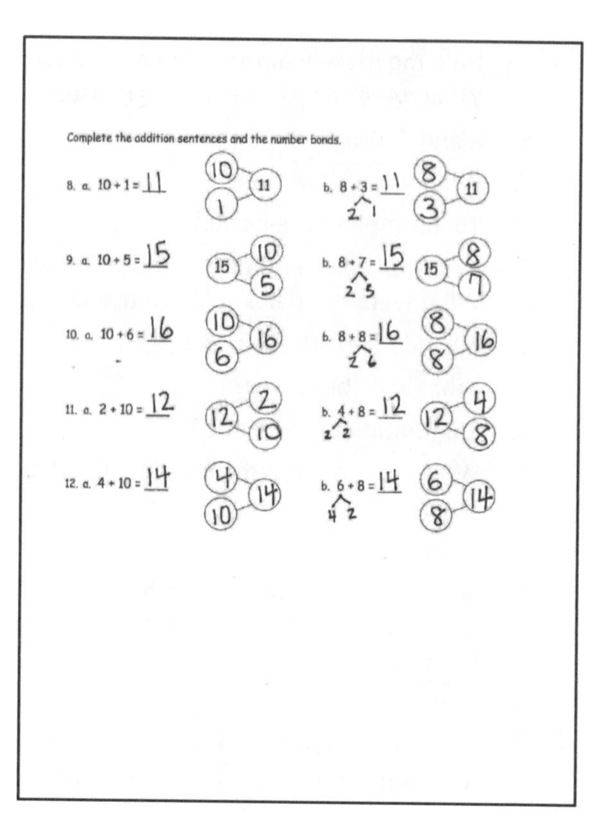
Complete the addition sentences and the number bonds.

8. a. 10 + 1 = 11 b. 8 + 3 = 11
9. a. 10 + 5 = 15 b. 8 + 7 = 15
10. a. 10 + 6 = 16 b. 8 + 8 = 16
11. a. 2 + 10 = 12 b. 4 + 8 = 12
12. a. 4 + 10 = 14 b. 6 + 8 = 14

- Look at Problem 1 and Problem 2. How are your bonds different? How can Problem 1 help you solve Problem 2?
- Look at Problems 8–10. What do you notice about the number bonds? How does knowing your 10+ facts help you with your 8+ facts?
- Look at Problem 5 and Problem 8. Do you think counting on or making ten was more efficient to solve these? Why?
- Look at your Problem Set from a few days ago. What do you notice about the answers when you have 9 as an addend compared to 8 as an addend? Why do you think this is?
- Look at your Application Problem. Would counting on or making ten help you solve this problem most efficiently? If you used making ten to solve this, share your work, and explain your thinking.
- One first grader I know makes ten for some of her 8+ facts and counts on to solve others. Sometimes she just knows the solution. Is that true for any of you? Which 8+ facts do you use a particular strategy to help you solve? Why?

Exit Ticket (3 minutes)

After the Student Debrief, instruct students to complete the Exit Ticket. A review of their work will help with assessing students' understanding of the concepts that were presented in today's lesson and planning more effectively for future lessons. The questions may be read aloud to the students.

Name ______________________________ Date ______________

Make ten to solve. Use a number bond to show how you took 2 out to make ten.

1. Ben has 8 green grapes and 3 purple grapes. How many grapes does he have?

8 + 3 = _____ 10 + _____ = _____

Ben has ___ grapes.

---

2. 8 + 4 = _____ 10 + _____ = _____

Use number bonds to show your thinking. Write the 10+ fact.

3. 8 + 5 = _____ _____ + _____ = _____

4. 8 + 7 = _____ _____ + _____ = _____

5. 4 + 8 = _____ _____ + _____ = _____

6. 7 + 8 = _____ _____ + _____ = _____

7. 8 + _____ = 17 _____ + _____ = _____

Complete the addition sentences and number bonds.

8. a. 10 + 1 = ____

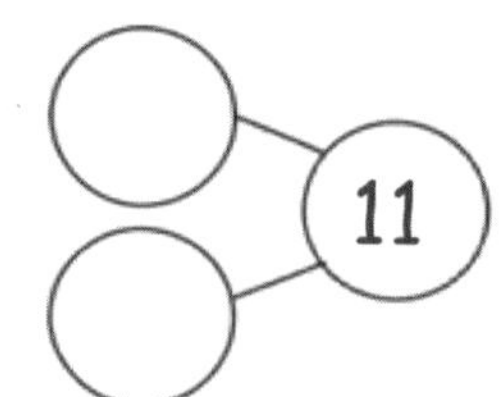

b. 8 + 3 = ____

11

9. a. 10 + 5 = ____

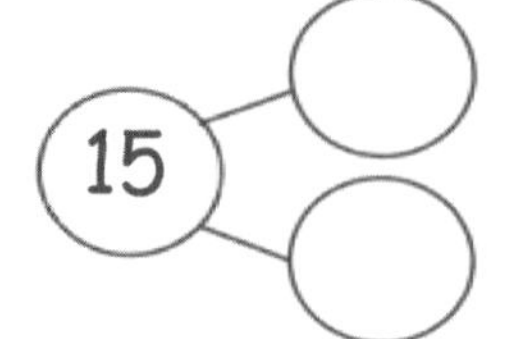

b. 8 + 7 = ____

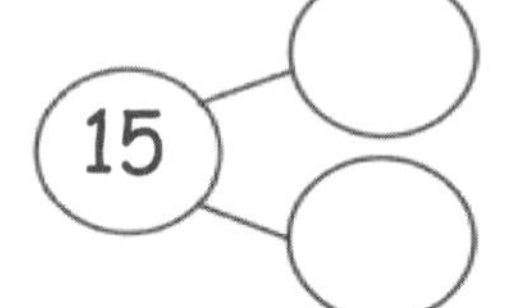

10. a. 10 + 6 = ____

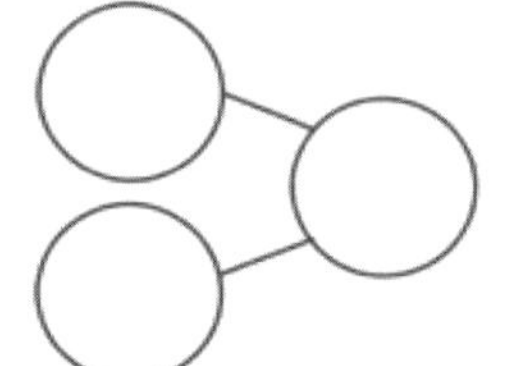

b. 8 + 8 = ____

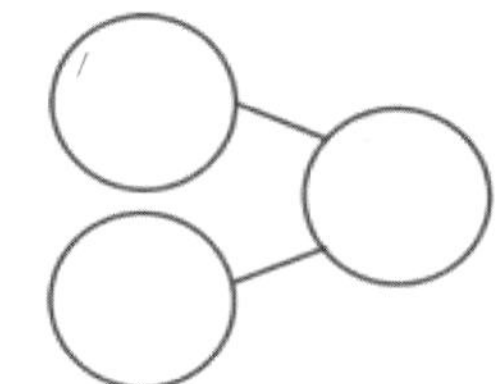

11. a. 2 + 10 = ____

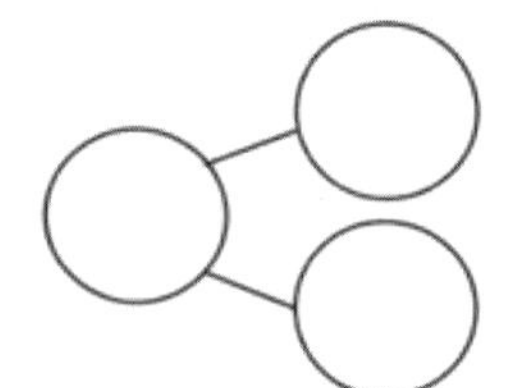

b. 4 + 8 = ____

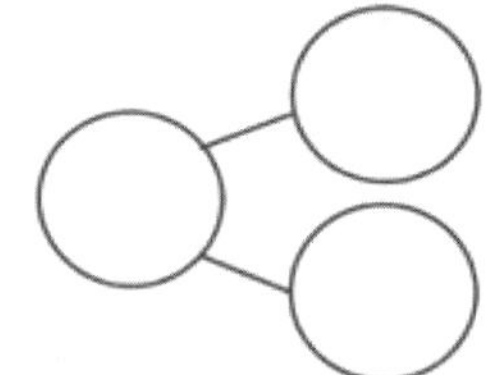

12. a. 4 + 10 = ____

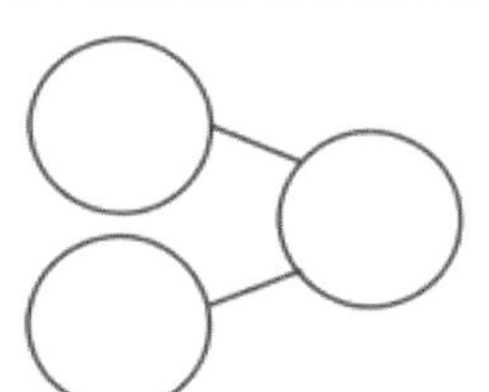

b. 6 + 8 = ____

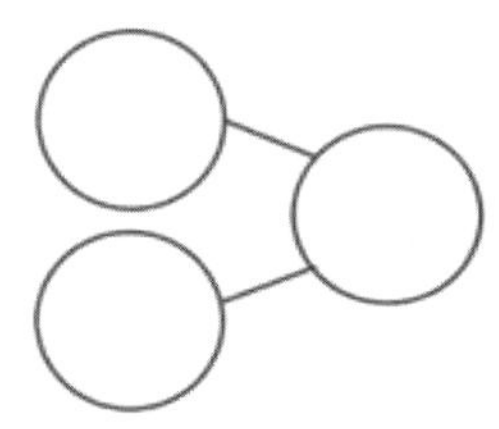

Name ______________________________ Date ______________

1. Seyla has 3 stamps in her collection. Her father gives her 8 more stamps. How many stamps does she have now? Show how you make ten, and write the 10+ fact.

$3 + 8 =$ _____ $10 +$ _____ $=$ _____

2. Complete the addition sentences and the number bonds.

a. $8 + 6 =$ ____

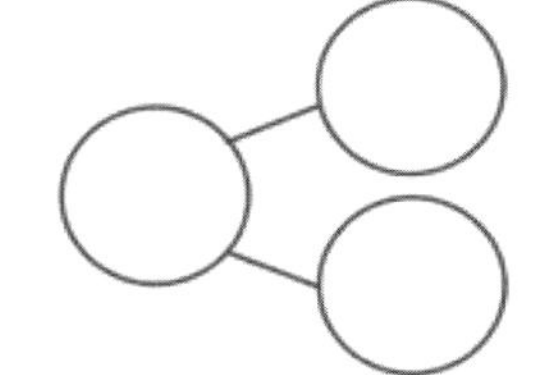

b. $10 +$ ____ $= 14$

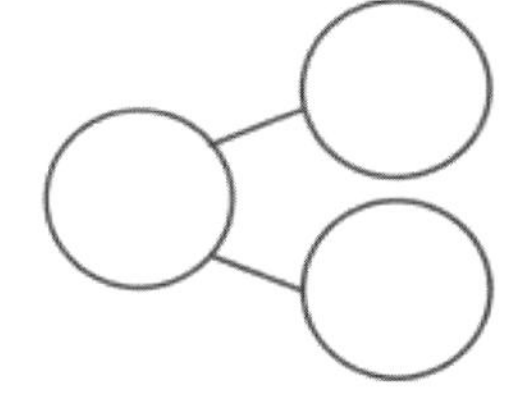

Name ______________________________ Date ______________

Use number bonds to show your thinking. Write the 10+ fact.

1. 8 + 3 = ____ 10 + ____ = ____

2. 6 + 8 = ____ ____ + 10 = ____

3. ____ = 8 + 8 ____ = 10 + ____

4. ____ = 5 + 8 ____ = 10 + ____

Complete the addition sentences and the number bonds.

5. a. 7 + 8 = ____

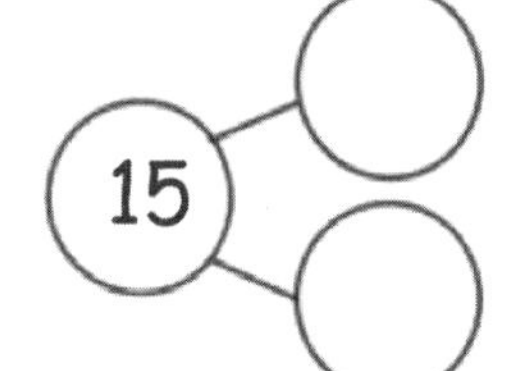

b. 10 + 5 = ____

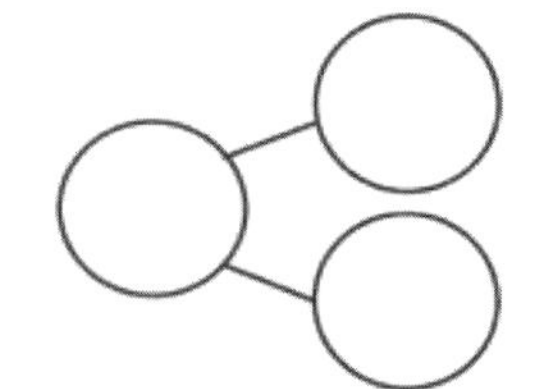

6. a. 16 = ____ + 8

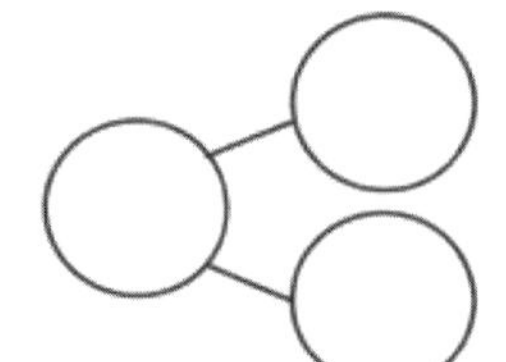

b. 10 + 6 = ____

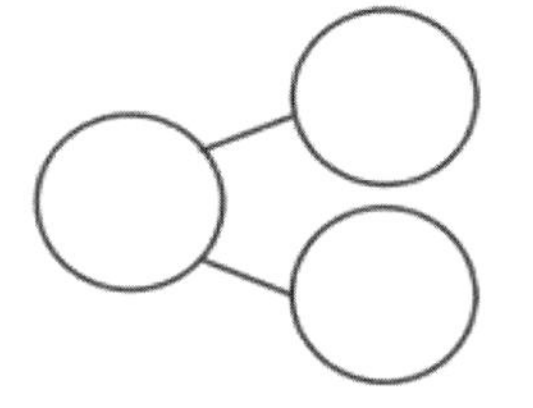

7. a. ____ = 9 + 8

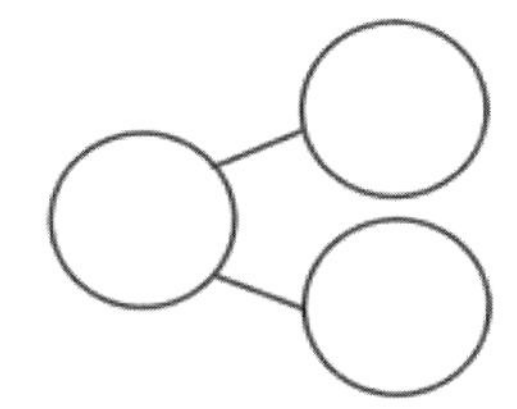

b. 10 + 7 = ____

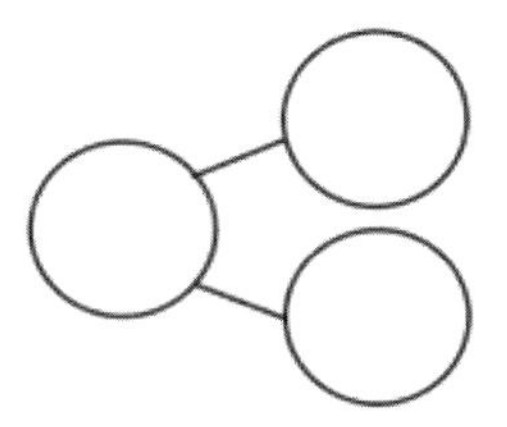

Draw a line to the matching number sentence. You may use a number bond or 5-group drawing to help you.

8. 11 = 8 + 3

8 + 6 = 14

9. Lisa had 5 red rocks and 8 white rocks.
   How many rocks did she have?

10 + 1 = 11

13 = 10 + 3

10.

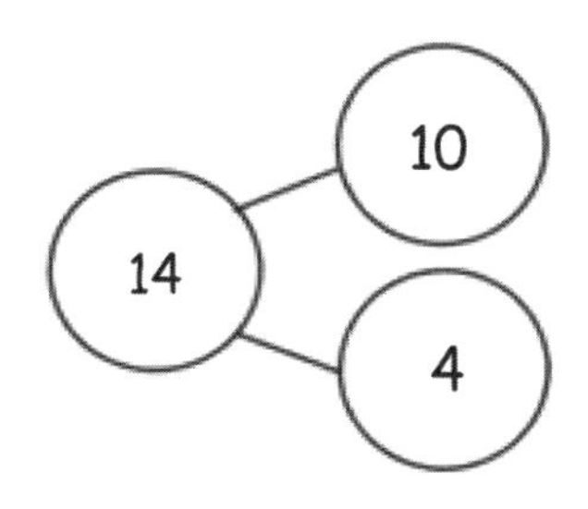

Lesson 10

Objective: Solve problems with addends of 7, 8, and 9.

Suggested Lesson Structure

■ Fluency Practice	(10 minutes)
■ Application Problem	(6 minutes)
■ Concept Development	(30 minutes)
■ Student Debrief	(14 minutes)
Total Time	**(60 minutes)**

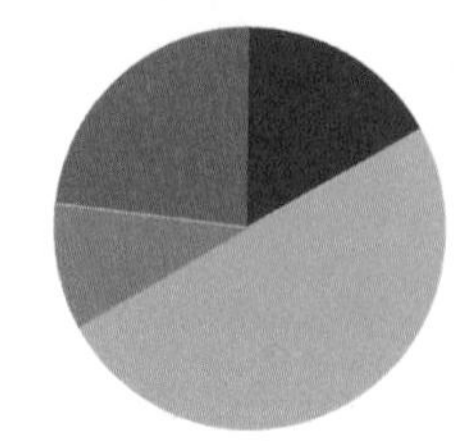

Fluency Practice (10 minutes)

- 1, 2, and 3 Less **1.OA.6** (3 minutes)
- Decomposing Addition Sentences **1.OA.6** (5 minutes)
- Happy Counting by Threes **1.OA.5** (2 minutes)

1, 2, and 3 Less (3 minutes)

Note: This fluency activity prepares students for today's lesson as students decompose numbers to make ten with addends of 7, 8, and 9.

T: On my signal, say the number that is 1 less.
T: 3.
S: 2.

Continue with all numbers within 10. Then repeat with 2 less and 3 less.

Decomposing Addition Sentences (5 minutes)

Note: This activity reviews how to decompose numbers to make ten, creating equivalent but easier number sentences.

Write number sentences on the board to model how to decompose number sentences into three addends.

9 + 5 = 14
9 + 1 + 4 = 14

T: (Write 9 + 5 = ___ on the board.) What does 9 need to make ten?
S: 1.
T: (Write 9 + 1 below 9 + 5 = ___.)
T: (Point to the 5.) If we take 1 from 5 to make ten, what part is left?
S: 4.

T: (Add *+ 4* after 9 + 1.) Say the number sentence with the answer.

S: 9 + 1 + 4 = 14.

T: (Write 14 to complete 9 + 1 + 4 = ____.) 9 + 1 + 4 = 14. 9 + 5 is...?

S: 14.

T: (Write 14 to complete 9 + 5 = ____.)

Continue with other 9 + *n* and 8 + *n* addition sentences. If students are ready, have them use their boards to independently decompose addition sentences into three parts.

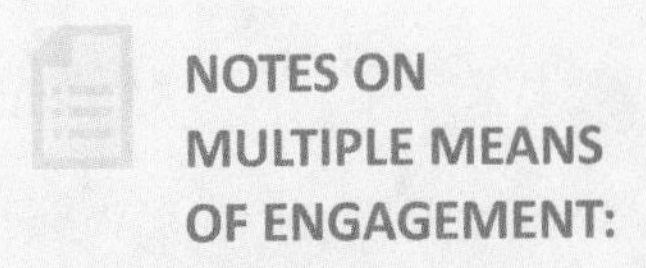

NOTES ON MULTIPLE MEANS OF ENGAGEMENT:

Maintain students' attention with short fluency games that are energetically presented.

Happy Counting by Threes (2 minutes)

Note: Review of counting on and back allows students to maintain fluency with adding and subtracting 3.

Repeat the Happy Counting activity from Lesson 4, counting by threes from 0 to 12 and back.

Application Problem (6 minutes)

There were 4 boots by the classroom door, 8 boots in the hallway, and 6 boots by the teacher's desk. How many boots were there altogether?

Extension: How many pairs of boots were there in all?

Note: In this problem, the numbers 4, 8, and 6 are used as addends. To solve, students may choose to make ten by adding (4 + 6) + 8, or they may choose to decompose either the 4 or 6 to make ten with 8. During the Student Debrief, students have the opportunity to share work and notice how peers are using Level 3 strategies such as making ten to solve.

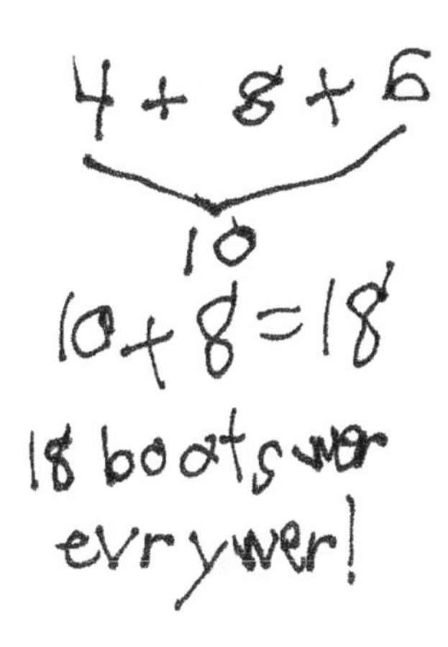

Concept Development (30 minutes)

Materials: (S) Personal white board, numeral cards or 5-group cards, one "+" card for each student, and one "=" card for each pair of students (Lesson 1 Fluency Template)

Have students come to the meeting area with their personal white boards and sit in a semicircle.

T: (Write 9 + 6 = ____ on the board.) Using an organized math drawing or a number bond, solve 9 + 6. Think about the equal ten-plus fact, and write a true number sentence using two expressions.

S: (Solve by drawing or using a number bond as the teacher circulates.)

T: (Choose one student to share the use of counting on and another student to share the use of making ten.) When there is a 9 as an addend, what could you do to the other addend?

S: Get the 1 out! Break apart 6 into 1 and 5 as parts.

Repeat the process with 4 + 8. Begin by asking students which number they should make ten with to solve more efficiently.

T: (Write 7 + 6 = ____ on the board.) Turn and talk to your artner. How might you solve this problem using what you already know about the make ten strategy?

T: Which number should we make ten with? Why?

S: Make ten with 7 because it's only 3 away from 10. → 6 is 4 away from 10. → It's easier for me to get the missing part from 7 than 6.

T: With your partner, use a number bond to solve this problem.

T: Look at your picture. What expression is 7 + 6 the same as?

S: 10 + 3.

T: Write that as a true number sentence.

S: (Write 7 + 6 = 10 + 3 or 10 + 3 = 7 + 6.)

T: What is 10 + 3?

S: 13.

T: So, what is 7 + 6? Say the number sentence.

S: 7 + 6 = 13.

NOTES ON MULTIPLE MEANS OF ACTION AND EXPRESSION:

Before sharing as a class, have students share their strategies with a partner. Hearing how problems were solved more than once helps those students who are still learning the process and English language learners.

Repeat the process with 4 + 7, 7 + 5, and 7 + 7.

MP.8

T: When 7 is the bigger addend, what could you do to the other addend?

S: Get the 3 out! → Make 3 as a part.

T: Now, we are going to play Simple Strategies! (Assign partners. Instruct each pair to combine their numeral cards and make two piles: digits 1–6 and digits 7–9, placing the 9 card on top of the second pile.) Here's how you play:

1. Partner A picks a card from the first pile (digits 1–6).
2. Using the 9 card from the second pile and the card picked by Partner A, Partner B writes an addition expression (e.g., 6 + 9).
3. Partners use counting on and then use making ten to solve the expression.
4. After using the make ten strategy, Partner A writes down the equal 10 + ___ fact.
5. Partners place the equal sign card between the boards to make a true number sentence.
6. Switch roles. Keep the 9 card up each time you begin a new expression.

As students play, the teacher circulates and moves students to replacing the 9 card with the 8 card and then the 7 card, as appropriate.

Problem Set (10 minutes)

Students should do their personal best to complete the Problem Set within the allotted 10 minutes. For some classes, it may be appropriate to modify the assignment by specifying which problems they work on first. Some problems do not specify a method for solving. Students should solve these problems using the RDW approach used for Application Problems.

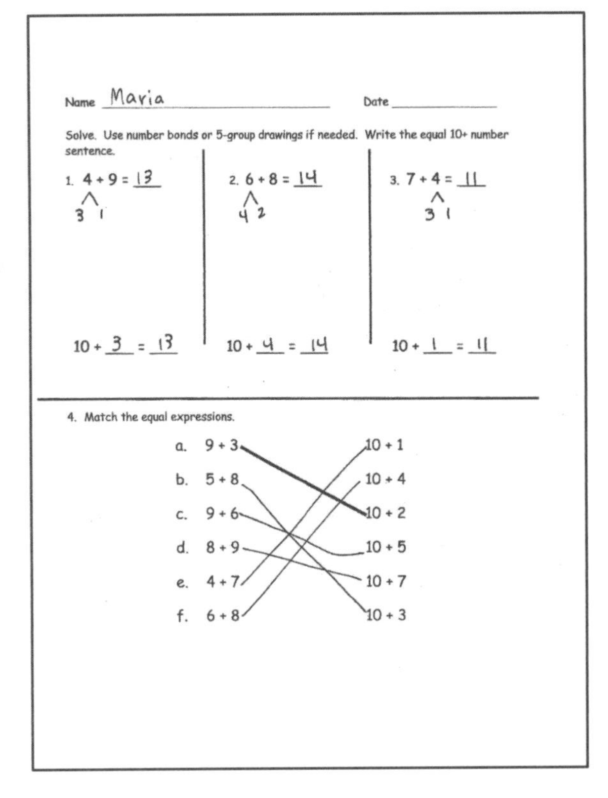
Name Maria Date

Solve. Use number bonds or 5-group drawings if needed. Write the equal 10+ number sentence.

1. 4 + 9 = 13 (3, 1) 10 + 3 = 13
2. 6 + 8 = 14 (4, 2) 10 + 4 = 14
3. 7 + 4 = 11 (3, 1) 10 + 1 = 11

4. Match the equal expressions.

a. 9 + 3 — 10 + 2
b. 5 + 8 — 10 + 3
c. 9 + 6 — 10 + 5
d. 8 + 9 — 10 + 7
e. 4 + 7 — 10 + 1
f. 6 + 8 — 10 + 4

Student Debrief (14 minutes)

Lesson Objective: Solve problems with addends of 7, 8, and 9.

The Student Debrief is intended to invite reflection and active processing of the total lesson experience.

Invite students to review their solutions for the Problem Set. They should check work by comparing answers with a partner before going over answers as a class. Look for misconceptions or misunderstandings that can be addressed in the Debrief. Guide students in a conversation to debrief the Problem Set and process the lesson.

Any combination of the questions below may be used to lead the discussion.

- Look at Problems 8–10. Can you find number sentences that have the same total? What are the number sentences? How are they related?
- Why is it efficient to start with a larger addend when you add? Give an example.
- Solve 9 + 6 = ___, 8 + 6 = ___, 7 + 6 = ___. What patterns do you notice? Look at how you broke apart the second addend. What patterns do you see there? How did this breaking apart affect your totals? (When you take out 1 more from the second addend, your total is 1 less.)
- Which is easiest for you to use? Counting on, making ten, or just knowing? Why?

Complete the addition sentences to make them true.

	(A)	(B)	(C)
5.	9 + 2 = 11	8 + 4 = 12	7 + 5 = 12
6.	9 + 5 = 14	8 + 3 = 11	7 + 6 = 13
7.	6 + 9 = 15	6 + 8 = 14	4 + 7 = 11
8.	7 + 9 = 16	5 + 8 = 13	7 + 7 = 14
9.	9 + 8 = 17	8 + 8 = 16	7 + 9 = 16
10.	6 + 9 = 15	7 + 8 = 15	10 + 7 = 17

Exit Ticket (3 minutes)

After the Student Debrief, instruct students to complete the Exit Ticket. A review of their work will help with assessing students' understanding of the concepts that were presented in today's lesson and planning more effectively for future lessons. The questions may be read aloud to the students.

Name ________________________ Date ____________

Solve. Use number bonds or 5-group drawings if needed. Write the equal ten-plus number sentence.

1. 4 + 9 = ____

10 + ____ = ____

2. 6 + 8 = ____

10 + ____ = ____

3. 7 + 4 = ____

10 + ____ = ____

4. Match the equal expressions.

a. 9 + 3	10 + 1
b. 5 + 8	10 + 4
c. 9 + 6	10 + 2
d. 8 + 9	10 + 5
e. 4 + 7	10 + 7
f. 6 + 8	10 + 3

Complete the addition sentences to make them true.

	a.	b.	c.
5.	9 + 2 = ____	8 + 4 = ____	7 + 5 = ____
6.	9 + 5 = ____	8 + 3 = ____	7 + 6 = ____
7.	6 + 9 = ____	6 + 8 = ____	4 + 7 = ____
8.	7 + 9 = ____	5 + 8 = ____	7 + 7 = ____
9.	9 + ____ = 17	8 + ____ = 16	7 + ____ = 16
10.	____ + 9 = 15	____ + 8 = 15	____ + 7 = 17

Name ______________________________ Date ______________

Solve. Use number bonds or 5-group drawings if needed. Write the equal ten-plus number sentence.

a.

9 + 5 = ___

10 + ___ = ___

b.

8 + 4 = ___

10 + ___ = ___

c.

7 + 6 = ___

10 + ___ = ___

Name ______________________ Date ____________

Solve. Match the number sentence to the ten-plus number bond that helped you solve the problem. Write the ten-plus number sentence.

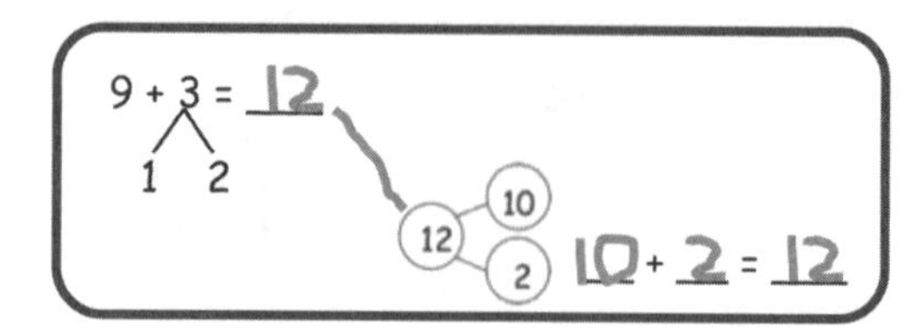

1. 8 + 6 = ___ (11: 10, 1) ___ + ___ = ____

2. 7 + 5 = ___ (15: 10, 5) ___ + ___ = ____

3. 5 + 8 = ___ (12: 10, 2) ___ + ___ = ____

4. 4 + 7 = ___ (14: 10, 4) ___ + ___ = ____

5. 6 + 9 = ___ (13: 10, 3) ___ + ___ = ____

Complete the number sentences so they equal the given number bond.

6.

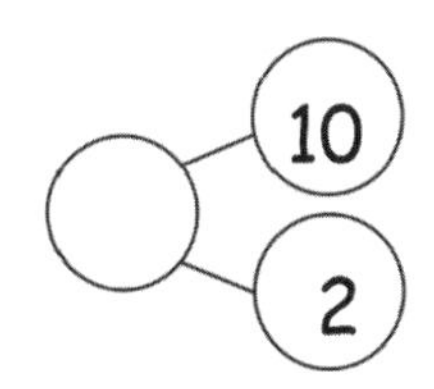

9 + ___ = 12

8 + ___ = 12

7 + ___ = 12

7.

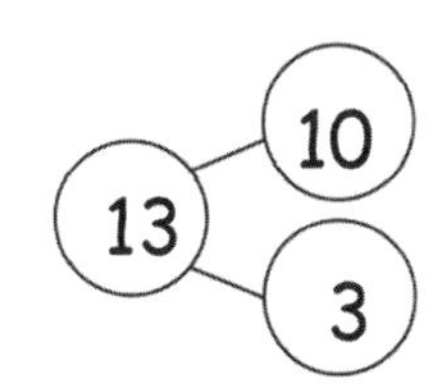

9 + ___ = 13

8 + ___ = 13

7 + ___ = 13

8.

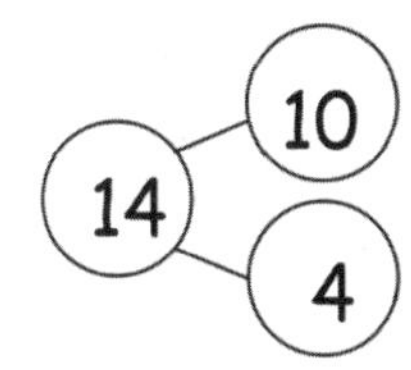

9 + ___ = 14

8 + ___ = 14

7 + ___ = 14

9.

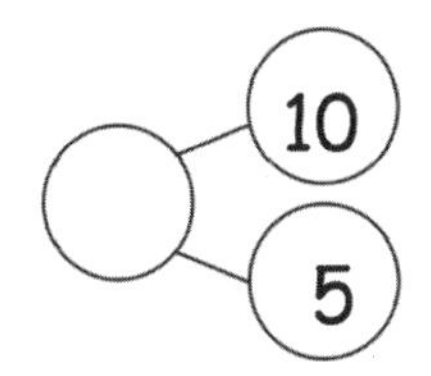

15 = 9 + ___

___ = 8 + ___

___ = 7 + ___

10.

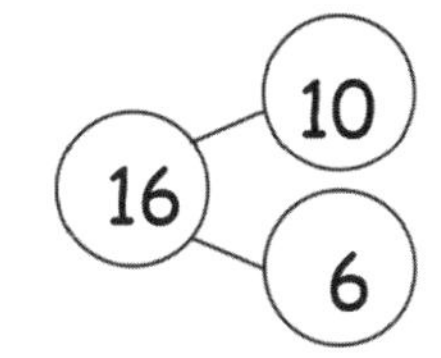

16 = 9 + ___

___ = 8 + ___

7 + ___ = ___

11.

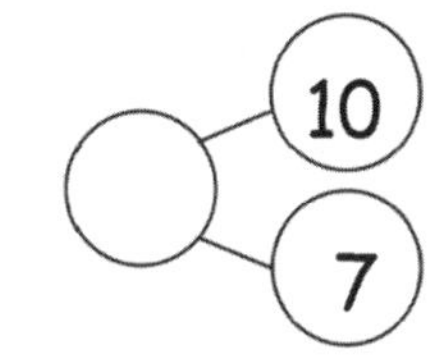

___ = 9 + 8

___ = 8 + ___

___ = 7 + ___

Lesson 11

Objective: Share and critique peer solution strategies for *put together with total unknown* word problems.

Suggested Lesson Structure

■ Fluency Practice	(13 minutes)
■ Application Problem	(6 minutes)
■ Concept Development	(31 minutes)
■ Student Debrief	(10 minutes)
Total Time	**(60 minutes)**

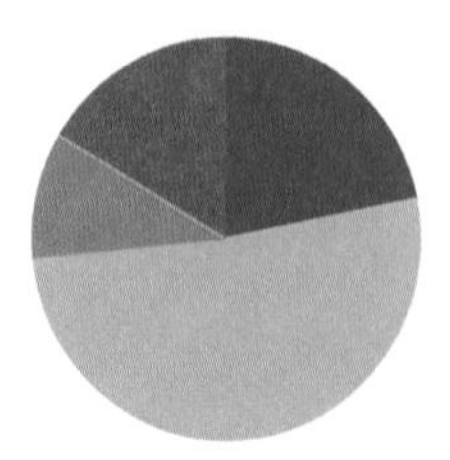

Fluency Practice (13 minutes)

- Sprint: Adding Across Ten **1.OA.6** (10 minutes)
- Rekenrek: Ten Less **1.NBT.5** (3 minutes)

Sprint: Adding Across Ten (10 minutes)

Materials: (S) Sprint: Adding Across Ten

Note: This Sprint reviews the make ten addition strategy.

Rekenrek: Ten Less (3 minutes)

Materials: (T) Rekenrek

Note: This is an anticipatory fluency activity for the take from ten subtraction strategy in Topic B where students need to decompose numbers by taking out a ten.

T: (Show 14 on the Rekenrek.) Say the number.
S: 14.
T: Say it the Say Ten way.
S: Ten 4.
T: What will my number be if I take out ten ones?
S: 4.
T: Let's check. (Take out ten.) Yes!

Continue with other teen numbers.

Application Problem (6 minutes)

Nicholas bought 9 green apples and 7 red apples. Sofia bought 10 red apples and 6 green apples. Sofia thinks she has more apples than Nicholas. Is she right? Choose a strategy you have learned to show your work. Then, write number sentences to show how many apples Nicholas and Sofia each have.

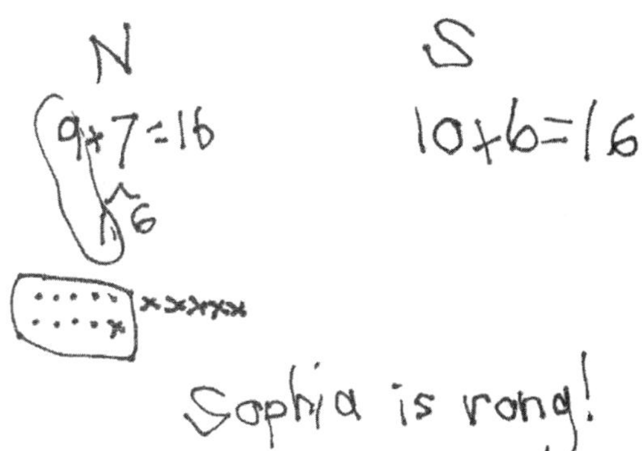

Note: This problem allows students to revisit equivalent expressions as they work with 9 + 7 and 10 + 6. The teacher can extend this thinking by either showing 9 + 7 = 10 + 6 or having students write the true number sentence themselves and then asking students to explain how they know.

Concept Development (31 minutes)

Materials: (T) Student work samples: make ten strategies (Template) (S) Personal white board

Have students come to the meeting area and sit in a semicircle.

T: (Project and read.) Louie made 7 puppets out of paper bags. Roberto made 6 puppets out of socks. How many puppets did the boys make? (Pause.) Turn and talk to your partner about how you would solve this problem.

S: (Discuss as the teacher circulates and listens.)

T: (Project the Student A sample.) How did Student A solve this problem? Explain to your partner what this student was thinking.

S: She counted all the circles starting with 1.
→ Maybe she used counting on. Seeeven, 8, 9, 10, 11, 12, 13.

T: (Project the Student B sample.) How did Student B solve this problem? Can you explain his thinking? Turn and talk to your partner.

S: He drew his shapes in 5-groups. When he made ten starting with 7, he drew a frame around it, so you can see 10 and 3. His strategy was to make ten from 7 by breaking 6 into 3 and 3.

T: (Project the Student C sample.) How did Student C solve this problem? How is it similar and different from Student B's work?

S: She didn't need to make a picture. She used the make ten strategy. But instead of making ten with 7, she made ten with 6 and broke apart 7 into 4 and 3.

T: (Project the Student D sample.) How did Student D solve the problem?

S: He drew a picture, but it's a little hard to count because the shapes are not organized. He probably had to count all of them, starting with 1. Or maybe he counted on from 7. Seeeven, 8, 9, 10, 11, 12, 13.

T: Do these all show ways to solve the problem? Which way seems like it's a better shortcut? Turn and talk to your partner.

S: (Discuss as the teacher circulates and listens.)

T: Oh, I found one more! Actually, I did this one. Ta-dah! Pretend you are my teacher, and take a look at my work. What are your thoughts? (Project the teacher work.)

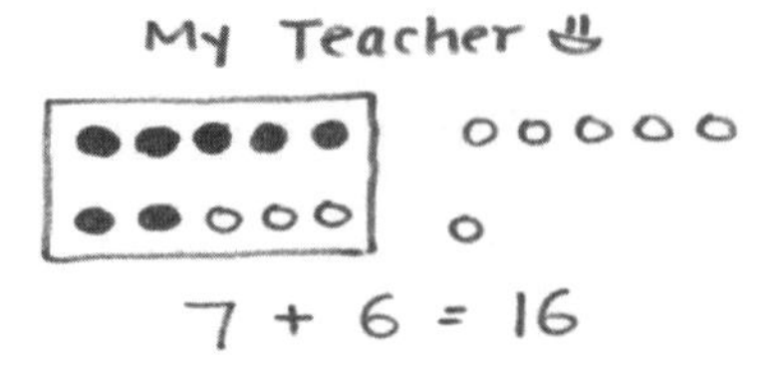

S: Your picture is organized. I like the way you drew your circles in a 5-group. But you didn't solve it right. The picture doesn't make sense.

T: What do you mean? With your partner, draw a picture that will help me see how I can make this better.

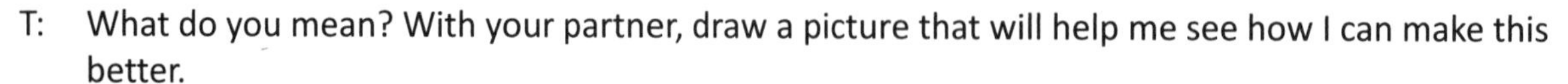

S: (Discuss as the teacher circulates and listens.)

T: How can you help me get the correct answer? What did I do wrong?

S: You need to make ten by taking apart 3 from 6. You just added 10 and 6 here. Not 10 and 3.

T: Good work! Let's try another problem!

T: (Project and read aloud.) Louie glued on 5 pieces of brown yarn for his puppet's hair. He then glued on 8 pieces of red yarn for more hair. How many pieces of yarn did Louie use? (Pause.) Solve this problem by showing your work clearly on your personal white board.

S: (Solve.)

NOTES ON MULTIPLE MEANS OF ENGAGEMENT:

Make sure to validate the different strategies students are using to solve so no students feel they have completed the work incorrectly. Be sensitive to students thinking in different ways, and encourage and cultivate healthy competition in the classroom.

Have students swap personal white boards with their partners, and discuss the following:

- Study what strategy your partner used.
- Did you get the same answer?
- Take turns to explain your partner's strategy.
- Are your strategies similar? How? Are they different? How?
- What did your partner do well?
- Which strategy is more efficient?

If time allows, repeat the partner work following the suggested sequence: 9 + 7, 8 + 6, and 7 + 7.

NOTES ON MULTIPLE MEANS OF ENGAGEMENT:

As students compare their strategies, be sure to listen to their conversations. Having these discussions with one another facilitates students' reflection and ability to actively process what they are learning.

Problem Set (10 minutes)

Students should do their personal best to complete the Problem Set within the allotted 10 minutes. For some classes, it may be appropriate to modify the assignment by specifying which problems they work on first. Some problems do not specify a method for solving. Students should solve these problems using the RDW approach used for Application Problems.

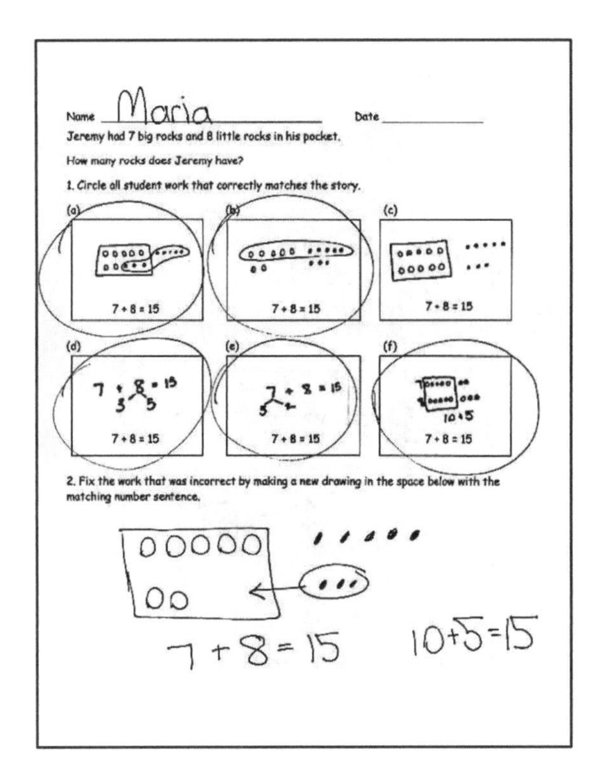
Name Maria Date

Jeremy had 7 big rocks and 8 little rocks in his pocket.

How many rocks does Jeremy have?

1. Circle all student work that correctly matches the story.

(a) 7 + 8 = 15 (b) 7 + 8 = 15 (c) 7 + 8 = 15

(d) 7 + 8 = 15 (e) 7 + 8 = 15 (f) 7 + 8 = 15

2. Fix the work that was incorrect by making a new drawing in the space below with the matching number sentence.

7 + 8 = 15 10+5=15

Student Debrief (10 minutes)

Lesson Objective: Share and critique peer solution strategies for *put together with total unknown* word problems.

The Student Debrief is intended to invite reflection and active processing of the total lesson experience.

Invite students to review their solutions for the Problem Set. They should check work by comparing answers with a partner before going over answers as a class. Look for misconceptions or misunderstandings that can be addressed in the Debrief. Guide students in a conversation to debrief the Problem Set and process the lesson.

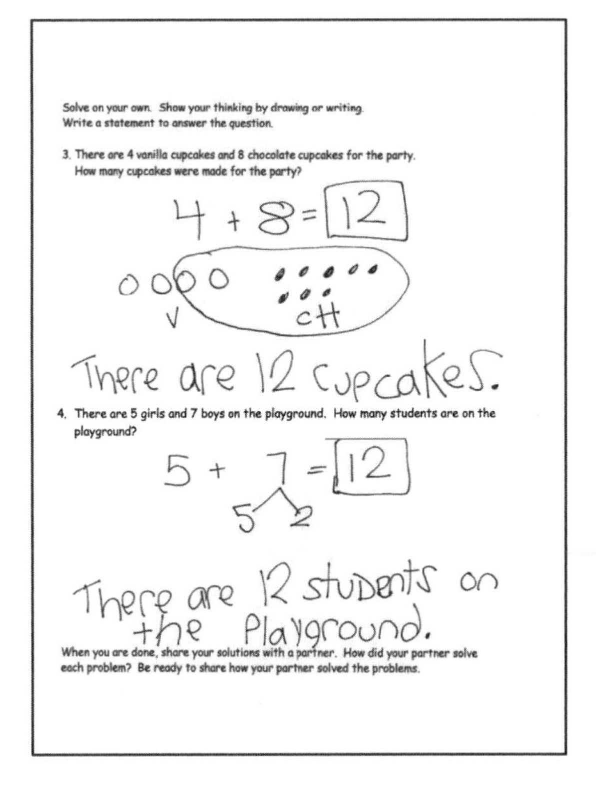
Solve on your own. Show your thinking by drawing or writing.
Write a statement to answer the question.

3. There are 4 vanilla cupcakes and 8 chocolate cupcakes for the party. How many cupcakes were made for the party?

4 + 8 = 12

There are 12 cupcakes.

4. There are 5 girls and 7 boys on the playground. How many students are on the playground?

5 + 7 = 12

There are 12 students on the playground.

When you are done, share your solutions with a partner. How did your partner solve each problem? Be ready to share how your partner solved the problems.

Any combination of the questions below may be used to lead the discussion.

- Compare Problem 3 to Problem 4 with your partner. How are your strategies similar and different?
- Look at Problem 1(b). How did this student solve his problem? How is it similar and different from the way we use the make ten strategy?
- Which samples use similar strategies? Explain your thinking.
- Which sample seems like it could be the most efficient strategy once you became an expert with it?

Exit Ticket (3 minutes)

After the Student Debrief, instruct students to complete the Exit Ticket. A review of their work will help with assessing students' understanding of the concepts that were presented in today's lesson and planning more effectively for future lessons. The questions may be read aloud to the students.

A

Name ______________________ Number Correct: Date ____________

*Write the missing number.

1.	$9 + 2 = \square$		16.	$4 + 8 = \square$	
2.	$9 + 3 = \square$		17.	$8 + 4 = \square$	
3.	$9 + 5 = \square$		18.	$7 + 4 = \square$	
4.	$9 + 4 = \square$		19.	$7 + 5 = \square$	
5.	$8 + 2 = \square$		20.	$7 + 6 = \square$	
6.	$8 + 3 = \square$		21.	$6 + 7 = \square$	
7.	$8 + 5 = \square$		22.	$9 + 9 = \square$	
8.	$8 + 4 = \square$		23.	$9 + \square = 18$	
9.	$9 + 4 = \square$		24.	$\square + 4 = 13$	
10.	$8 + 5 = \square$		25.	$\square + 4 = 12$	
11.	$9 + 5 = \square$		26.	$12 = 3 + \square$	
12.	$8 + 6 = \square$		27.	$16 = 8 + \square$	
13.	$9 + 6 = \square$		28.	$9 + 4 = 8 + \square$	
14.	$6 + 9 = \square$		29.	$9 + 3 = 5 + \square$	
15.	$9 + 6 = \square$		30.	$\square + 7 = 8 + 6$	

B

Number Correct:

Name ______________________ Date ____________

*Write the missing number.

1.	9 + 1 = □		16.	3 + 8 = □	
2.	9 + 2 = □		17.	8 + 3 = □	
3.	9 + 4 = □		18.	7 + 3 = □	
4.	9 + 3 = □		19.	7 + 4 = □	
5.	8 + 2 = □		20.	7 + 5 = □	
6.	8 + 3 = □		21.	5 + 7 = □	
7.	8 + 5 = □		22.	8 + 8 = □	
8.	8 + 4 = □		23.	8 + □ = 16	
9.	9 + 4 = □		24.	□ + 3 = 12	
10.	8 + 5 = □		25.	□ + 4 = 12	
11.	9 + 5 = □		26.	12 = 3 + □	
12.	8 + 7 = □		27.	14 = 7 + □	
13.	9 + 7 = □		28.	9 + 3 = 8 + □	
14.	7 + 9 = □		29.	9 + 3 = 5 + □	
15.	9 + 7 = □		30.	□ + 7 = 8 + 5	

Name ______________________________ Date ______________

Jeremy had 7 big rocks and 8 little rocks in his pocket.

How many rocks does Jeremy have?

1. Circle all student work that correctly matches the story.

a.

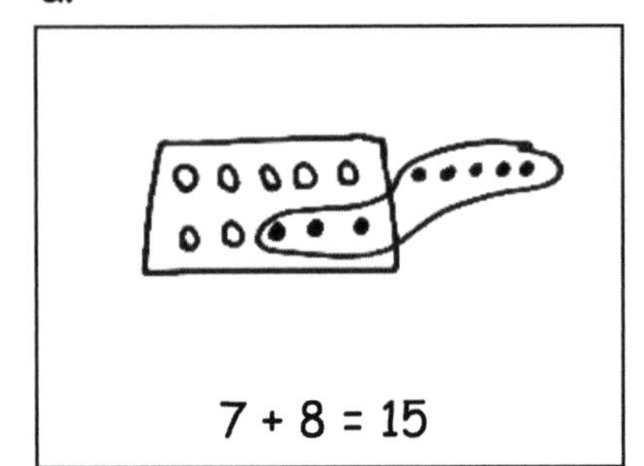

b.

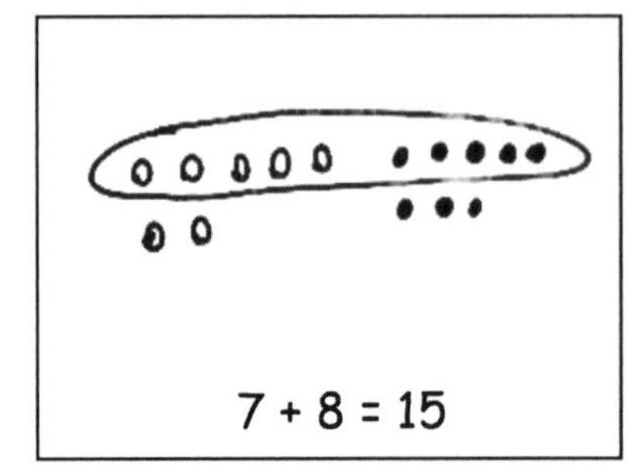

c.

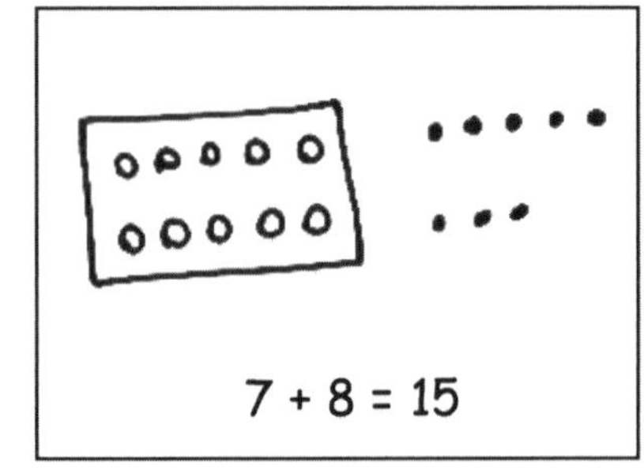

d.

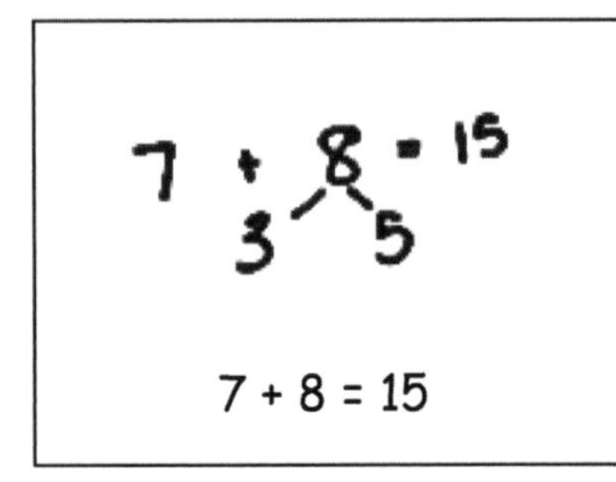

e.

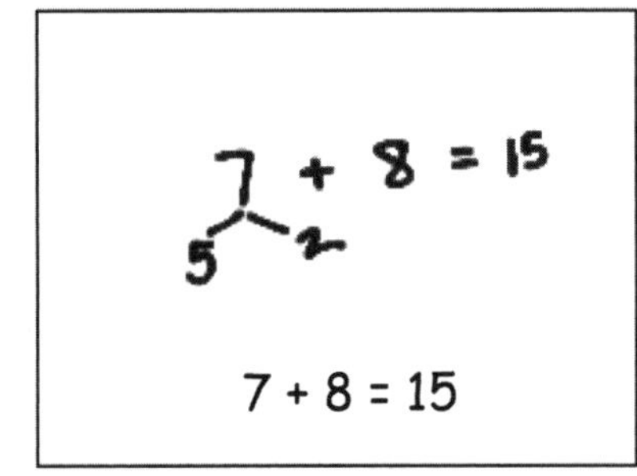

f.

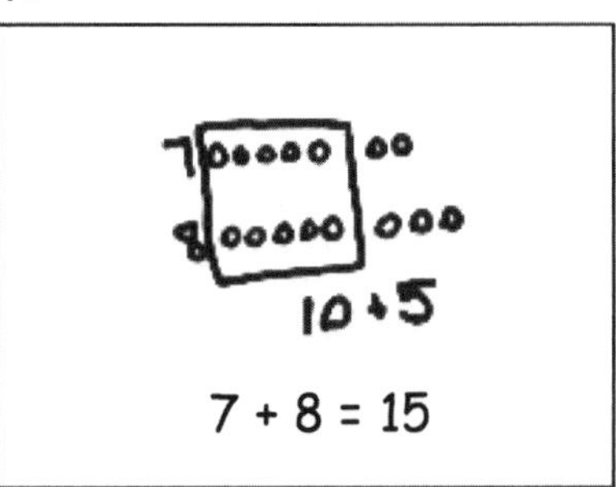

2. Fix the work that was incorrect by making a new drawing in the space below with the matching number sentence.

Solve on your own. Show your thinking by drawing or writing. Write a statement to answer the question.

3. There are 4 vanilla cupcakes and 8 chocolate cupcakes for the party. How many cupcakes were made for the party?

---

4. There are 5 girls and 7 boys on the playground. How many students are on the playground?

When you are done, share your solutions with a partner. How did your partner solve each problem? Be ready to share how your partner solved the problems.

Name ______________________________ Date ______________

John thinks the problem below should be solved using 5-group drawings, and Sue thinks it should be solved using a number bond. Solve both ways, and circle the strategy you think is the more efficient.

Kim scores 5 goals in her soccer game and 8 runs in her softball game. How many points does she score altogether?

John's Work

Sue's Work

Name ______________________________ Date ______________

Look at the student work. Correct the work. If the answer is incorrect, show a correct solution in the space below the student work.

1. Todd has 9 red cars and 7 blue cars. How many cars does he have altogether?

Mary's Work

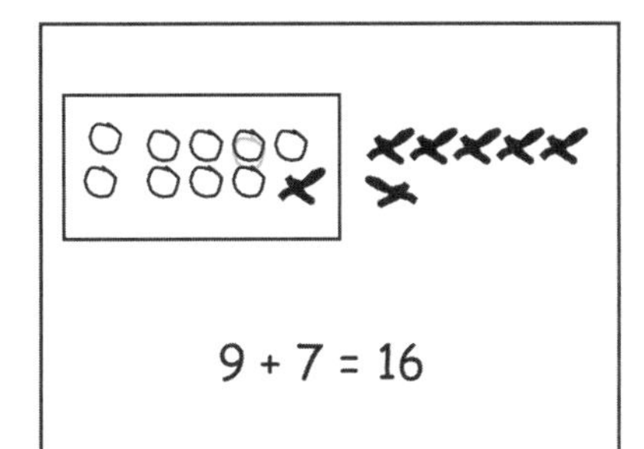

Joe's Work

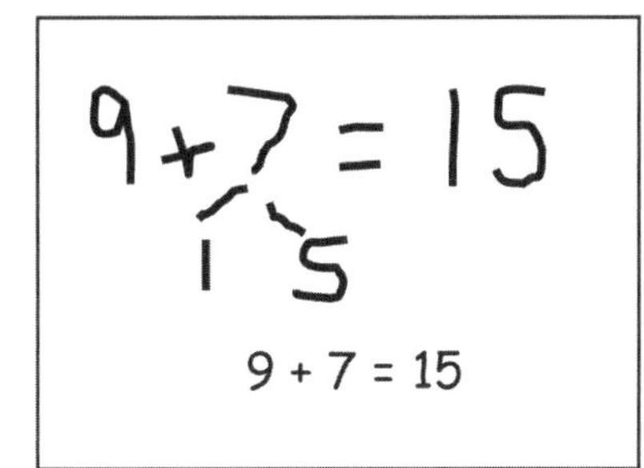

Len's Work

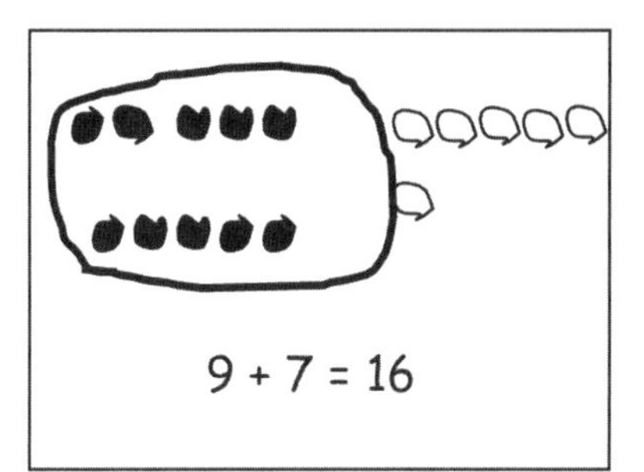

2. Jill has 8 beta fish and 5 goldfish. How many fish does she have in total?

Frank's Work

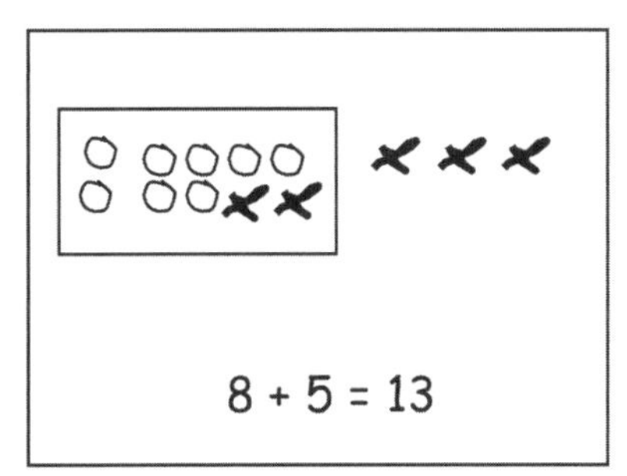

Lori's Work

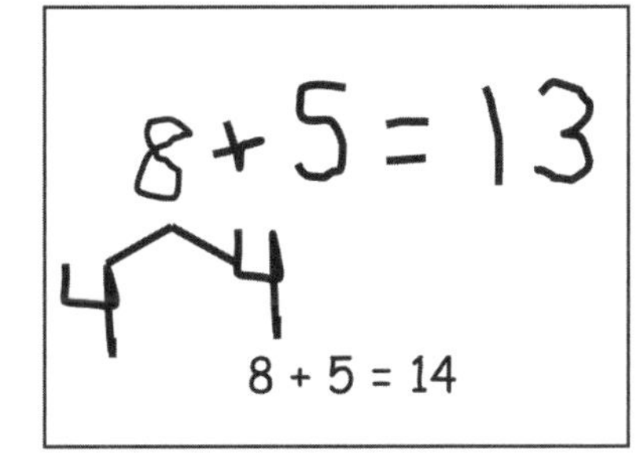

Mike's Work

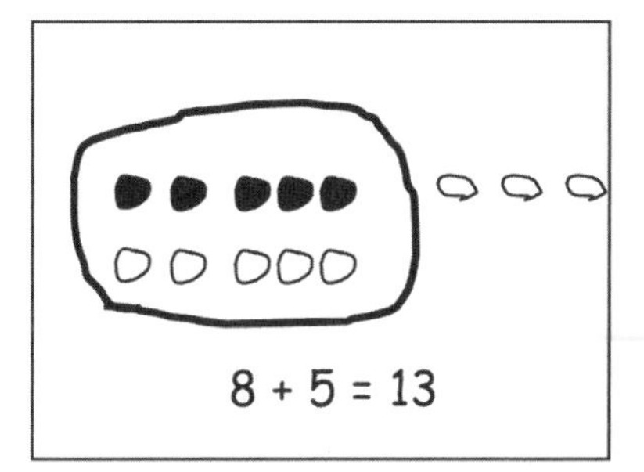

3. Dad baked 7 chocolate and 6 vanilla cupcakes. How many cupcakes did he bake in all?

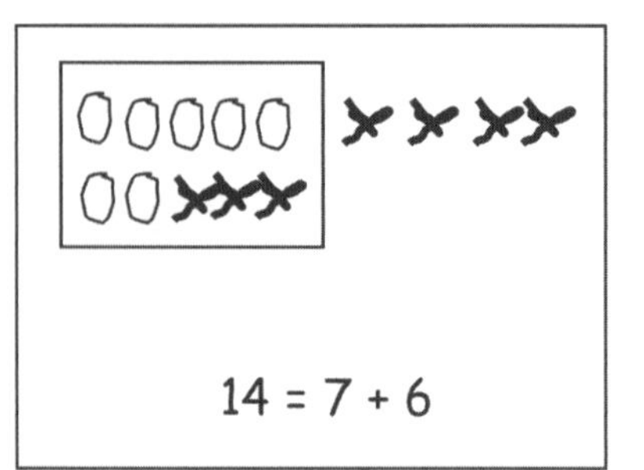

Joe's Work

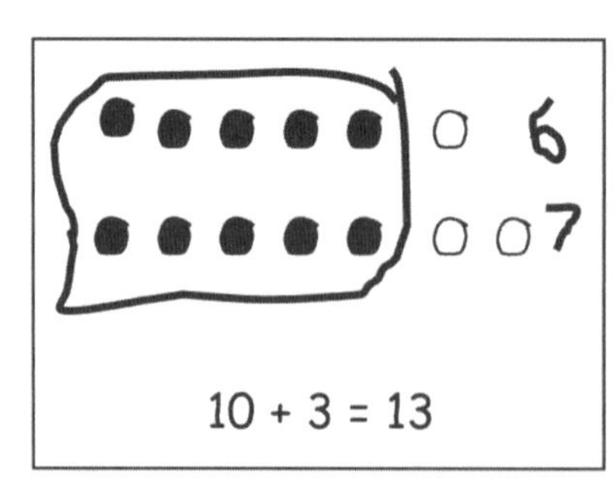

Lori's Work

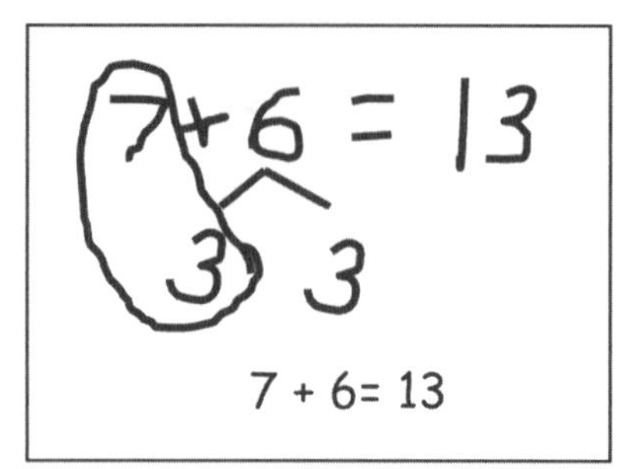

4. Mom caught 9 fireflies, and Sue caught 8 fireflies. How many fireflies did they catch altogether?

Mike's Work

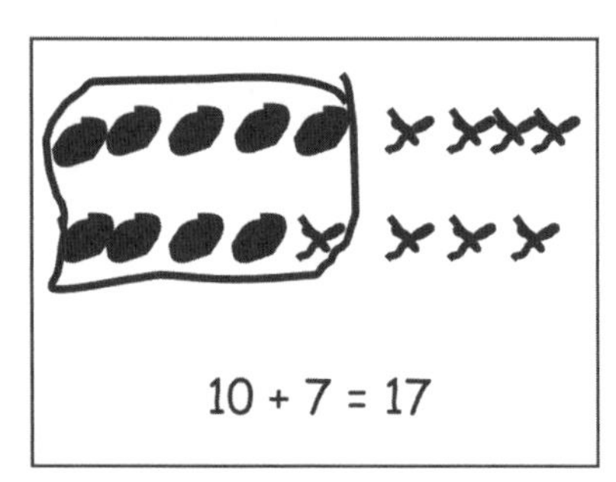

Len's Work

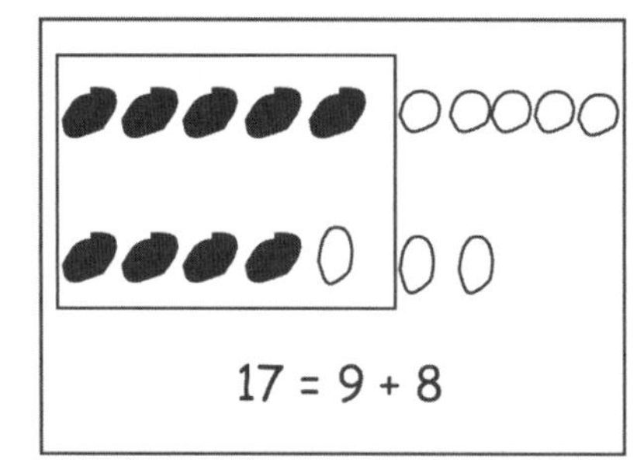

Frank's Work

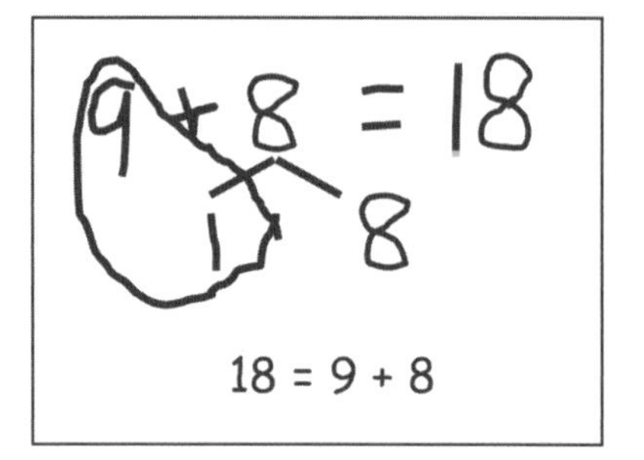

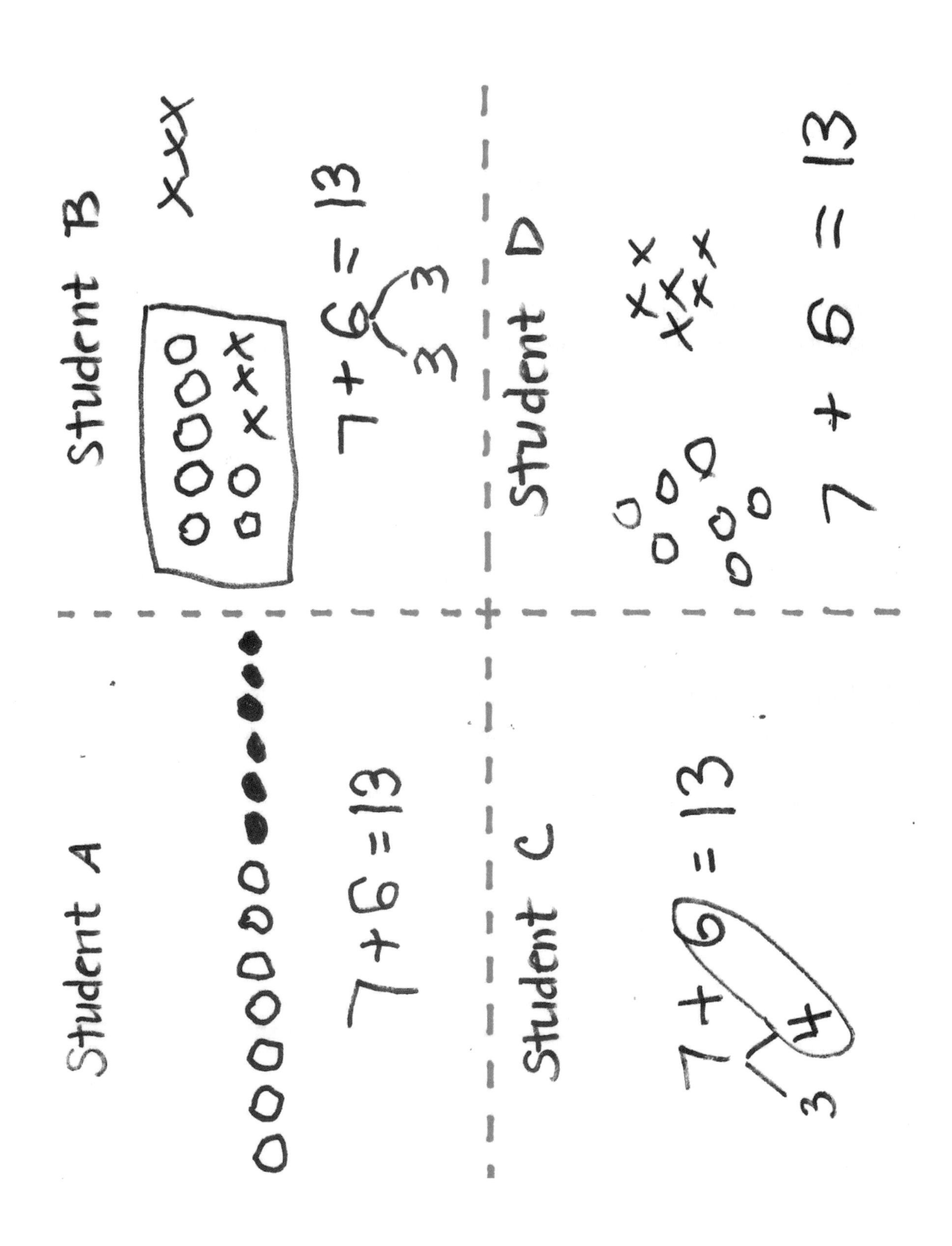

student work samples: make ten strategies

Name ______________________________ Date ______________

1. Pedro has 8 pennies. Anita has 4 pennies. Olga has 2 pennies.

   a. Whose pennies together make ten?

   b. How many pennies do Pedro, Anita, and Olga have in all? Explain your thinking using a math drawing and a number sentence. Complete the statement.

   Pedro, Anita, and Olga have ______ pennies in all.

2. Circle the pairs of numbers that make ten in each problem. Then, write the numbers that make the number sentences true. The first one is done for you.

   a. (9) + 5 + (1) = 15   2 + 6 + 8 = _____   4 + 3 + 7 = ______

   b. 8 + 2 + ____ = 15   9 + _____ + 1 = 16   1 + 7 + 9 = 10 + _____

3. Hakop has 6 pennies in a bowl. Nine pennies are in his drawer. How many pennies does Hakop have in all? Explain how you know with a labeled math drawing and number sentence. Complete the statement.

Hakop has _____ pennies in all.

4. Write a number bond in each number sentence to show how to make ten.

a. $9 + 5 = 14$

b. $8 + 5 = 13$

c. $6 + 9 = 15$

d. $17 = 8 + 9$

5. Eva has 6 marbles in her hand and 8 in her pocket.

   a. Two students drew the pictures below to find out how many marbles Eva has. Label their drawings with P and H for Pocket and Hand. Write a number sentence to go with each drawing.

   b. True or false: You have to start with 6 marbles and then add the 8 marbles.
   (Circle one.) **True False**
   Use pictures or words to explain how you know.

   c. Show two ways to find the number of Eva's marbles that show how to make ten. Write a number sentence for each.

   d. Jerry has 4 marbles in his pocket and 10 in his hand. Explain how it is that Jerry and Eva have the same number of marbles. Use words, math drawings, and numbers.

Mid-Module Assessment Task Standards Addressed	Topic A

Represent and solve problems involving addition and subtraction.

1.OA.1 Use addition and subtraction within 20 to solve word problems involving situations of adding to, taking from, putting together, taking apart, and comparing, with unknowns in all positions, e.g., by using objects, drawings, and equations with a symbol for the unknown number to represent the problem.

1.OA.2 Solve word problems that call for addition of three whole numbers whose sum is less than or equal to 20, e.g., by using objects, drawings, and equations with a symbol for the unknown number to represent the problem.

Understand and apply properties of operations and the relationship between addition and subtraction.

1.OA.3 Apply properties of operations as strategies to add and subtract. (Students need not use formal terms for these properties.) *Examples: If 8 + 3 = 11 is known, then 3 + 8 = 11 is also known. (Commutative property of addition.) To add 2 + 6 + 4, the second two numbers can be added to make a ten, so 2 + 6 + 4 = 2 + 10 = 12. (Associative property of addition.)*

Add and subtract within 20.

1.OA.6 Add and subtract within 20, demonstrating fluency for addition and subtraction within 10. Use strategies such as counting on; making ten (e.g., 8 + 6 = 8 + 2 + 4 = 10 + 4 = 14); decomposing a number leading to a ten (e.g., 13 – 4 = 13 – 3 – 1 = 10 – 1 = 9); using the relationship between addition and subtraction (e.g., knowing that 8 + 4 = 12, one knows 12 – 8 = 4); and creating equivalent but easier or known sums (e.g., adding 6 + 7 by creating the known equivalent 6 + 6 + 1 = 12 + 1 = 13).

Evaluating Student Learning Outcomes

A Progression Toward Mastery is provided to describe steps that illuminate the gradually increasing understandings that students develop *on their way to proficiency*. In this chart, this progress is presented from left (Step 1) to right (Step 4). The learning goal for students is to achieve Step 4 mastery. These steps are meant to help teachers and students identify and celebrate what the students CAN do now and what they need to work on next.

A Progression Toward Mastery				
Assessment Task Item	**STEP 1 Little evidence of reasoning without a correct answer. (1 Point)**	**STEP 2 Evidence of some reasoning without a correct answer. (2 Points)**	**STEP 3 Evidence of some reasoning with a correct answer or evidence of solid reasoning with an incorrect answer. (3 Points)**	**STEP 4 Evidence of solid reasoning with a correct answer. (4 Points)**
1 **1.OA.1** **1.OA.2**	Student is unable to complete either question accurately.	Student correctly answers one question but may not explain his thinking adequately.	Student correctly answers both questions but fails to explain using a math drawing, number sentence, and complete statement. OR Student explains her thinking using a math drawing, number sentence, and complete statement but answers one or both questions incorrectly.	Student correctly: ▪ Identifies that Olga and Pedro's pennies together make ten. ▪ Solves for 14 pennies in total. ▪ Explains his thinking using a math drawing, number sentence, and complete statement.
2 **1.OA.3** **1.OA.6**	Student solves for one unknown correctly or is unable to complete the task.	Student solves one or two unknowns correctly and circles the pairs of ten for at least two problems.	Student may solve for the unknown in each equation but fails to circle the pairs that make ten or solves for one unknown incorrectly.	Student correctly circles the pairs that make ten and solves as follows: a. 15, 16, 14 b. 5, 6, 7.
3 **1.OA.1**	Student's answer is incorrect, and there is no evidence of reasoning.	Student's answer is incorrect, but there is evidence of reasoning. For example, the student is able to write a number sentence or draw 5-groups.	Student's answer is correct but his response is incomplete, possibly missing labels for the drawing or an addition sentence, but the work is essentially strong.	Student correctly: ▪ Finds there are 15. ▪ Correctly draws and labels. ▪ Writes a corresponding number sentence.

A Progression Toward Mastery				
4 **1.OA.3** **1.OA.6**	Student is unable to draw number bonds that demonstrate the make ten strategy.	Student draws one or two of the number bonds correctly, showing how to make ten.	Student draws three out of the four number bonds correctly, showing how to make ten.	Student correctly draws a number bond for each of the four problems, showing how to make ten for each.
5 **1.OA.1** **1.OA.2** **1.OA.3** **1.OA.6**	Student's answers are incorrect and there is no evidence of reasoning.	Student's answers are incorrect, but there is evidence of reasoning. For example, the student is able to write a number sentence.	Student's answers are correct, but the responses are incomplete (e.g., may be missing labels for the drawing, an addition sentence, or may lack explanation). The student's work is essentially strong.	Student correctly: ▪ Labels the student drawings and writes a number sentence for each. ▪ Identifies the statement as false, and explains why, citing the commutative property with pictures or words (no formal terms necessary). ▪ Shows how to make ten to solve the problem. ▪ Explains how they have the same number of marbles.

Name Maria Date ________

1. Pedro has 8 pennies. Anita has 4 pennies. Olga has 2 pennies.

   a. Whose pennies together make ten?

      Pedro Olga

   b. How many pennies do Pedro, Anita, and Olga have in all? Explain your thinking using a math drawing and a number sentence. Complete the statement.

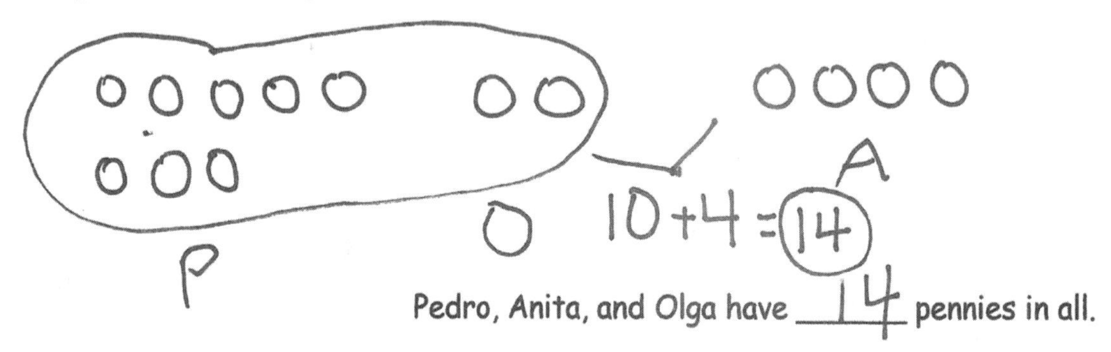

Pedro, Anita, and Olga have 14 pennies in all.

2. Circle the pairs of numbers that make ten in each problem. Then, write the numbers that make the number sentences true. The first one is done for you.

   a. 9 + 5 + 1 = 15   2 + 6 + 8 = 16   4 + 3 + 7 = 14

   b. 8 + 2 + 5 = 15   9 + 6 + 1 = 16   1 + 7 + 9 = 10 + 7

3. Hakop has 6 pennies in a bowl. Nine pennies are in his drawer. How many pennies does Hakop have in all? Explain how you know with a labeled math drawing and number sentence. Complete the statement.

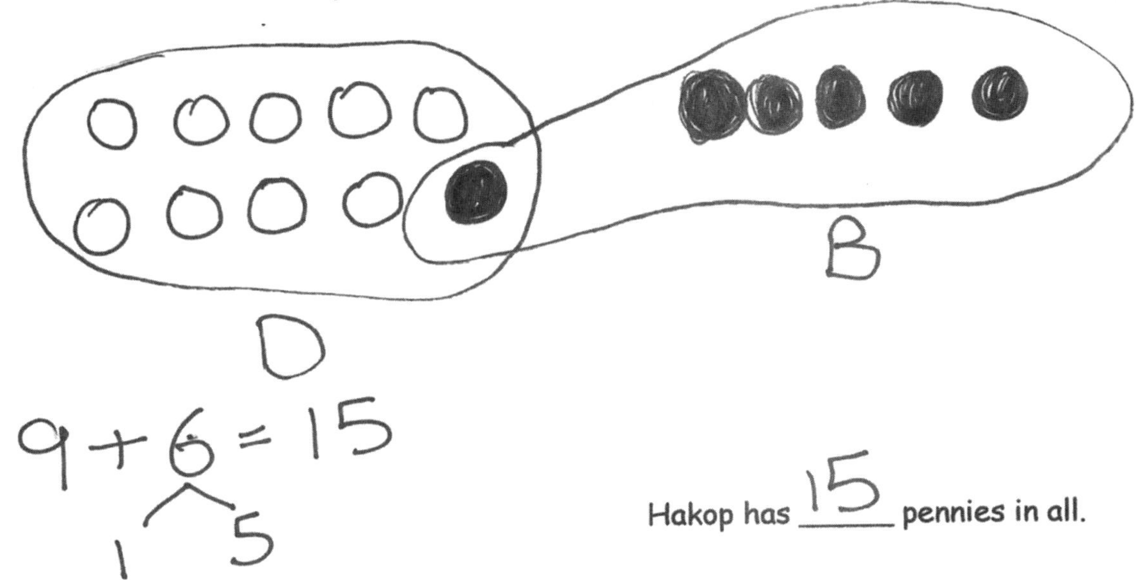

Hakop has 15 pennies in all.

4. Write a number bond in each number sentence to show how to make ten

a. 9 + 5 = 14 (1 4)

b. 8 + 5 = 13 (2 3)

c. 6 + 9 = 15 (5 1)

d. 17 = 8 + 9 (7 1)

5. Eva has 6 marbles in her hand and 8 in her pocket.

   a. Two students drew the pictures below to find out how many marbles Eva has. Label their drawings with P and H for Pocket and Hand. Write a number sentence to go with each drawing.

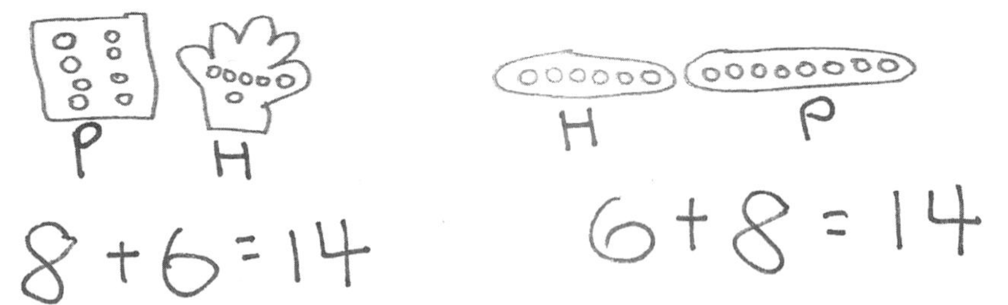

   b. True or false: You have to start with 6 marbles and then add the 8 marbles.
   (Circle one.) **True** **False**
   Use pictures or words to explain how you know.

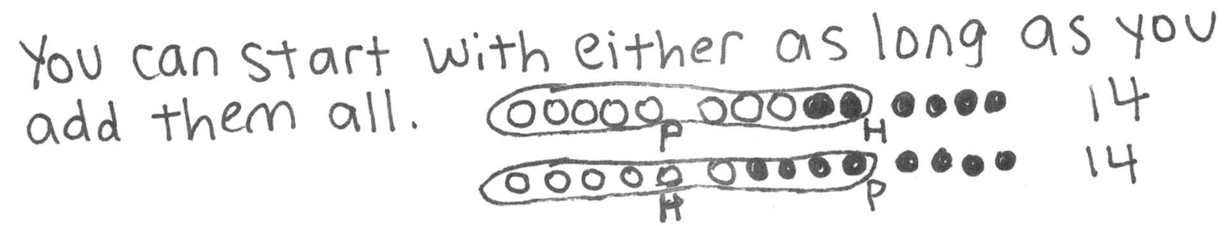

   c. Show two ways to find the number of Eva's marbles that show how to make ten. Write a number sentence for each.

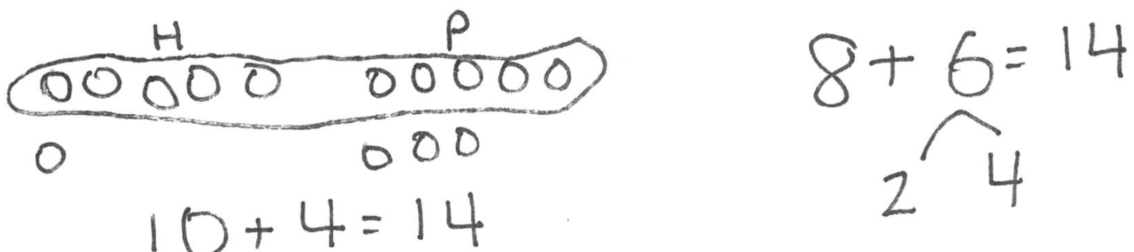

   d. Jerry has 4 marbles in his pocket and 10 in his hand. Explain how it is that Jerry and Eva have the same number of marbles. Use words, math drawings, and numbers.

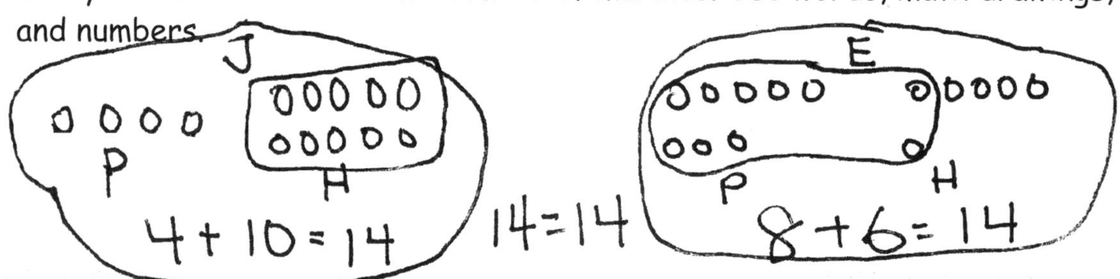

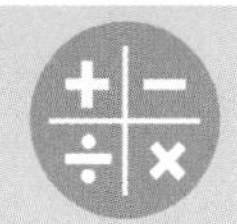

Topic B

Counting On or Taking from Ten to Solve *Result Unknown* and *Total Unknown* Problems

1.OA.1, 1.OA.3, 1.OA.4, 1.OA.6, 1.OA.5, 1.OA.7

Focus Standards:	1.OA.1	Use addition and subtraction within 20 to solve word problems involving situations of adding to, taking from, putting together, taking apart, and comparing, with unknowns in all positions, e.g., by using objects, drawings, and equations with a symbol for the unknown number to represent the problem.
	1.OA.3	Apply properties of operations as strategies to add and subtract. (Students need not use formal terms for these properties.) *Examples: If 8 + 3 = 11 is known, then 3 + 8 = 11 is also known. (Commutative property of addition.) To add 2 + 6 + 4, the second two numbers can be added to make a ten, so 2 + 6 + 4 = 2 + 10 = 12. (Associative property of addition.)*
	1.OA.4	Understand subtraction as an unknown-addend problem. *For example, subtract 10 – 8 by finding the number that makes 10 when added to 8.*
	1.OA.6	Add and subtract within 20, demonstrating fluency for addition and subtraction within 10. Use mental strategies such as counting on; making ten (e.g., 8 + 6 = 8 + 2 + 4 = 10 + 4 = 14); decomposing a number leading to a ten (e.g., 13 – 4 = 13 – 3 – 1 = 10 – 1 = 9); using the relationship between addition and subtraction (e.g., knowing that 8 + 4 = 12, one knows 12 – 8 = 4); and creating equivalent but easier or known sums (e.g., adding 6 + 7 by creating the known equivalent 6 + 6 + 1 = 12 + 1 = 13).
Instructional Days:	10	
Coherence -Links from:	GK–M4	Number Pairs, Addition and Subtraction to 10
-Links to:	G2–M3	Place Value, Counting, and Comparison of Numbers to 1,000
	G2–M5	Addition and Subtraction Within 1,000 with Word Problems to 100

Topic B focuses on the take from ten Level 3 strategy (**1.OA.6**). Students begin with word problems calling on them to subtract 9 from 10 in Lessons 12 and 13, first with concrete objects, then with drawings, and then with number bonds. The problems students solve are similar to this one: "Mary has two plates of cookies, one with 10 and one with 2. At the party, 9 cookies were eaten from the plate with 10 cookies. How many cookies were left after the party?" (**1.OA.1**) 10 – 9 = 1 and 1 + 2 = 3. This allows students to use this take from ten strategy when the ten is already separated for them and in a variety of contexts (concrete, pictorial, and abstract), which sets them up for the work of the later lessons of the topic where they must decompose teen numbers on their own to take from ten.

Lessons 14, 15, and 16 focus students on modeling subtraction of 9 from teen numbers, first with manipulatives, then 5-groups drawings, and finally number bonds. Students relate counting on to subtraction in a couple of ways (pictured below) (**1.OA.4**). Students begin to realize that there is both simplicity and efficiency when they decompose the teen number into 10 and some ones, subtract the 9 from 10, and finally add the 1 left over with some ones; this is key in Lesson 16 as students share their thinking and compare efficiency.

S: To solve 12 – 9, I count on from 9 to 12. Niiiine, 10, 11, 12, three counts. → To solve 12 – 9, I make 12 into 10 and 2 and subtract 9 from ten. 1 + 2 = 3.

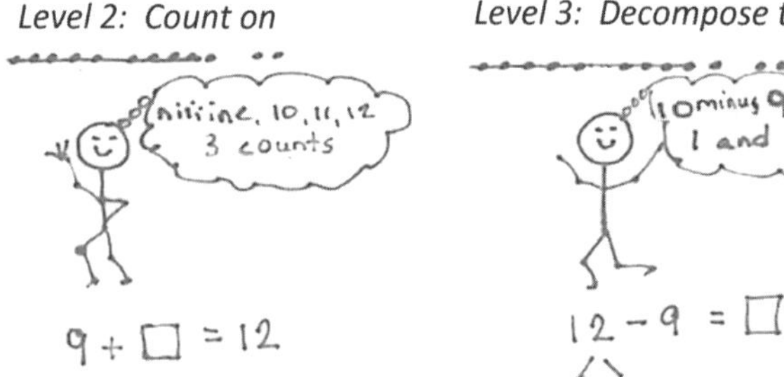

This same progression that occurred with subtracting 9 from teen numbers repeats itself in Lessons 17, 18, and 19 as students subtract 8 from teen numbers in concrete, pictorial, and abstract contexts. Students practice a pattern of action, take from ten and add the ones, as they face different contexts in word problems (MP.8). For example, "Maria has 12 snowballs. She threw 8 of them. How many does she have left?" (**1.OA.3**).

Lesson 20 both broadens and solidifies students' strategy use as they are faced with a combination of 7, 8, and 9 as subtrahends being taken away from teen numbers in both story problems and abstract equations. Lesson 21 closes Topic B with a student-centered discussion about solution strategies as they solve both action-oriented (*take from with result unknown*) and relationship (*take apart with addend unknown*) problems. Students ask each other, "How and why did you solve it this way?" and then discuss which strategies are the most efficient.

A Teaching Sequence Toward Mastery of Counting On or Taking from Ten to Solve *Result Unknown* and *Total Unknown* Problems	
Objective 1:	**Solve word problems with subtraction of 9 from 10.** **(Lessons 12–13)**
Objective 2:	**Model subtraction of 9 from teen numbers.** **(Lessons 14–15)**
Objective 3:	**Relate counting on to making ten and taking from ten.** **(Lesson 16)**
Objective 4:	**Model subtraction of 8 from teen numbers.** **(Lessons 17–18)**
Objective 5:	**Compare efficiency of counting on and taking from ten.** **(Lesson 19)**
Objective 6:	**Subtract 7, 8, and 9 from teen numbers.** **(Lesson 20)**
Objective 7:	**Share and critique peer solution strategies for *take from with result unknown* and *take apart with addend unknown* word problems from the teens.** **(Lesson 21)**

Lesson 12

Objective: Solve word problems with subtraction of 9 from 10.

Suggested Lesson Structure

Fluency Practice	(11 minutes)
Application Problem	(6 minutes)
Concept Development	(33 minutes)
Student Debrief	(10 minutes)
Total Time	**(60 minutes)**

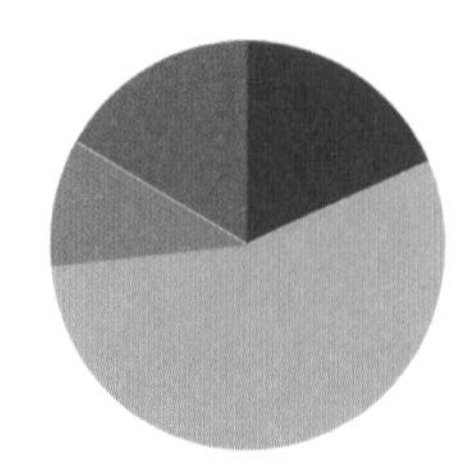

Fluency Practice (11 minutes)

- Rewrite Expressions as 10+ Sentences **1.OA.6** (5 minutes)
- 5-Group Flash: Partners to Ten **1.OA.6** (2 minutes)
- Teen Number Bonds **1.NBT.2** (4 minutes)

Rewrite Expressions as 10+ Sentences (5 minutes)

Materials: (S) Personal white board

Note: This review fluency activity reinforces the make ten addition strategy where students mentally decompose numbers to create equivalent but easier number sentences.

Write addition sentences with 9, 8, or 7 as an addend. Tell students to rewrite the sentence with 10 as an addend (e.g., write 9 + 2, and students write 10 + 1 = 11). Suggested sequence: 9 + 1, 9 + 2, 9 + 3, 9 + 5, 9 + 6, 8 + 2, 8 + 3, 8 + 5, 8 + 6, 7 + 3.

5-Group Flash: Partners to Ten (2 minutes)

Materials: (T) 5-group row cards (Fluency Template 1)

Note: This activity supports Grade 1's core fluency standard of adding and subtracting within 10. Notice the shift in the visual representation of ten, which transitions students into seeing ten as a single unit by the module's end.

This fluency activity focuses on the partners to ten.

Flash a card for two to three seconds. Snap. Students say the number. Snap again. Students say the partner to ten.

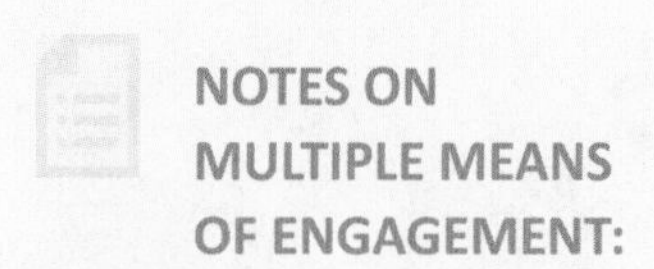

NOTES ON MULTIPLE MEANS OF ENGAGEMENT:

Certain fluency games that are played in class are good to play at home. Send home directions to the games so that parents can play with their children. Suggest ways parents can use different numbers to challenge their children and extend their learning during the games.

Teen Number Bonds (4 minutes)

Materials: (S) Personal white board with 5-group row insert (Fluency Template 2)

Note: Composing teen numbers as 10 ones and some more ones prepares students for the take from ten subtraction strategy.

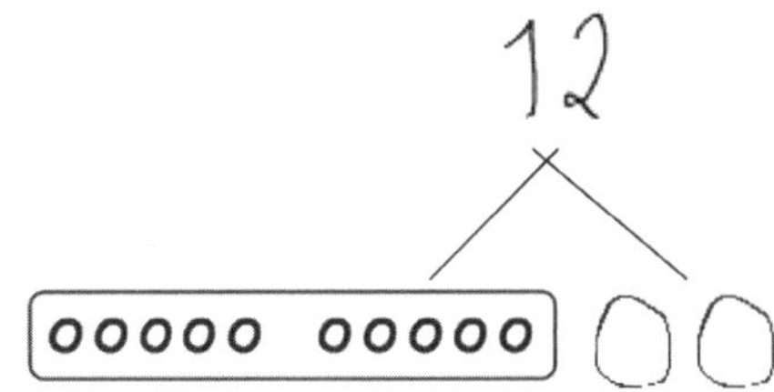

T: Draw more circles to show a total of 12.

S: (Draw 2 more circles.)

T: Say 12 as a number bond with 10 as a part.

S: 10 and 2 make 12.

T: Draw lines to show the total of 12 from your circles.

S: (Draw lines to make a number bond with the numeral 12 on top.)

Continue with other numbers between 11 and 20.

Application Problem (6 minutes)

Claudia bought 8 red apples and 9 green apples. How many apples does Claudia have altogether? Make a math drawing, number sentence, and statement to show your thinking.

Extension: Claudia ate 3 red apples, and her friend ate 4 green apples. How many apples does Claudia have now?

Note: This problem revisits the make ten strategy introduced in Topic A. It provides a foundation for today's work of solving word problems with subtraction of 9 from 10 using the same numbers and story problem character.

R G

$10 + 7 = 17$

Claudia has 17 apples altogether.

Concept Development (33 minutes)

Materials: (T) Chart paper (S) Personal white board

Have students sit at their tables with their personal white boards.

T: (Project and read aloud.) When Claudia brought home her 17 apples, she put 10 in a bowl and 7 on the table. Then, she decided to give 9 apples to her babysitter. How many apples did Claudia have left? (Pause.) Solve the problem on your personal board, and talk with your partner about how you solved it.

S: (Solve the problem, and discuss strategies as the teacher circulates.)

T: What strategies did you use?

S: I drew all of the apples and then crossed off the ones on the table and 2 more. I counted the ones that were left. 8. → I drew 10 circles for the bowl and 7 for the table. Then, I took 9 from the 10 in the bowl. 7 and 1 is 8.

T: Let's all try another. (Project and read aloud: "Bailey Bunny had 10 carrots in a basket and 5 on a plate. She ate 9 carrots from the basket. How many carrots were left?")
T: On your personal white board, draw how many carrots Bailey Bunny had in the basket, and label it.
S: (Draw 10 circles, and write B or basket.)
T: In the next row, draw the carrots that were on the plate, and label it.
S: (Draw 5 circles, and write P or plate.)
T: The problem says that she ate 9 carrots from the basket. What should we do?
S: Cross off 9.
T: From where?
S: From the basket, from 10.
T: Show on your personal white board.
S: (Cross off 9 circles from 10.)
T: How many carrots are left in the basket?
S: 1 carrot!
T: How many are left on the plate?
S: 5 carrots.
T: Then, how many carrots are left in all?
S: 6 carrots.

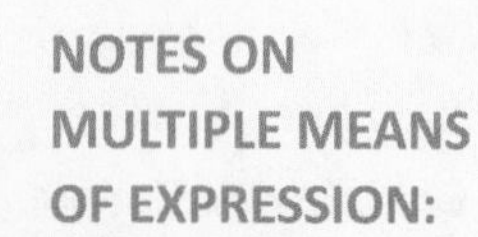

NOTES ON MULTIPLE MEANS OF EXPRESSION:

At this time, students may be working at varying stages of subtracting, as depicted in the image below. While praising students for accurate solutions, encourage them to move to the next level strategy. If they are counting all, they should be encouraged to make the connection to counting on. If students are counting on, they should be encouraged toward taking from ten.

Repeat the process using the suggested sequence: 11 – 9, 12 – 9, and 14 – 9, recording the work on a chart paper for the Student Debrief.

T: Let's record how we solved our story problem with a number bond. (Read the story again.) Draw a number bond to show Bailey Bunny's total number of carrots, the part in the basket, and the part on the plate.
S: (Draw.)
T: Draw circles to show the different parts.
S: (Draw.)
T: What did we do next? Show in your picture.
S: (Cross off 9 circles.) We took away 9 carrots from the basket. → We took away 9 from 10.
T: Turn and talk to your partner about how you can find how many carrots are left.
S: I counted 1, 2, 3, 4, 5, 6. → I didn't use the picture. I counted on. Niiine, 10, 11, 12, 13, 14, 15. That is 6 counts. → I added 1 and 5. That's 6 carrots.

Repeat the process using the following suggested sequence: 16 – 9, 17 – 9, and 18 – 9, recording the work on chart paper for the Student Debrief.

Problem Set (10 minutes)

Students should do their personal best to complete the Problem Set within the allotted 10 minutes. For some classes, it may be appropriate to modify the assignment by specifying which problems they work on first. Some problems do not specify a method for solving. Students should solve these problems using the RDW approach used for Application Problems.

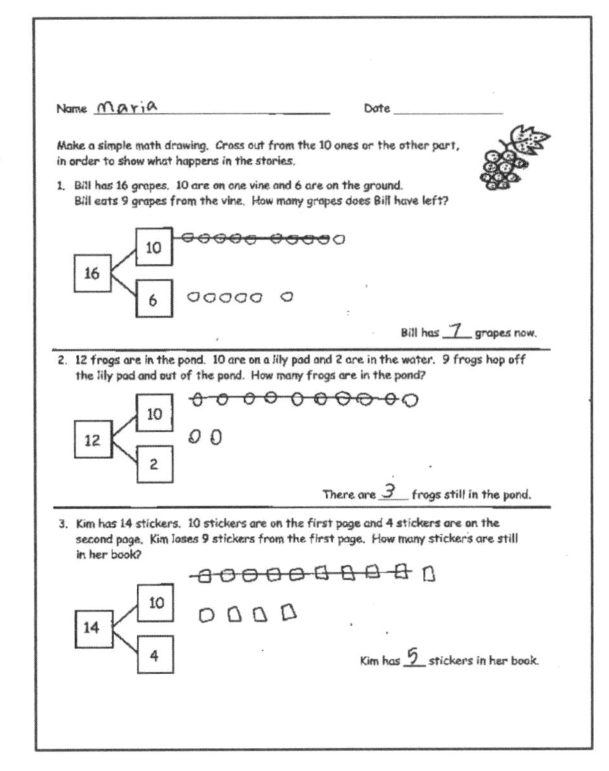
Name Maria Date

Make a simple math drawing. Cross out from the 10 ones or the other part, in order to show what happens in the stories.

1. Bill has 16 grapes. 10 are on one vine and 6 are on the ground. Bill eats 9 grapes from the vine. How many grapes does Bill have left?

16 — 10, 6

Bill has 7 grapes now.

2. 12 frogs are in the pond. 10 are on a lily pad and 2 are in the water. 9 frogs hop off the lily pad and out of the pond. How many frogs are in the pond?

12 — 10, 2

There are 3 frogs still in the pond.

3. Kim has 14 stickers. 10 stickers are on the first page and 4 stickers are on the second page. Kim loses 9 stickers from the first page. How many stickers are still in her book?

14 — 10, 4

Kim has 5 stickers in her book.

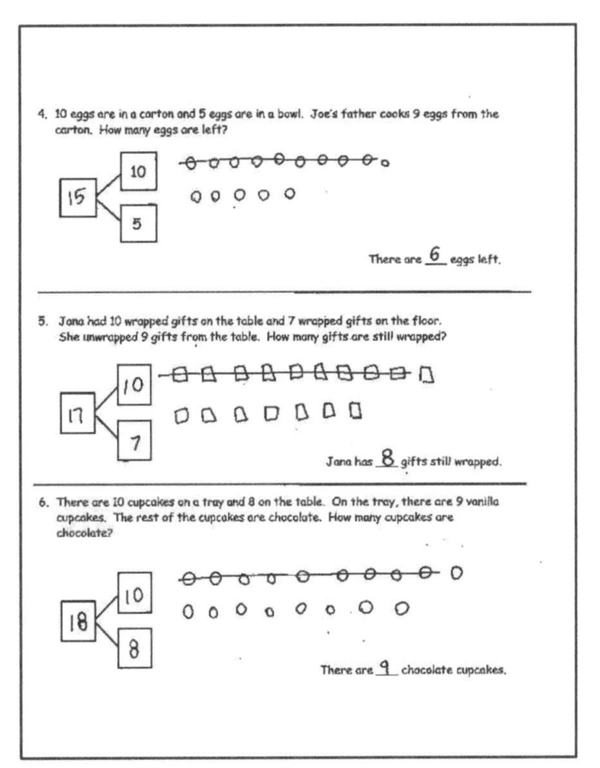
4. 10 eggs are in a carton and 5 eggs are in a bowl. Joe's father cooks 9 eggs from the carton. How many eggs are left?

15 — 10, 5

There are 6 eggs left.

5. Jana had 10 wrapped gifts on the table and 7 wrapped gifts on the floor. She unwrapped 9 gifts from the table. How many gifts are still wrapped?

17 — 10, 7

Jana has 8 gifts still wrapped.

6. There are 10 cupcakes on a tray and 8 on the table. On the tray, there are 9 vanilla cupcakes. The rest of the cupcakes are chocolate. How many cupcakes are chocolate?

18 — 10, 8

There are 9 chocolate cupcakes.

Student Debrief (10 minutes)

Lesson Objective: Solve word problems with subtraction of 9 from 10.

The Student Debrief is intended to invite reflection and active processing of the total lesson experience.

Invite students to review their solutions for the Problem Set. They should check work by comparing answers with a partner before going over answers as a class. Look for misconceptions or misunderstandings that can be addressed in the Debrief. Guide students in a conversation to debrief the Problem Set and process the lesson.

Any combination of the questions below may be used to lead the discussion.

- Look at your drawings on your Problem Set. What did you notice when we took away 9 for each problem?
- Look at the chart of work from the Concept Development. What do you notice about the answers to each of these questions? (The answer is always 1 more than the second part of the number bond.) Why do you think this is?
- How can solving Problem 3 help you solve Problem 4?
  - After taking 9 from 10, how did you find the total amount left over? Which is the most efficient way to find out how many are left? Explain your thinking.
  - Look at your Application Problem, and think about what Claudia did with the apples once she got home (model this problem again). How are these problems similar? How are they different?

Exit Ticket (3 minutes)

After the Student Debrief, instruct students to complete the Exit Ticket. A review of their work will help with assessing students' understanding of the concepts that were presented in today's lesson and planning more effectively for future lessons. The questions may be read aloud to the students.

Name ______________________________ Date ______________

Make a simple math drawing. Cross out from the 10 ones or the other part in order to show what happens in the stories.

1. Bill has 16 grapes. 10 are on one vine, and 6 are on the ground. Bill eats 9 grapes from the vine. How many grapes does Bill have left?

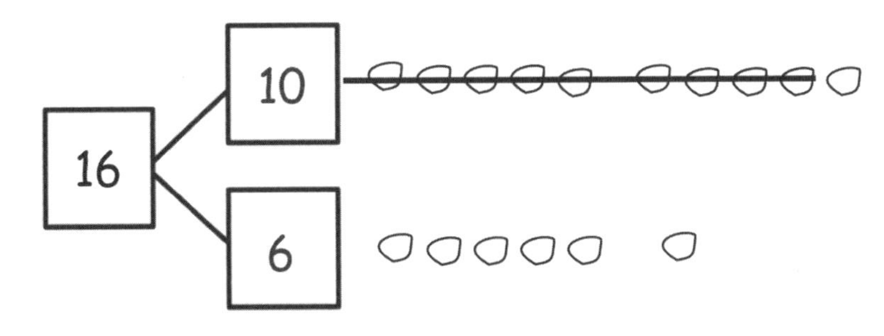

Bill has _____ grapes now.

---

2. 12 frogs are in the pond. 10 are on a lily pad, and 2 are in the water. 9 frogs hop off the lily pad and out of the pond. How many frogs are in the pond?

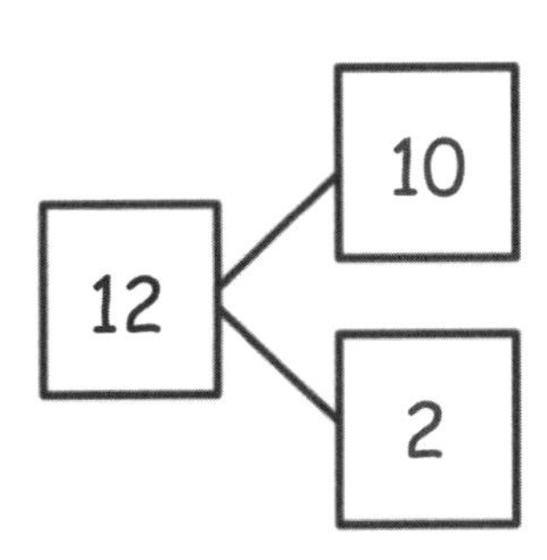

There are _____ frogs still in the pond.

---

3. Kim has 14 stickers. 10 stickers are on the first page, and 4 stickers are on the second page. Kim loses 9 stickers from the first page. How many stickers are still in her book?

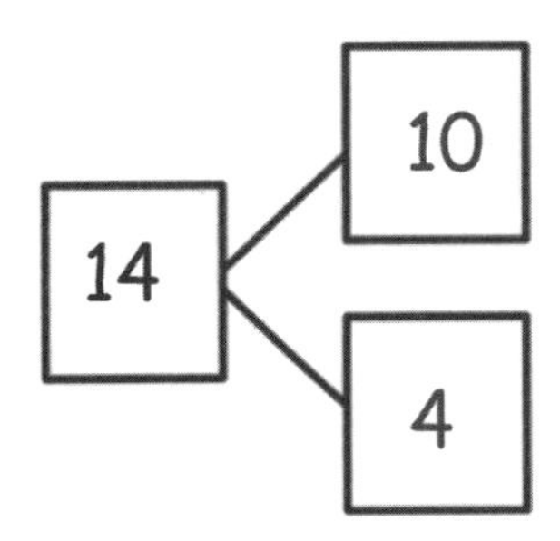

Kim has _____ stickers in her book.

4. 10 eggs are in a carton, and 5 eggs are in a bowl. Joe's father cooks 9 eggs from the carton. How many eggs are left?

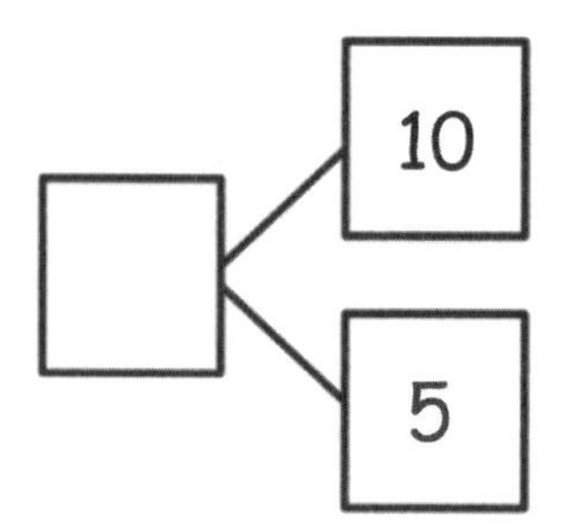

There are ____ eggs left.

---

5. Jana had 10 wrapped gifts on the table and 7 wrapped gifts on the floor. She unwrapped 9 gifts from the table. How many gifts are still wrapped?

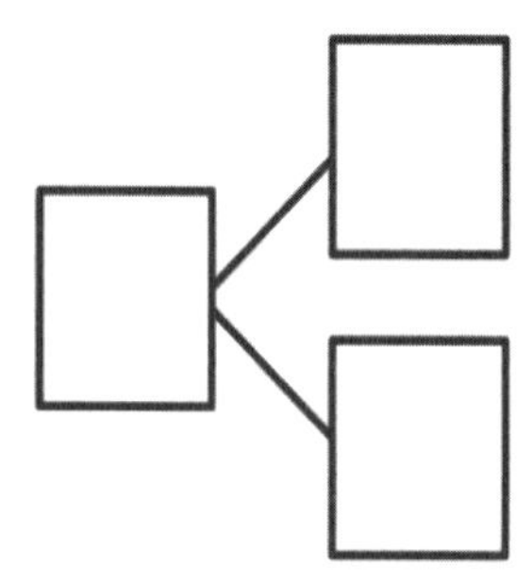

Jana has ____ gifts still wrapped.

---

6. There are 10 cupcakes on a tray and 8 on the table. On the tray, there are 9 vanilla cupcakes. The rest of the cupcakes are chocolate. How many cupcakes are chocolate?

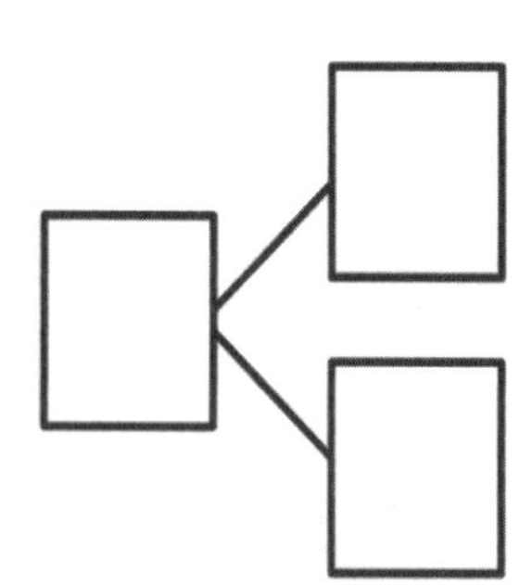

There are ____ chocolate cupcakes.

Name ________________________________ Date _______________

Make a simple math drawing. Cross out from the 10 ones to show what happens in the story.

There were 16 books on the table. 10 books were about dinosaurs. 6 books were about fish. A student took 9 of the dinosaur books. How many books were left on the table?

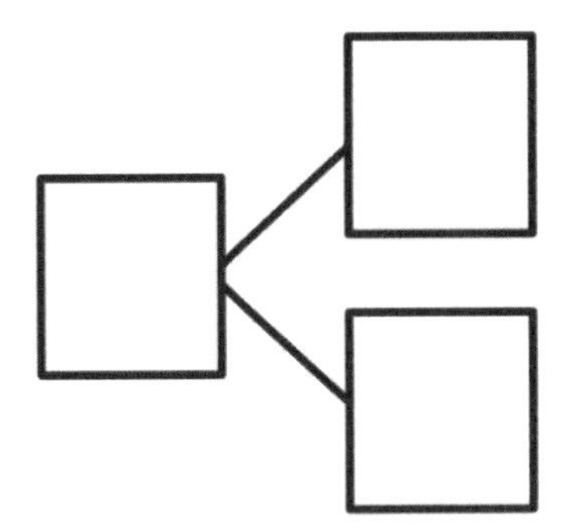

There were ____ books left on the table.

Name ______________________________ Date ______________

Make a simple math drawing. Cross out from the 10 ones to show what happens in the stories.

I had 16 grapes.
10 of them were red,
and 6 were green.
I ate 9 red grapes.
How many grapes do
I have now?

16
10
6

Now I have 7 grapes.

1. There were 15 squirrels by a tree. 10 of them were eating nuts. 5 squirrels were playing. A loud noise scared away 9 of the squirrels eating nuts. How many squirrels were left by the tree?

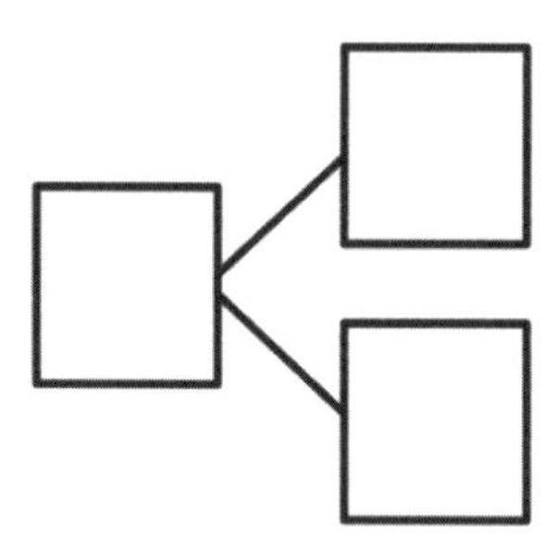

There were ___ squirrels left by the tree.

2. There are 17 ladybugs on the plant. 10 of them are on a leaf, and 7 of them are on the stem. 9 of the ladybugs on the leaf crawled away. How many ladybugs are still on the plant?

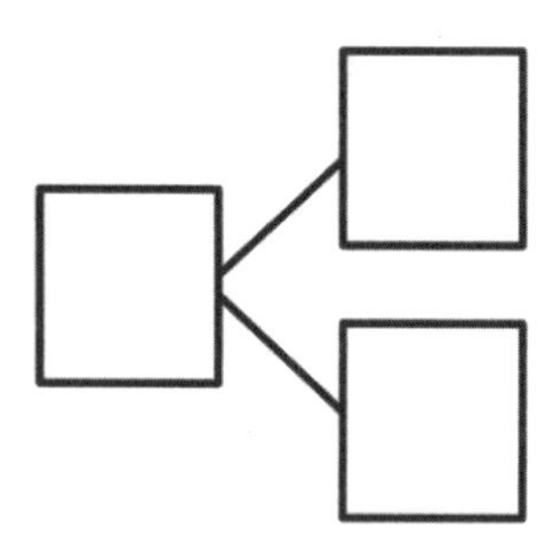

There are ___ ladybugs on the plant.

3. Use the number bond to fill in the math story. Make a simple math drawing. Cross out from 10 ones or some ones to show what happens in the stories.

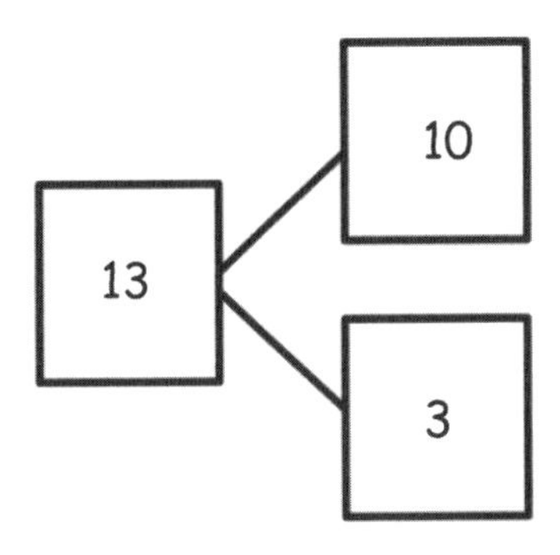

There were 13 ants in the anthill.

10 of the ants are sleeping, and 3 of them are awake.

9 of the sleeping ants woke up and crawled away.

How many ants are left in the anthill?

Math drawing:

_____ ants are left in the anthill.

---

4. Use the number bond below to come up with your own math story. Include a simple math drawing. Cross out from 10 ones to show what happens.

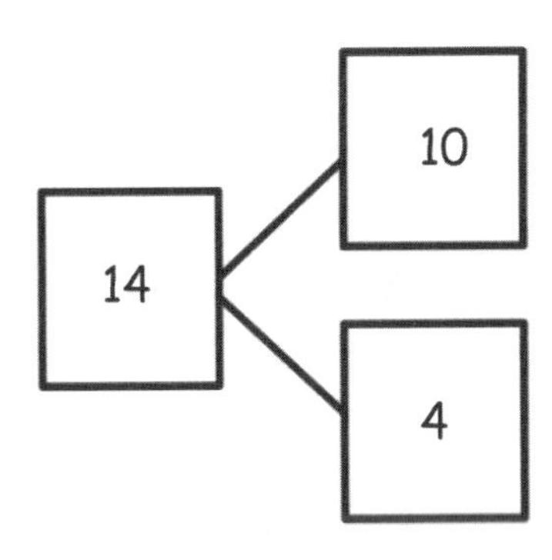

Math drawing:

Number sentences:

Statement:

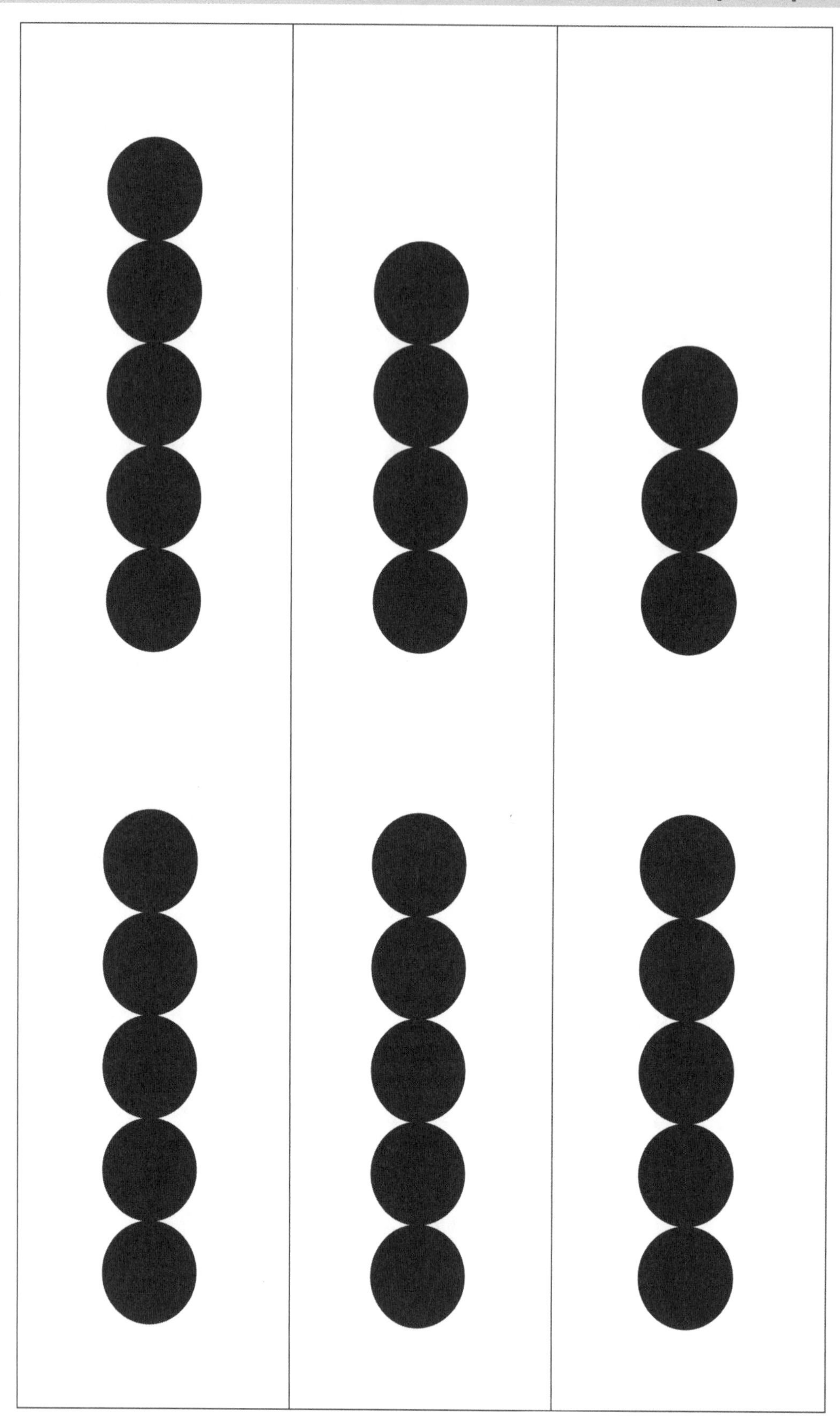

5-group row cards

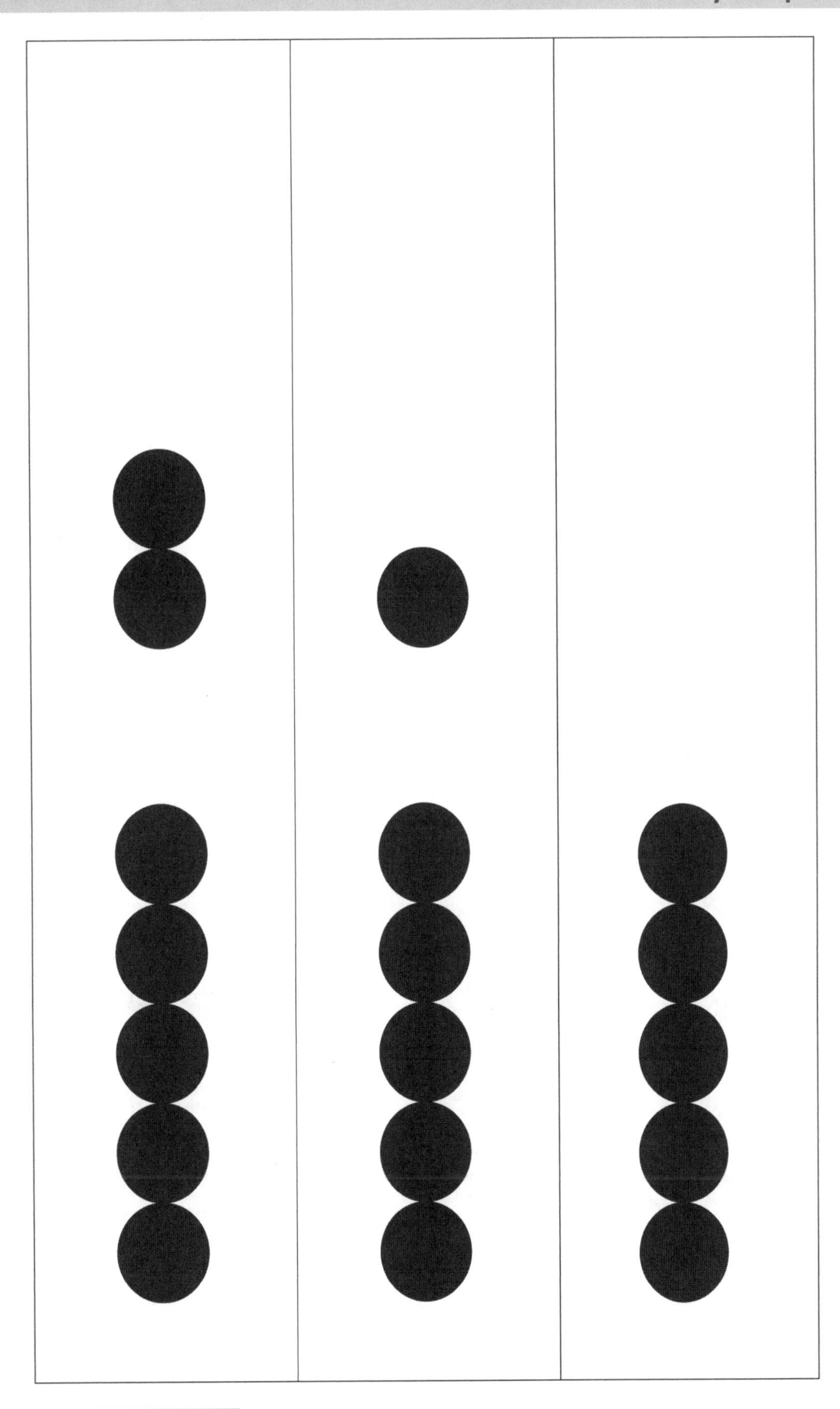

5-group row cards

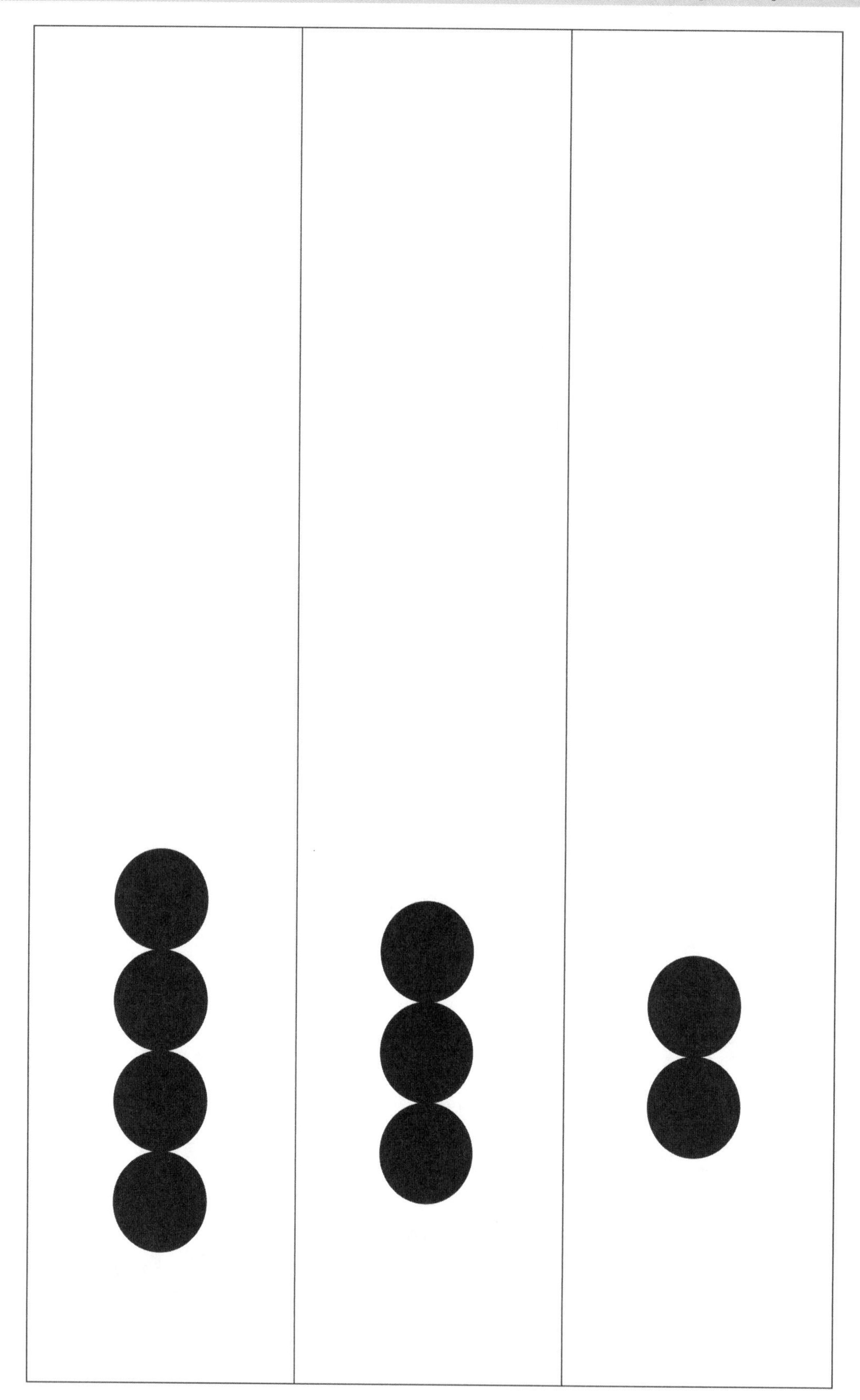

5-group row cards

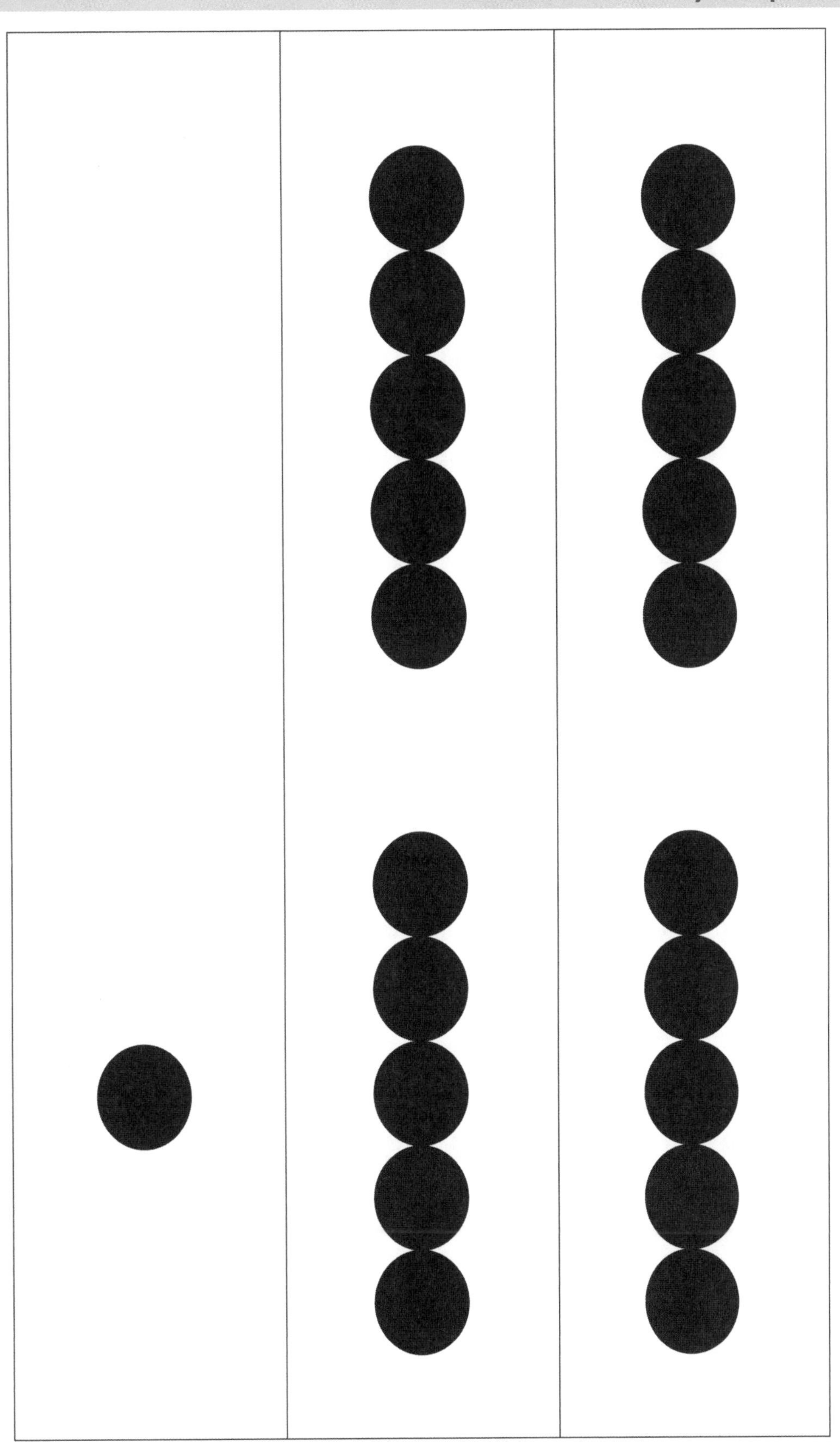

5-group row cards

ooooo ooooo

5-group row insert

Lesson 13

Objective: Solve word problems with subtraction of 9 from 10.

Suggested Lesson Structure

■ Fluency Practice	(13 minutes)
■ Application Problem	(7 minutes)
■ Concept Development	(30 minutes)
■ Student Debrief	(10 minutes)
Total Time	**(60 minutes)**

Fluency Practice (13 minutes)

- 2, 3, 5 Less **1.OA.6** (3 minutes)
- Subtraction with Cards **1.OA.6** (5 minutes)
- 5-Group Flash: Take from Ten **1.OA.6** (5 minutes)

2, 3, 5 Less (3 minutes)

Note: This fluency activity supports Grade 1's core fluency standard of adding and subtracting within 10.

T: On my signal, say the number that is 2 less.
T: 5.
S: 3.

Continue with numbers between 4 and 10. Then, review 3 less and 5 less.

Subtraction with Cards (5 minutes)

Materials: (S) 1 deck of numeral cards with 2 extra tens for each pair of students (Lesson 1 Fluency Template, numeral side only), counters (if needed)

Note: Reviewing subtraction facts supports Grade 1's core fluency standard of adding and subtracting within 10. Provide the number bond template for students who need extra support. Students can place the larger number as the whole and the smaller as a part to figure out the missing part.

Students place the deck of cards facedown between them. Each partner flips over two cards and subtracts the smaller number from the larger number. The partner with the smaller difference keeps the cards played by both players that round. The player with the most cards at the end of the game wins.

5-Group Flash: Take from Ten (5 minutes)

Materials: (T) 5-group row cards (Lesson 12 Fluency Template 1) (S) Personal white board with 5-group row insert (Lesson 12 Fluency Template 2)

Note: This maintenance fluency activity with partners to ten facilitates the take from ten subtraction strategy that students are learning.

Flash a card (e.g., 9) for one to three seconds. Students cross off the number flashed from the 5-group row insert and write the corresponding subtraction sentence.

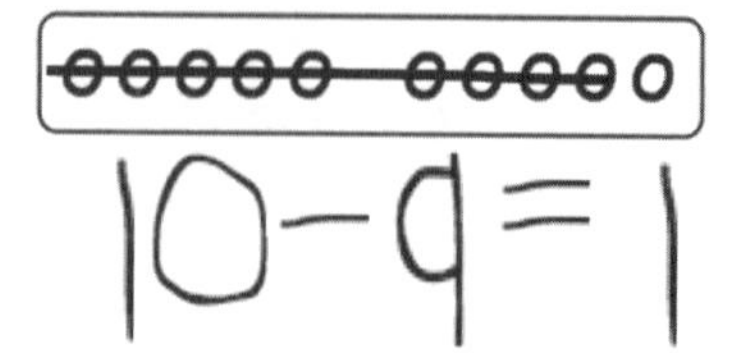

Application Problem (7 minutes)

Ten snowflakes fell on Sam's mitten, and 6 fell on his coat. Nine of the snowflakes on Sam's mitten melted. How many snowflakes are left? Write a subtraction sentence to show how many snowflakes are left.

Note: This problem continues the work started in Lesson 12, asking students to subtract 9 from 10.

M C

10 - 9 = 1 1 + 6 = 7

7 snowflakes are left.

Concept Development (30 minutes)

Materials: (T) Image of 5-group rows (Lesson 12 Fluency Template 1) (S) Personal white board with 5-group rows insert (Lesson 12 Fluency Template 2)

Have students come to the meeting area with their personal white boards and sit in a semicircle.

T: (Project and read aloud.) There were 10 ants on the picnic blanket and 4 ants on the grass. Nine ants from the picnic blanket went into the anthill with a bread crumb. How many ants are not in the anthill?

T: Show me a number bond that shows how many ants were around at the beginning of the story.

S: (Write 14, 10, and 4.)

T: Using the picture from our fluency activity, I'll make a math drawing to show the parts. (Model drawing a 5-group row of 10 that is framed and labeled as 10 and 4 dark circles to the right, labeled as 4.)

T: Talk with a partner. If 9 ants left the blanket to go into the anthill, how many ants are not in the anthill?

S: (Discuss with a partner and solve.)

T: How many ants are not in the anthill?

S: 5.

T: Use my math drawing to show me how you know.

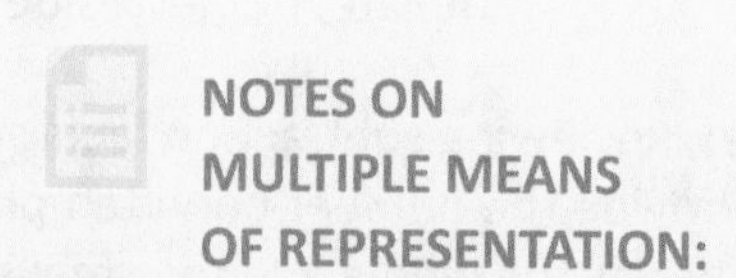

NOTES ON MULTIPLE MEANS OF REPRESENTATION:

Reading aloud word problems facilitates problem solving for those students who have difficulty reading the text within the problems. Hearing the word problem also helps students who are auditory learners.

S: These 10 circles are the ants from the blanket. If I cross off 9 of them, I have 1 here (point to the framed 5-group row) and 4 more here (point to the 4 dark circles next to the frame). → If we start from the 9 we had, we can count up. (Point to the 5-group picture, starting at the last circle in the framed 5-group row.) 1 more to get to 10, and then 4 more to get to 14. → I knew that we had 4 black circles, and I added 1 more. That's 5.

T: Which strategy is more efficient?

S: Adding 1 to the other part.

T: Turn and talk to you partner, and write the number sentence that shows how we solved this problem. Explain your thinking.

S: We took away 9 ants from the 10 ants on the blanket. There was 1 ant left, plus there were 4 ants still on the grass. So, 10 – 9 = 1, and then 4 + 1 = 5. → I can write 14 – 9 = 5. In the beginning, there were 14 ants. Then, 9 ants went into the anthill, so I took 9 away. There are 5 ants left.

T: Let's take a look at the math drawing. Do these 10 open circles remind you of any other drawings?

S: They look like 5-groups, except they are all in a line. We used to make them with 5 on top and 5 on the bottom.

T: You are right! Since these are all in a row, we'll call them a **5-group row**. There is a space to separate 5 circles from the other 5.

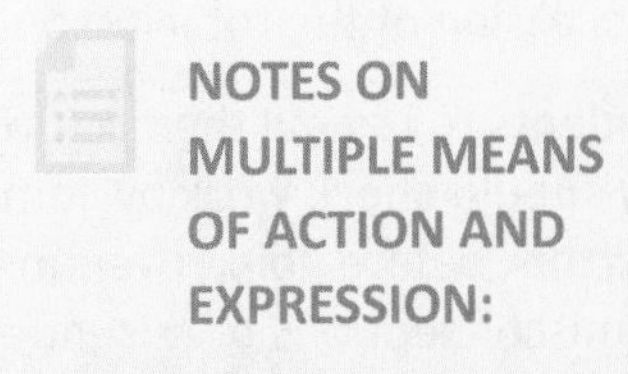

NOTES ON MULTIPLE MEANS OF ACTION AND EXPRESSION:

In this lesson, students are transitioning from drawing 5-groups to drawing 5-group rows. Some students may need some time to make the transition and complete the drawings the new way.

Repeat the process by having students write the number bond, draw the picture, and write the number sentence using the following suggested sequence: 13 – 9, 15 – 9, 16 – 9, 17 – 9, and 18 – 9. For the first few problems, use the 5-group rows template (with the group of 10 framed), revisiting the fluency activities from yesterday's and today's lessons. Then, leave the last couple of problems for students to draw their 5-group rows (with or without frames) independently.

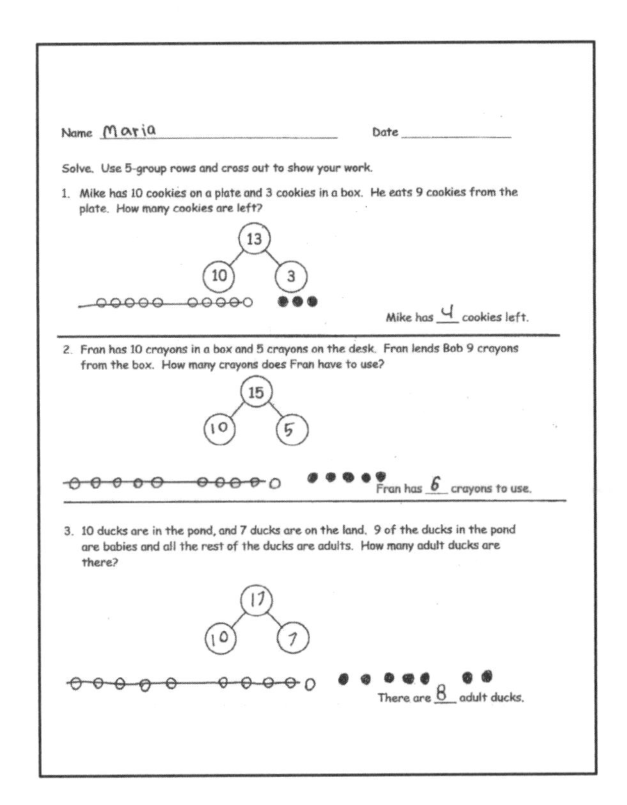

Name Maria Date

Solve. Use 5-group rows and cross out to show your work.

1. Mike has 10 cookies on a plate and 3 cookies in a box. He eats 9 cookies from the plate. How many cookies are left?

13, 10, 3

Mike has 4 cookies left.

2. Fran has 10 crayons in a box and 5 crayons on the desk. Fran lends Bob 9 crayons from the box. How many crayons does Fran have to use?

15, 10, 5

Fran has 6 crayons to use.

3. 10 ducks are in the pond, and 7 ducks are on the land. 9 of the ducks in the pond are babies and all the rest of the ducks are adults. How many adult ducks are there?

17, 10, 7

There are 8 adult ducks.

Problem Set (10 minutes)

Students should do their personal best to complete the Problem Set within the allotted 10 minutes. For some classes, it may be appropriate to modify the assignment by specifying which problems they work on first. Some problems do not specify a method for solving. Students should solve these problems using the RDW approach used for Application Problems.

Student Debrief (10 minutes)

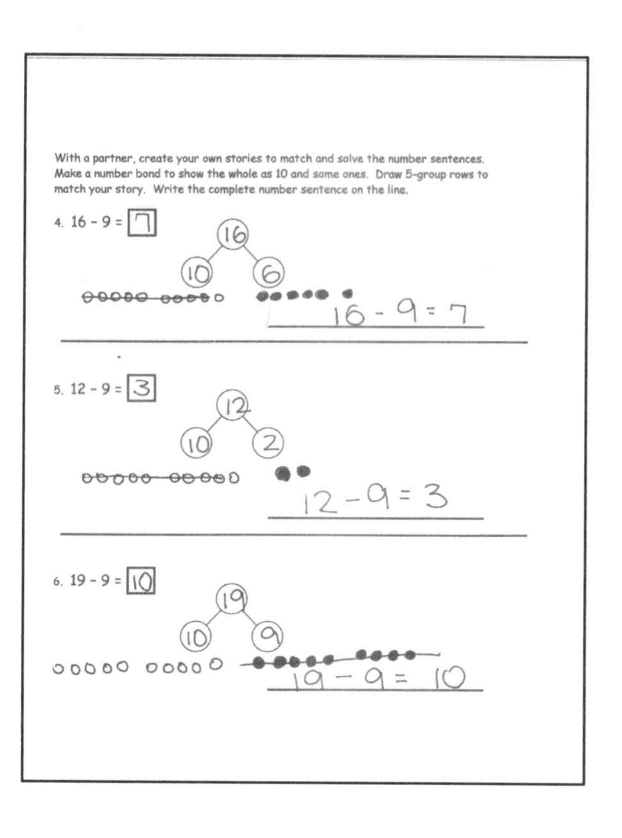
With a partner, create your own stories to match and solve the number sentences. Make a number bond to show the whole as 10 and some ones. Draw 5-group rows to match your story. Write the complete number sentence on the line.

4. 16 - 9 = 7

16 - 9 = 7

5. 12 - 9 = 3

12 - 9 = 3

6. 19 - 9 = 10

19 - 9 = 10

Lesson Objective: Solve word problems with subtraction of 9 from 10.

The Student Debrief is intended to invite reflection and active processing of the total lesson experience.

Invite students to review their solutions for the Problem Set. They should check work by comparing answers with a partner before going over answers as a class. Look for misconceptions or misunderstandings that can be addressed in the Debrief. Guide students in a conversation to debrief the Problem Set and process the lesson.

Any combination of the questions below may be used to lead the discussion.

- What pattern did you notice about how we solved – 9 problems? (We always took away 9 from 10. The answer is always the other part plus 1 because taking away 9 from 10 always leaves you with 1.)
- How can Problem 2 help you solve Problem 4?
- Look at Problem 6. Which part did you take the 9 from? Why? Explain your thinking.
- What new math drawing did we use today to solve subtraction problems? (5-group rows.) How is this drawing helpful?
- Look at your Application Problem. Where did you take your 9 from? Share your strategy.
- How can we use what we learned today to solve the Application Problem?

Exit Ticket (3 minutes)

After the Student Debrief, instruct students to complete the Exit Ticket. A review of their work will help with assessing students' understanding of the concepts that were presented in today's lesson and planning more effectively for future lessons. The questions may be read aloud to the students.

Name ____________________ Date ____________

Solve. Use 5-group rows, and cross out to show your work.

1. Mike has 10 cookies on a plate and 3 cookies in a box. He eats 9 cookies from the plate. How many cookies are left?

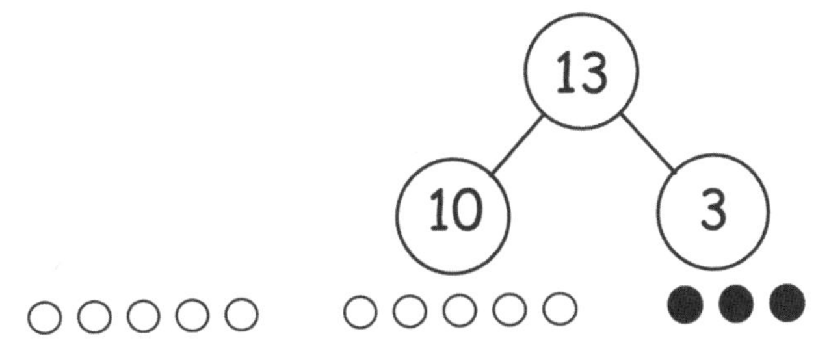

Mike has ___ cookies left.

2. Fran has 10 crayons in a box and 5 crayons on the desk. Fran lends Bob 9 crayons from the box. How many crayons does Fran have to use?

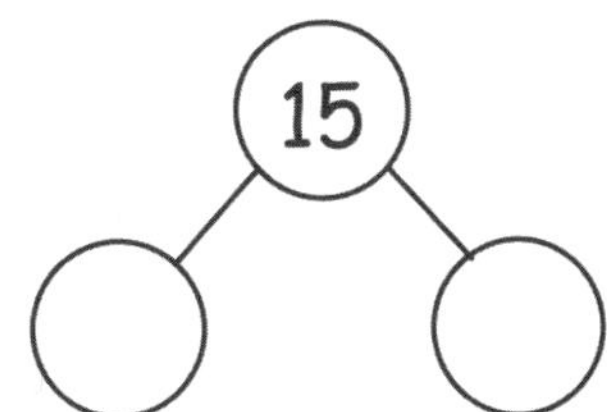

Fran has ___ crayons to use.

3. 10 ducks are in the pond, and 7 ducks are on the land. 9 of the ducks in the pond are babies, and all the rest of the ducks are adults. How many adult ducks are there?

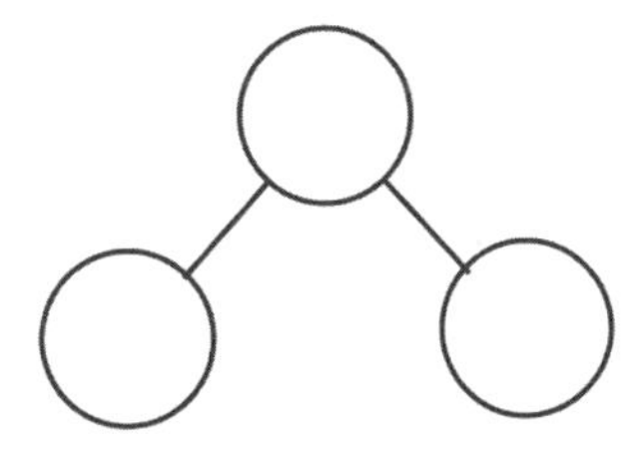

There are ___ adult ducks.

With a partner, create your own stories to match, and solve the number sentences. Make a number bond to show the whole as 10 and some ones. Draw 5-group rows to match your story. Write the complete number sentence on the line.

4. 16 - 9 = ☐

____

5. 12 - 9 = ☐

____

6. 19 - 9 = ☐

____

Name ______________________________ Date ______________

Solve. Fill in the number bond. Use 5-group rows, and cross out to show your work.

Gabriela has 4 hair clips in her hair and 10 hair clips in her bedroom. She gives 9 of the hair clips in her room to her sister. How many hair clips does Gabriela have now?

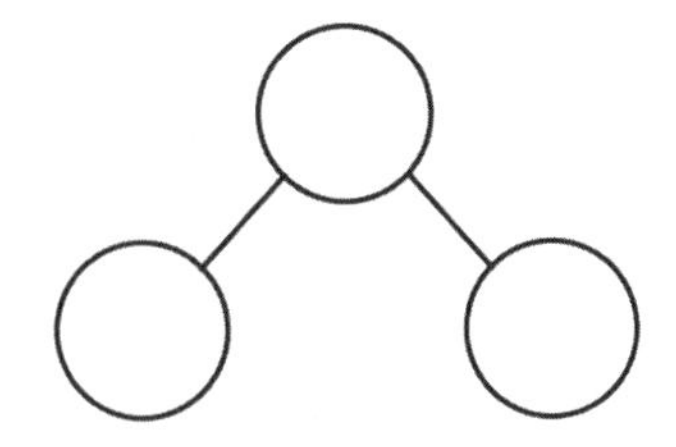

Gabriela has ___ hair clips.

Name ______________________________ Date ______________

Solve. Use 5-group rows, and cross out to show your work. Write number sentences.

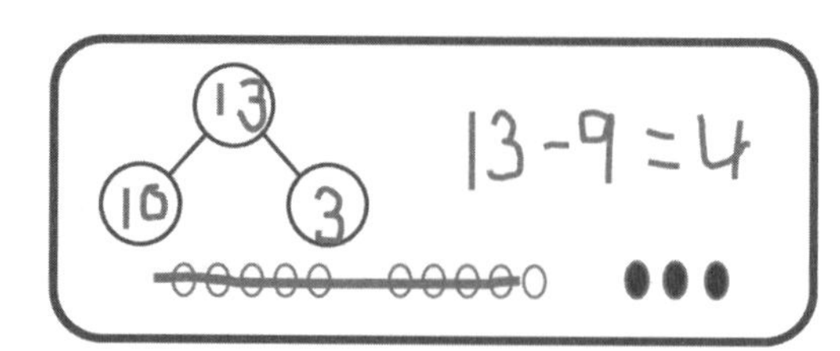

1. In a park, 10 dogs are running on the grass, and 1 dog is sleeping under the tree. 9 of the running dogs leave the park. How many dogs are left in the park?

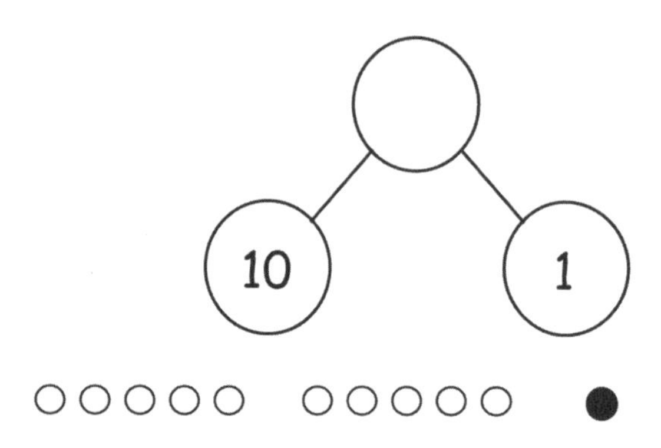

There are ___ dogs left in the park.

---

2. Alejandro had 9 rocks in his yard and 10 rocks in his room. 9 of the rocks in his room are gray rocks, and the rest of the rocks are white. How many white rocks does Alejandro have?

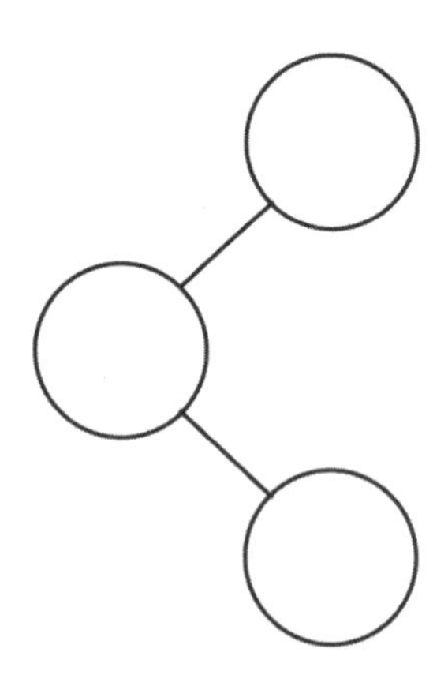

Alejandro has ___ white rocks.

3. Sophia has 8 toy cars in the kitchen and 10 toy cars in her bedroom. 9 of the toy cars in the bedroom are blue. The rest of her cars are red. How many red cars does Sophia have?

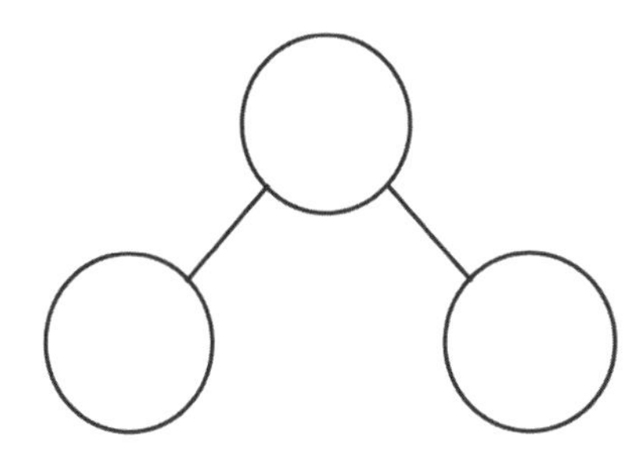

Sophia has ___ red cars.

---

4. Complete the number bond, and fill in the math story. Use 5-group rows, and cross out to show your work. Write number sentences.

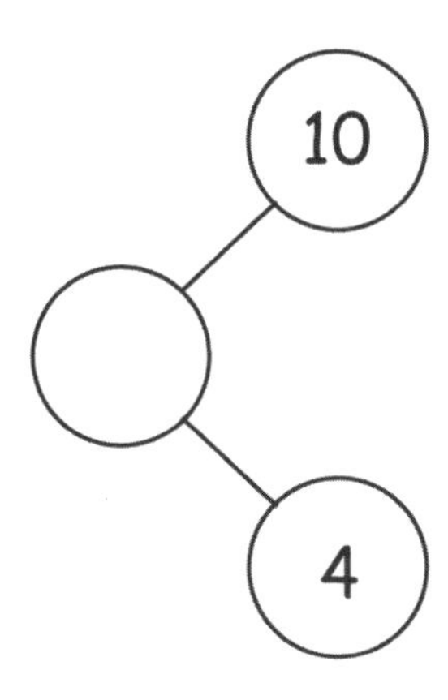

There were _____ birds splashing in a puddle and _____ birds walking on the dry grass. 9 of the splashing birds flew away. How many birds are left?

There are ___ birds left.

Lesson 14

Objective: Model subtraction of 9 from teen numbers.

Suggested Lesson Structure

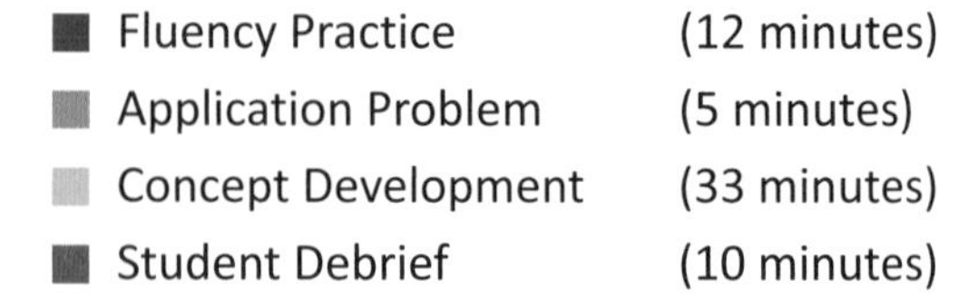

Fluency Practice	(12 minutes)
Application Problem	(5 minutes)
Concept Development	(33 minutes)
Student Debrief	(10 minutes)
Total Time	**(60 minutes)**

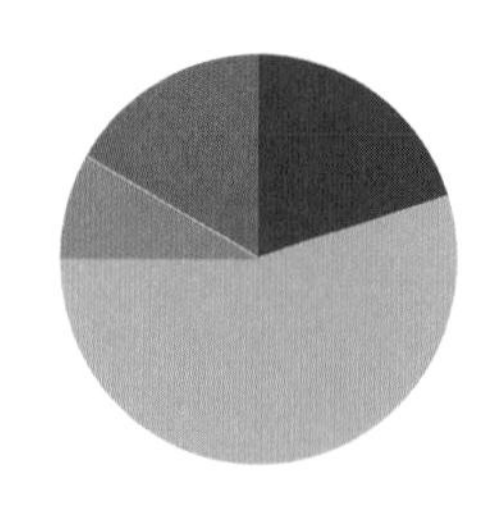

Fluency Practice (12 minutes)

- 5-Group Flash: Partners to Ten **1.OA.6** (2 minutes)
- Sprint: Subtraction Within 10 **1.OA.6** (10 minutes)

5-Group Flash: Partners to Ten (2 minutes)

Materials: (T) 5-group row cards (Lesson 12 Fluency Template 1)

Note: This activity supports Grade 1's core fluency standard of adding and subtracting within 10.

Flash a card for two to three seconds. Signal students to say the number. Signal again for students to say the partner to ten.

Sprint: Subtraction Within 10 (10 minutes)

Materials: (S) Subtraction Within 10 Sprint

Note: This Sprint reviews subtracting from ten, along with other subtraction facts within the Grade 1 core fluency objective of adding and subtracting within 10.

Application Problem (5 minutes)

Sarah has 6 blue beads in her bag and 4 green beads in her pocket. She gives away the 6 blue beads and 3 green beads. How many beads does she have left?

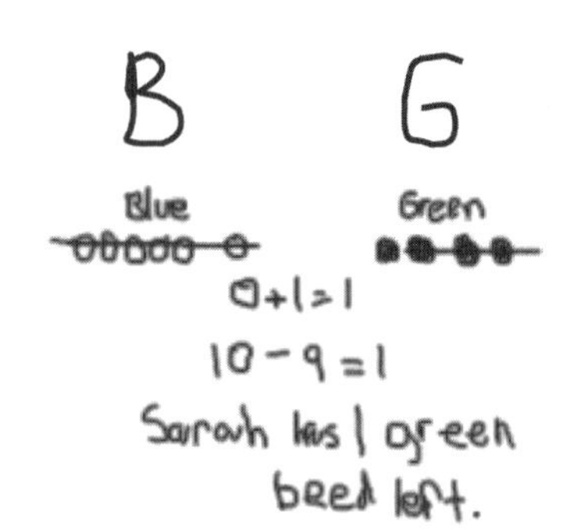

Note: This problem again asks students to subtract 9 from 10 but from two different places: some from the green bead group and some from the blue bead group. Using numbers within 10, students can explore how it is sometimes more efficient to take from a particular group(s) when subtracting. During the Student Debrief, students have the opportunity to share their strategies.

Concept Development (33 minutes)

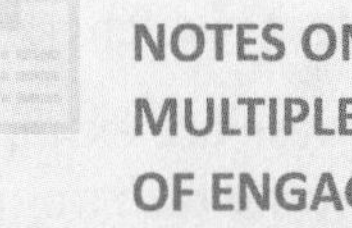

NOTES ON MULTIPLE MEANS OF ENGAGEMENT:

Be aware of the different learning needs in the class, and adjust the lesson as necessary. Since some students may need to work at the concrete level for a longer period of time, allow access to manipulatives. Other students may grasp the take from ten strategy quickly and be able to do mental math for some number sentences.

Materials: (T) Linking cubes (S) Personal white board, linking cubes (optional)

Students sit in a semicircle in the meeting area with their personal white boards.

T: (Project and read aloud.) Shayan has 12 eggs. He uses nine of them to make breakfast for his family. How many eggs are left?

T: How would you solve this problem? Use your personal white board to show your work.

S: (Solve as the teacher circulates.)

T: How did you solve this problem?

S: I drew 12 eggs. I crossed off 9, and I had 3 eggs left. → I counted on from 9 (9, 10, 11, 12). I have 3 fingers up, too. → I used the strategy from yesterday. I saw that I can take apart 12 as 10 and 2. I took away 9 from 10 and did 1 + 2 = 3. Three eggs.

T: No matter which strategies these students used, did they get the same answer?

S: Yes!

T: Here is a stick of 12 linking cubes to show how many eggs Shayan had in the beginning. Just like what we practiced yesterday, let's break it off into 10 and 2. (Break off and separate into two sticks.) We need to take away...?

S: 9 eggs.

MP.7

T: Where should I take 9 from? Turn and talk to your partner.

S: Take from 2 and then more from 10. → Take 9 from 10.

T: (Model taking away from 2.) Do I have enough? I need to take away more from 10. Help me count until we take away 9. (Count and take away 7 more.) How many do we have left?

S: 3.

T: (Model taking away from 10.) Taking away 9 from 10 will first leave us with...? (Break off 9 and show 1.)

S: 1.

T: 1 and 2 make...?

S: 3.

T: Turn and talk to your partner. Which was more efficient, simpler? Taking 9 from 10 or taking away the 2 and then some more from 10?

S: Taking 9 from 10.

T: I agree. Let's try more.

NOTES ON MULTIPLE MEANS OF ACTION AND EXPRESSION:

It is important to guide students to evaluate their thinking, as well as their partners', during the turn and talks. This provides students an opportunity to evaluate their process and analyze errors.

Repeat the process using the following suggested sequence: 11 – 9, 14 – 9, and 17 – 9. For each story problem, ask students which number 9 should be taken from.

T: Most of these are examples of 10 being a friendly number. When we take a number away from 10, we'll call it the take from ten strategy.

T: On your personal white board, draw a picture to show how we took 9 away from 10 to solve 17 – 9.

S: (Draw as the teacher circulates and supports students.)

T: Let's do just a few more. This time, you can use drawings or the linking cubes to show how we use the take from ten strategy to solve.

Repeat with 15 – 9, 18 – 9, and 19 – 9.

Problem Set (10 minutes)

Students should do their personal best to complete the Problem Set within the allotted 10 minutes. For some classes, it may be appropriate to modify the assignment by specifying which problems they work on first. Some problems do not specify a method for solving. Students should solve these problems using the RDW approach used for Application Problems.

Student Debrief (10 minutes)

Lesson Objective: Model subtraction of 9 from teen numbers.

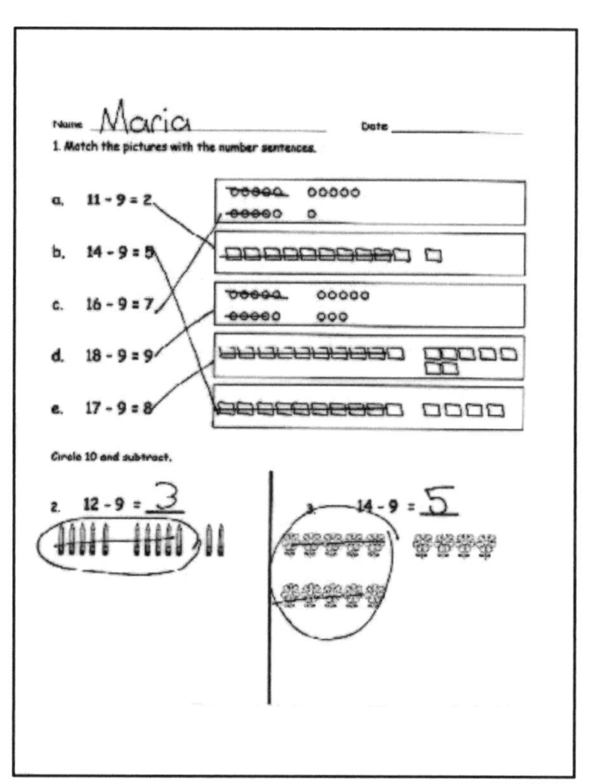
Name Maria Date

1. Match the pictures with the number sentences.

a. 11 - 9 = 2

b. 14 - 9 = 5

c. 16 - 9 = 7

d. 18 - 9 = 9

e. 17 - 9 = 8

Circle 10 and subtract.

2. 12 - 9 = 3

3. 14 - 9 = 5

The Student Debrief is intended to invite reflection and active processing of the total lesson experience.

Invite students to review their solutions for the Problem Set. They should check work by comparing answers with a partner before going over answers as a class. Look for misconceptions or misunderstandings that can be addressed in the Debrief. Guide students in a conversation to debrief the Problem Set and process the lesson.

Any combination of the questions below may be used to lead the discussion.

- Look at Problems 8–10. What is happening with the difference in each of these problems? If the pattern continued, what would be the next problem? What problem would come before the first problem?
- When solving 19 – 9, where can you take 9 from? Explain your answer.
- A student says, "Taking away 9 is like adding 1 to the part that is not 10 from the number bond. To solve 17 – 9, you can do 1 + 7." Is she correct? Explain your answer.

- What new strategy did we learn to solve our problems today? (Take from ten strategy.) Explain to your partner why it's an efficient strategy.
- Look at your Application Problem. How did you solve it? Do you have to add the blue beads and the green beads together to solve this problem? Why or why not? How is it like our lesson today? How is it different?

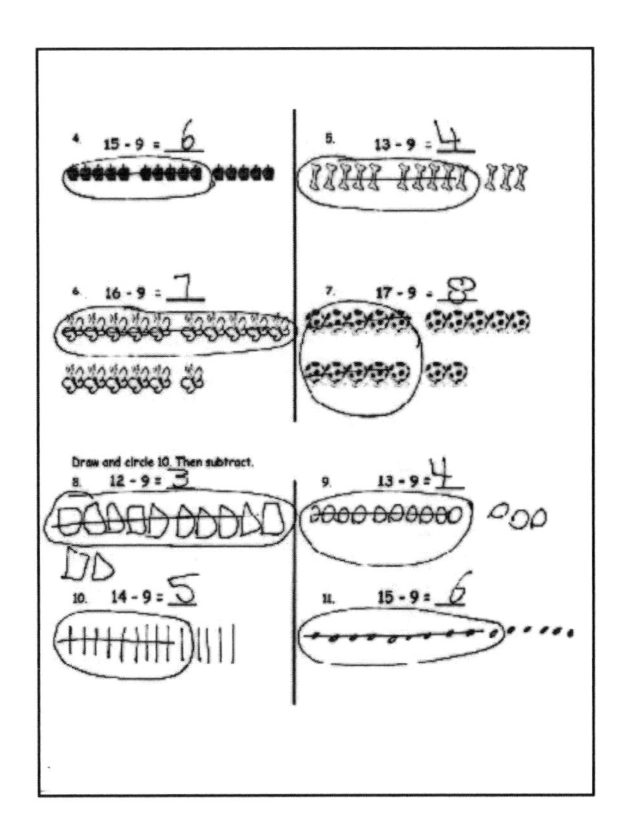

Exit Ticket (3 minutes)

After the Student Debrief, instruct students to complete the Exit Ticket. A review of their work will help with assessing students' understanding of the concepts that were presented in today's lesson and planning more effectively for future lessons. The questions may be read aloud to the students.

A

Number Correct:

Name ______________________ Date ____________

*Write the missing number.

1.	10 - 9 = □		16.	10 - □ = 5	
2.	10 - 8 = □		17.	9 - □ = 5	
3.	10 - 6 = □		18.	8 - □ = 5	
4.	10 - 7 = □		19.	10 - □ = 3	
5.	10 - 6 = □		20.	9 - □ = 3	
6.	10 - 5 = □		21.	8 - □ = 3	
7.	10 - 6 = □		22.	□ - 6 = 4	
8.	10 - 4 = □		23.	□ - 6 = 3	
9.	10 - 3 = □		24.	□ - 6 = 2	
10.	10 - 7 = □		25.	10 - 4 = 9 - □	
11.	10 - 8 = □		26.	8 - 2 = 10 - □	
12.	10 - 2 = □		27.	8 - □ = 10 - 3	
13.	10 - 1 = □		28.	9 - □ = 10 - 3	
14.	10 - 9 = □		29.	10 - 4 = 9 - □	
15.	10 - 10 = □		30.	□ - 2 = 10 - 4	

B

Number Correct:

Name ____________________ Date ____________

*Write the missing number.

1.	10 - 8 = □		16.	10 - □ = 0	
2.	10 - 9 = □		17.	9 - □ = 0	
3.	10 - 8 = □		18.	8 - □ = 0	
4.	10 - 9 = □		19.	10 - □ = 1	
5.	10 - 7 = □		20.	9 - □ = 1	
6.	10 - 9 = □		21.	8 - □ = 1	
7.	10 - 8 = □		22.	□ - 5 = 5	
8.	10 - 7 = □		23.	□ - 5 = 4	
9.	10 - 3 = □		24.	□ - 5 = 3	
10.	10 - 7 = □		25.	10 - 8 = 9 - □	
11.	10 - 6 = □		26.	8 - 6 = 10 - □	
12.	10 - 4 = □		27.	8 - □ = 10 - 2	
13.	10 - 3 = □		28.	9 - □ = 10 - 2	
14.	10 - 7 = □		29.	10 - 3 = 9 - □	
15.	10 - 5 = □		30.	□ - 1 = 10 - 3	

Name ______________________________ Date ______________

1. Match the pictures with the number sentences.

a. 11 - 9 = 2

b. 14 - 9 = 5

c. 16 - 9 = 7

d. 18 - 9 = 9

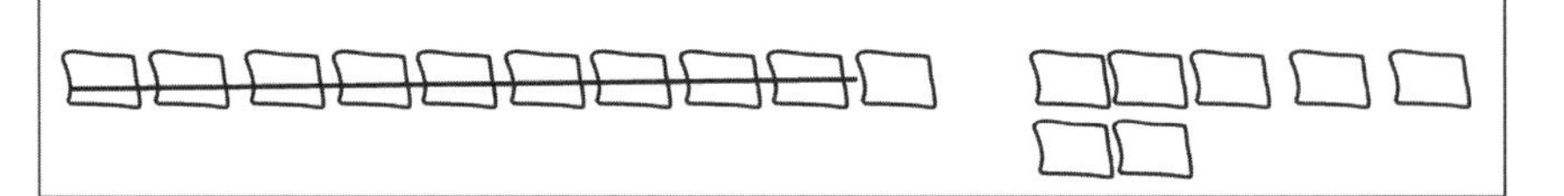

e. 17 - 9 = 8

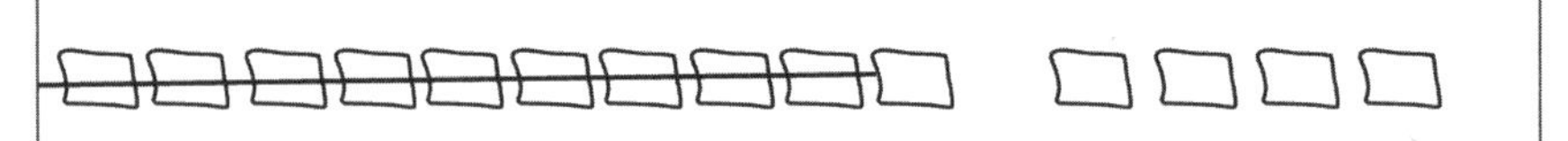

Circle 10 and subtract.

2. 12 - 9 = _____

3. 14 - 9 = _____

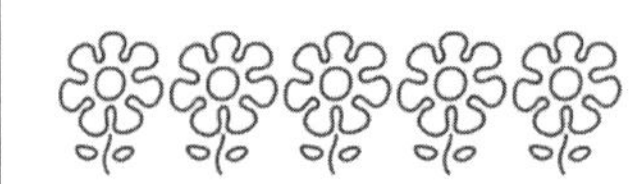
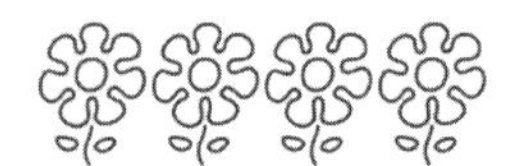
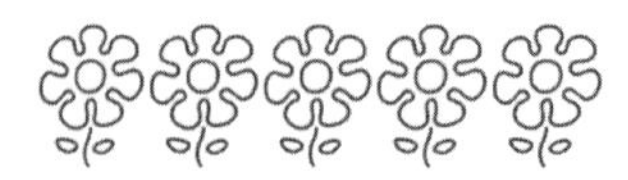

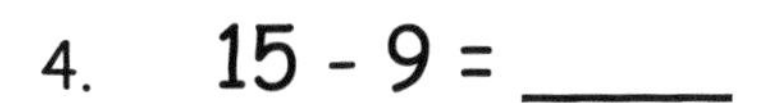

4. 15 - 9 = _____

5. 13 - 9 = _____

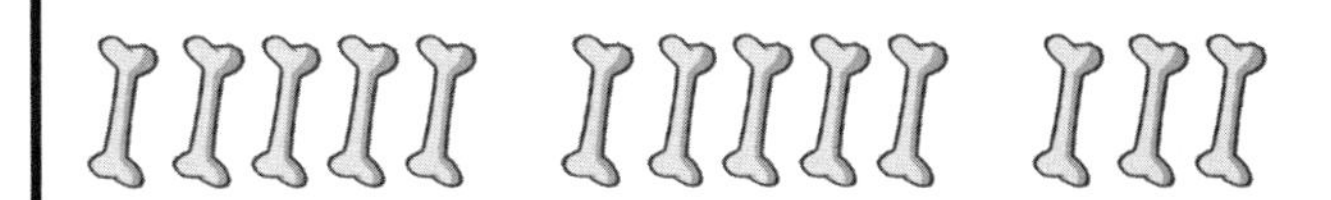

6. 16 - 9 = _____

7. 17 - 9 = _____

Draw and circle 10. Then subtract.

8. 12 - 9 = _____

9. 13 - 9 = _____

10. 14 - 9 = _____

11. 15 - 9 = _____

Name ______________________________ Date ______________

Draw and circle 10. Solve and make a number bond.

1. 17 - 9 = _____

2. 14 - 9 = _____

3. 15 - 9 = _____

4. 18 - 9 = _____

Name ______________________________ Date ________________

Circle 10 and subtract. Make a number bond.

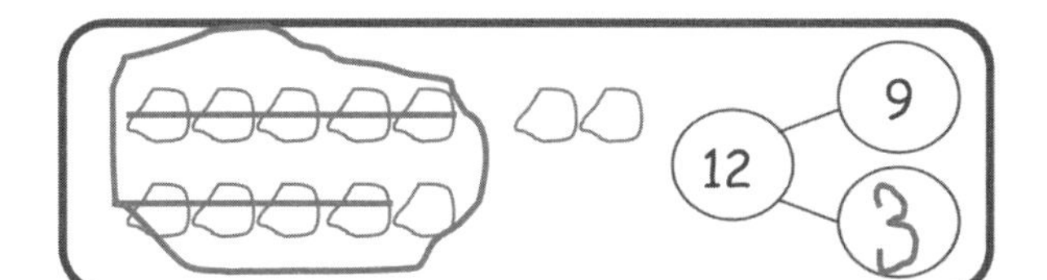

1. 15 - 9 = _____

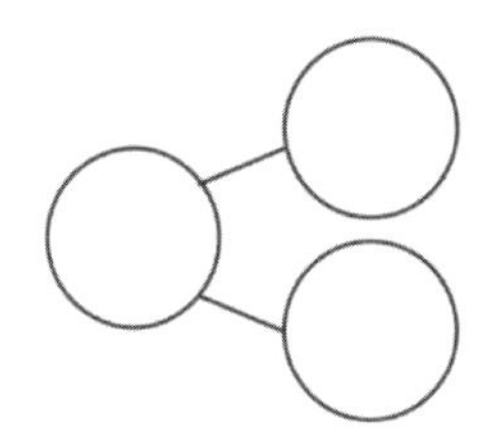

Draw and circle 10. Subtract and make a number bond.

2. 14 - 9 = _____

3. 12 - 9 = _____

4. 13 - 9 = _____

5. 16 - 9 = _____

6. Complete the number bond, and write the number sentence that helped you.

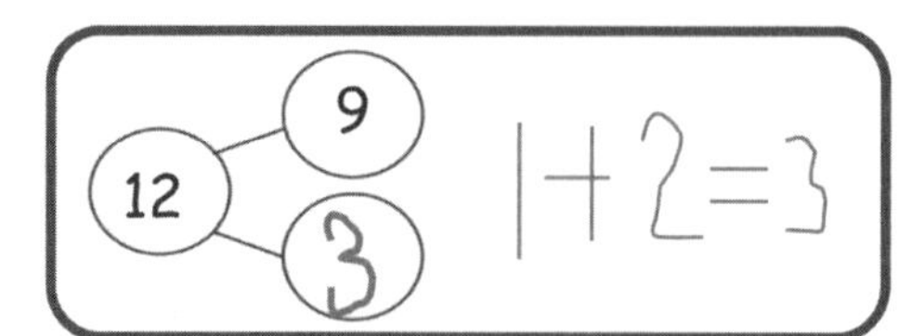

a. 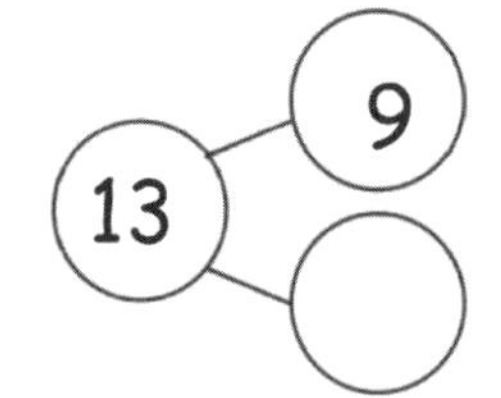

____

b. 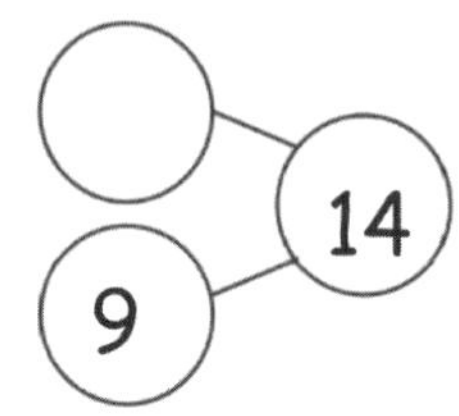

____

c. 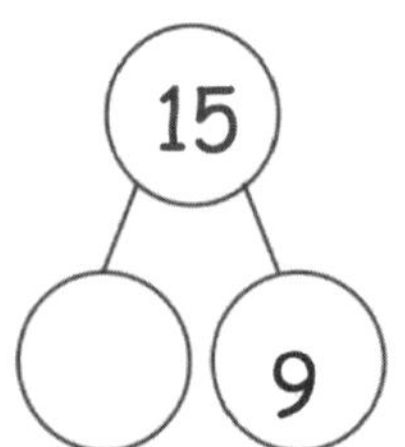

____

d. 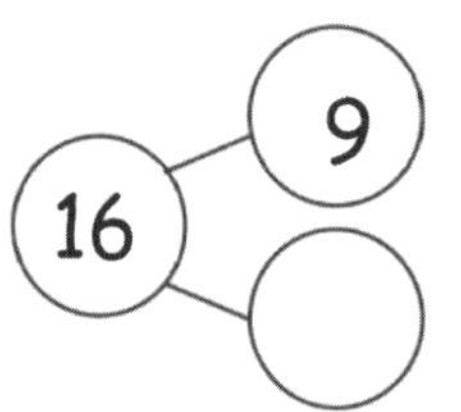

____

7. Make the number bond that would come next, and write a number sentence that matches.

Lesson 15

Objective: Model subtraction of 9 from teen numbers.

Suggested Lesson Structure

■ Fluency Practice	(10 minutes)
■ Application Problem	(7 minutes)
■ Concept Development	(33 minutes)
■ Student Debrief	(10 minutes)
Total Time	**(60 minutes)**

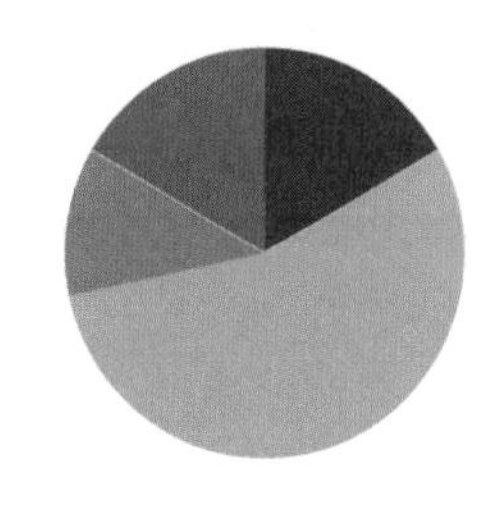

Fluency Practice (10 minutes)

- 5-Group Flash: 5 Less and 4 Less **1.OA.6** (2 minutes)
- Make It Equal: Subtraction Expressions **1.OA.6** (5 minutes)
- Partners to Ten **1.OA.6** (3 minutes)

5-Group Flash: 5 Less and 4 Less (2 minutes)

Materials: (T) 5-group row cards (Lesson 12 Fluency Template 1)

Note: This activity supports Grade 1's core fluency standard of adding and subtracting within 10 and helps students to see the relationship with 5 less (easy, one 5-group less) to 4 less (take out the five except for 1). For struggling students, lead them to visualize 5 less by hiding a 5-group. Make the connection to seeing the number on their fingers and hiding one hand.

Flash a card for two to three seconds. Students say the number that is 5 less and then 4 less.

Make It Equal: Subtraction Expressions (5 minutes)

Materials: (S) 5-group cards (Lesson 1 Fluency Template), minus and equal symbol cards, one "=" card and two "–" cards (Fluency Template) per set of partners

Note: This activity builds fluency for subtraction within 10 and promotes an understanding of equality.

Assign students partners of similar skill level. Students arrange 5-group cards from 0 to 10, including the extra 5, and place the "=" card between them. Write four numbers on the board (e.g., 10, 9, 2, 1). Partners take the 5-group cards that match the numbers written to make two equivalent subtraction expressions (e.g., 10 – 9 = 2 – 1). Students can be encouraged to make another sentence of equivalent expressions for the same set of cards as well.

Suggested sequence: 10, 9, 2, 1; 10, 3, 9, 2; 10, 4, 5, 9; 10, 8, 7, 9; 10, 7, 9, 6; 10, 8, 4, 2.

Partners to Ten (3 minutes)

Materials: (S) Personal white board with 5-group row insert (Lesson 12 Fluency Template 2)

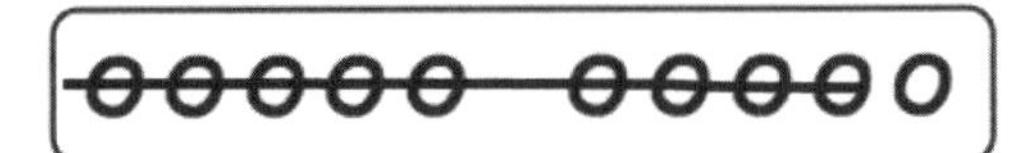

Note: This maintenance fluency activity with partners to ten facilitates the take from ten subtraction strategy.

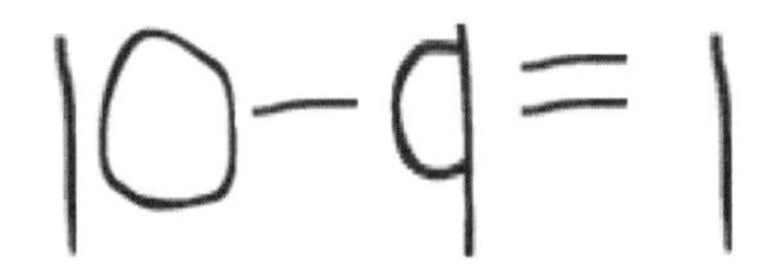

Say a number between 0 and 10 (e.g., 9). Students cross off the number from the 5-group row insert and write the corresponding subtraction sentence.

Application Problem (7 minutes)

Julian has 7 markers. His mother gives him 8 more. He loses 9 markers. How many does he have left?

Note: In the Student Debrief, students can discuss their drawings and number sentences and share various strategies, one of which may be decomposing 15 into 10 and 5 and taking 9 from 10. Though it is covered formally in a later lesson, teachers might also choose to encourage students to see that the expressions 15 – 9 and 1 + 5 are equivalent.

J M

10 − 9 = 1

1 + 5 = 6

15 − 9 = 6

Concept Development (33 minutes)

Materials: (S) Personal white board

T: (Project 15 – 9 = ___.) With a partner, solve this on your personal white board. Use words or a drawing to show how you know.

S: (Discuss and solve with partners as the teacher circulates and notices the solution strategies students are using independently.)

T: What is the unknown number in this number sentence?

S: 6.

T: How did you solve that?

S: I started at 9 and counted on until I got to 15. That took 6 fingers. → I took 9 away from 15 and had 6 left. → I know 15 is made of 10 and 5, so I took 9 from 10 and then saw that I had 6 left.

NOTES ON MULTIPLE MEANS OF ACTION AND EXPRESSION:

Some students may have trouble organizing their dots in a row with spaces in the correct places. Be sure to accommodate these students and set up a way for them to be successful with their drawings. Avoiding a frustration like drawing keeps students focused on the math or task at hand.

T: I noticed that many of you used drawings on your personal white boards. How can we draw 15 so that we can tell how many we have when we look quickly?

S: Use 5-group pictures!

T: Let's use 5-groups in one long row, like we did during Fluency Practice today. (On the board, draw a 5-group row to show 15. Leave extra space between the first 10 circles and the last 5 circles.)

T: Let's frame the 10 circles we have so we can see 10 and 5 more easily. (Draw a rectangle around the first 10 circles.)

T: Now we can see 15 as 10 and 5. (Add the number 15 and the bond lines above as shown.)

T: If we want to take 9 out of 15, how can this drawing help us find a quick and easy place to take the 9 from?

S: The group of 10 inside the frame!

T: Hmm. If I take 9 out of 10, how much would that leave me in the frame?

S: Just one!

T: How much do we have when we take 9 out of 15?

S: 6. There is 1 left in the frame and 5 left on the other side, so that's 6.

T: (Project 14 – 9 = ___.) Let's all make 5-group drawings like that last one as we solve for the unknown number.

Repeat the process above with the following sequence: 16 – 9, 13 – 9, 17 – 9. Support students in drawing 5-group rows so that they can see the ten and the additional circles easily. Circulate and encourage students to share where they can find 9 quickly and easily.

Suggest students cover the 9 to help them move away from counting all and move toward visualizing and mental math.

> **NOTES ON MULTIPLE MEANS OF ENGAGEMENT:**
>
> For those students who can fluently solve math facts within 20, cultivate excitement by connecting on-level math to higher math, presenting numbers to 100. If they can solve 15 – 9 with ease, present problems such as 25 – 9 or 35 – 9.

After two problems, ask students to close their eyes and see if they can visualize or see in their minds' eyes what is happening in the story when they subtract 9.

Have students draw the problem using 5-group drawings. Before they cross out the 9, ask them to visualize what the picture will look like once it is crossed out and determine how many will not be crossed out. Then, have students cross out the 9 and see if their pictures match what they visualized.

Problem Set (10 minutes)

Students should do their personal best to complete the Problem Set within the allotted 10 minutes. For some classes, it may be appropriate to modify the assignment by specifying which problems they work on first. Some problems do not specify a method for solving. Students should solve these problems using the RDW approach used for Application Problems.

Student Debrief (10 minutes)

Lesson Objective: Model subtraction of 9 from teen numbers.

The Student Debrief is intended to invite reflection and active processing of the total lesson experience.

Invite students to review their solutions for the Problem Set. They should check work by comparing answers with a partner before going over answers as a class. Look for misconceptions or misunderstandings that can be addressed in the Debrief. Guide students in a conversation to debrief the Problem Set and process the lesson.

Any combination of the questions below may be used to lead the discussion.

- Look at your Problem Set. How did you find an easy way to take 9 out of the teen numbers?
- Look at Problems 6–8. What do you notice is similar about the pictures in these problems? What do you notice about the numbers in these problems? If this pattern continued, what problem would come next? How can the problems help us solve 11 – 9?
- Look at Problem 10. How are the two number sentences related? What was the same or different about your drawings?
- Look at your Application Problem. How does the problem connect to today's lesson? How would you change or add to your work?

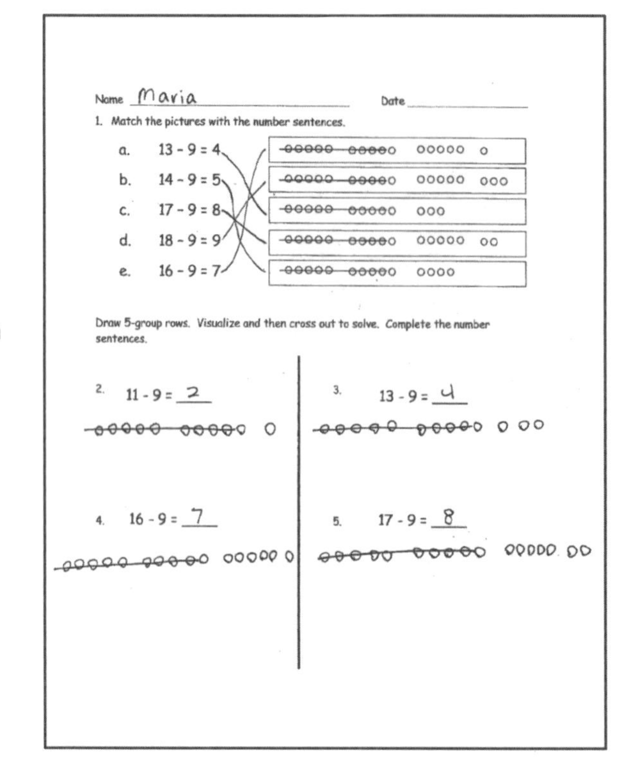
Name Maria Date

1. Match the pictures with the number sentences.

a. 13 - 9 = 4
b. 14 - 9 = 5
c. 17 - 9 = 8
d. 18 - 9 = 9
e. 16 - 9 = 7

Draw 5-group rows. Visualize and then cross out to solve. Complete the number sentences.

2. 11 - 9 = 2
3. 13 - 9 = 4
4. 16 - 9 = 7
5. 17 - 9 = 8

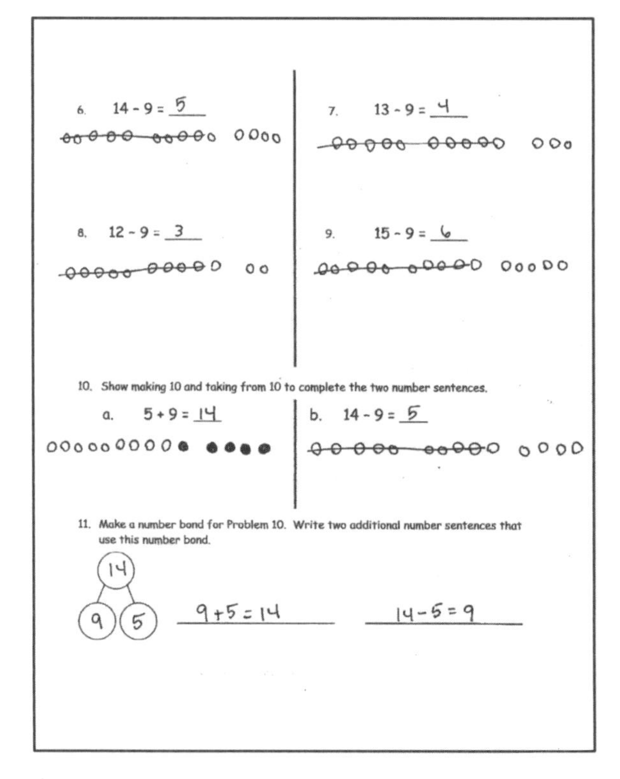
6. 14 - 9 = 5
7. 13 - 9 = 4
8. 12 - 9 = 3
9. 15 - 9 = 6

10. Show making 10 and taking from 10 to complete the two number sentences.

a. 5 + 9 = 14
b. 14 - 9 = 5

11. Make a number bond for Problem 10. Write two additional number sentences that use this number bond.

14, 9, 5
9 + 5 = 14
14 - 5 = 9

Exit Ticket (3 minutes)

After the Student Debrief, instruct students to complete the Exit Ticket. A review of their work will help with assessing students' understanding of the concepts that were presented in today's lesson and planning more effectively for future lessons. The questions may be read aloud to the students.

G1-M2-TE-BK2-1.3.1-01.2016

Name ____________________ Date __________

1. Match the pictures with the number sentences.

a. 13 - 9 = 4

b. 14 - 9 = 5

c. 17 - 9 = 8

d. 18 - 9 = 9

e. 16 - 9 = 7

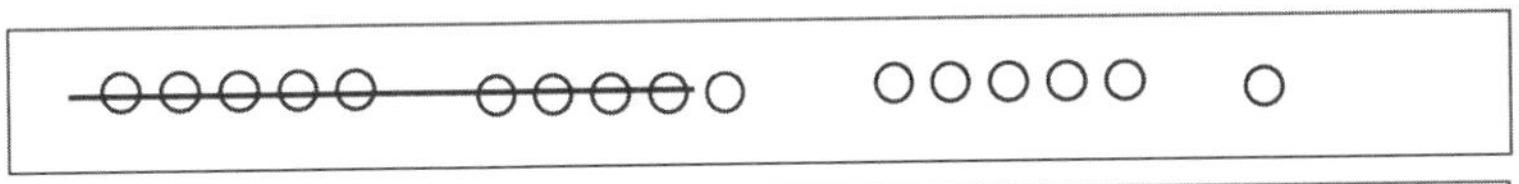

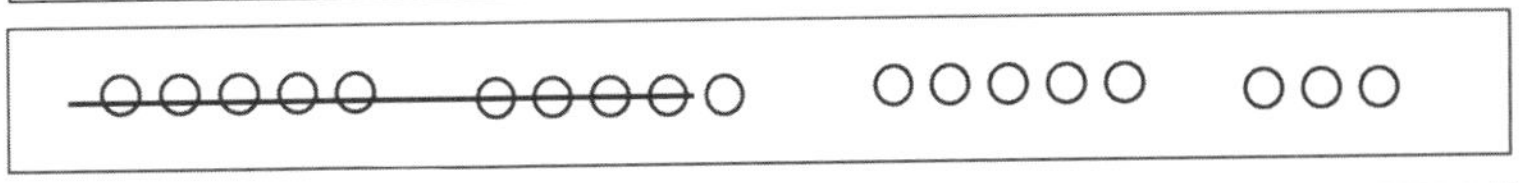

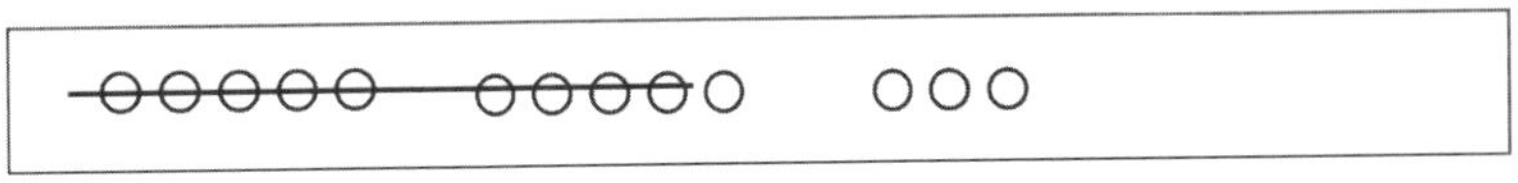

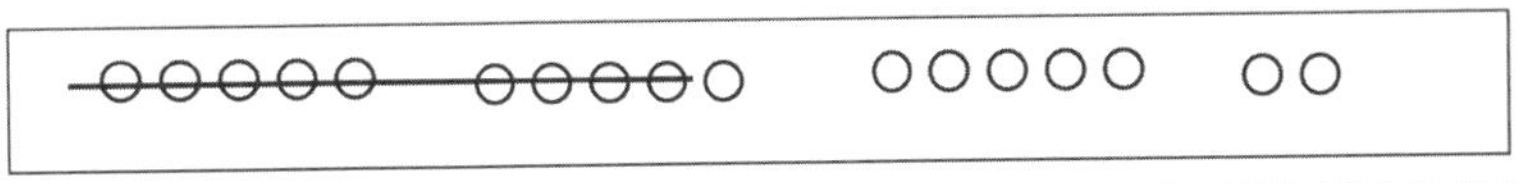

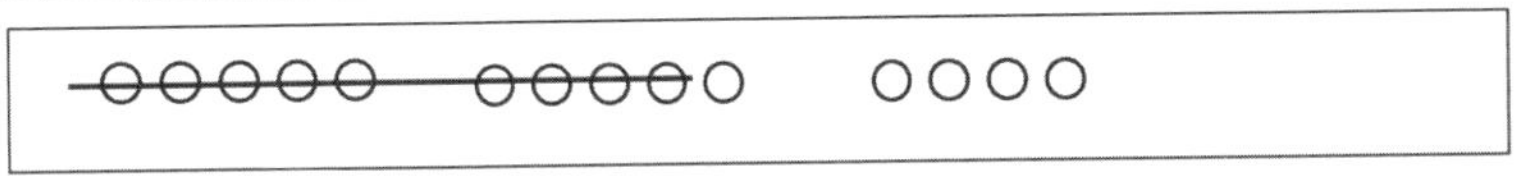

Draw 5-group rows. Visualize and then cross out to solve. Complete the number sentences.

2. 11 - 9 = ______

3. 13 - 9 = ______

4. 16 - 9 = ______

5. 17 - 9 = ______

6. 14 - 9 = ____

7. 13 - 9 = ____

8. 12 - 9 = ____

9. 15 - 9 = ____

10. Show making 10 and taking from 10 to complete the two number sentences.

a. 5 + 9 = ____

b. 14 - 9 = ____

11. Make a number bond for Problem 10. Write two additional number sentences that use this number bond.

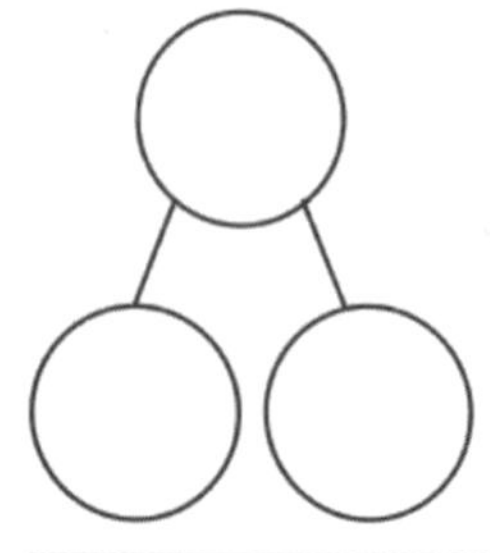

____________________ ____________________

Name ________________________ Date ____________

Draw 5-group rows, and cross out to solve. Complete the number sentences.

1. 17 - 9 = _____

2. 19 - 9 = _____

Name ______________________ Date ____________

Write the number sentence for each 5-group row drawing.

1.

13 - 9 = 4

____

Draw 5-groups to complete the number bond, and write the 9- number sentence.

2.

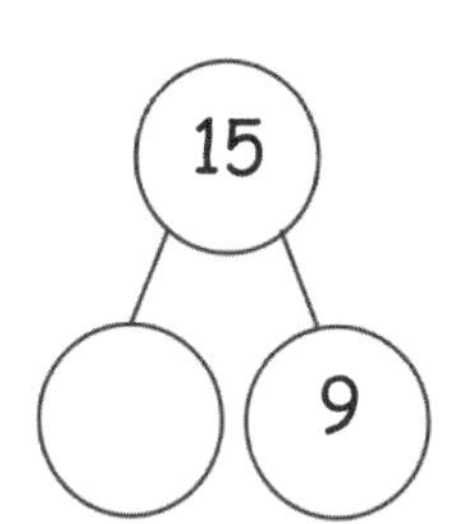

3.

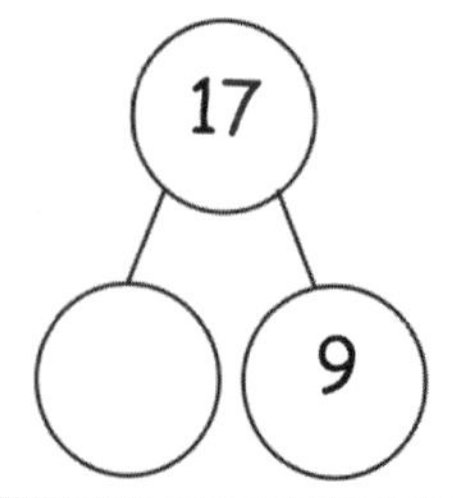

Draw 5-groups to complete the number bond, and write the 9– number sentence.

4.

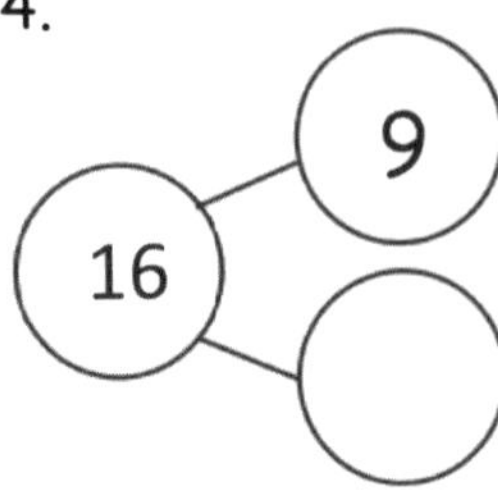

Draw 5-groups to show making ten and taking from ten to solve the two number sentences. Make a number bond, and write two additional number sentences that would have this number bond.

5. $8 + 9 =$ _____

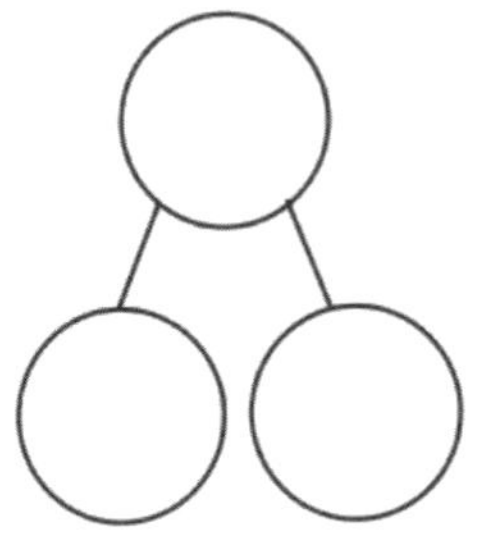

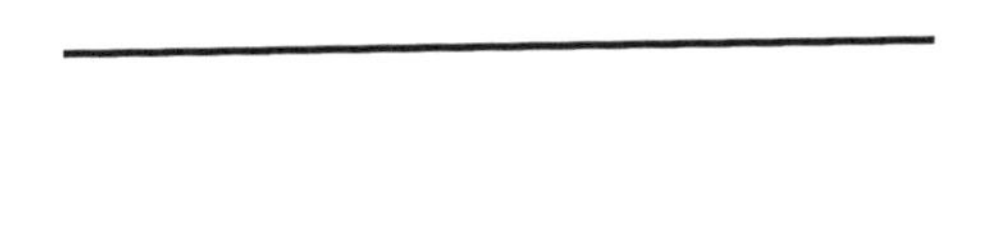

6. $17 - 9 =$ _____

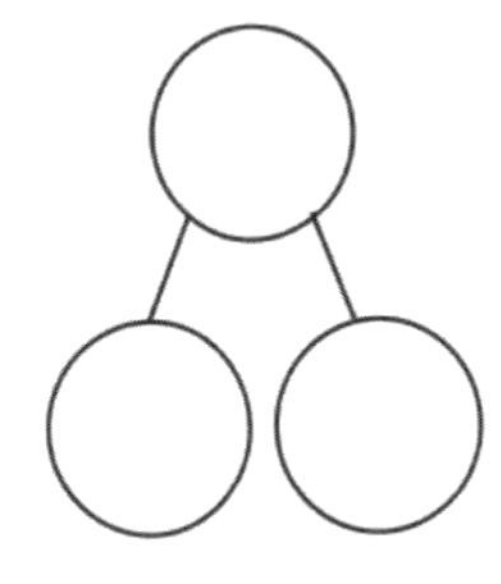

____

____

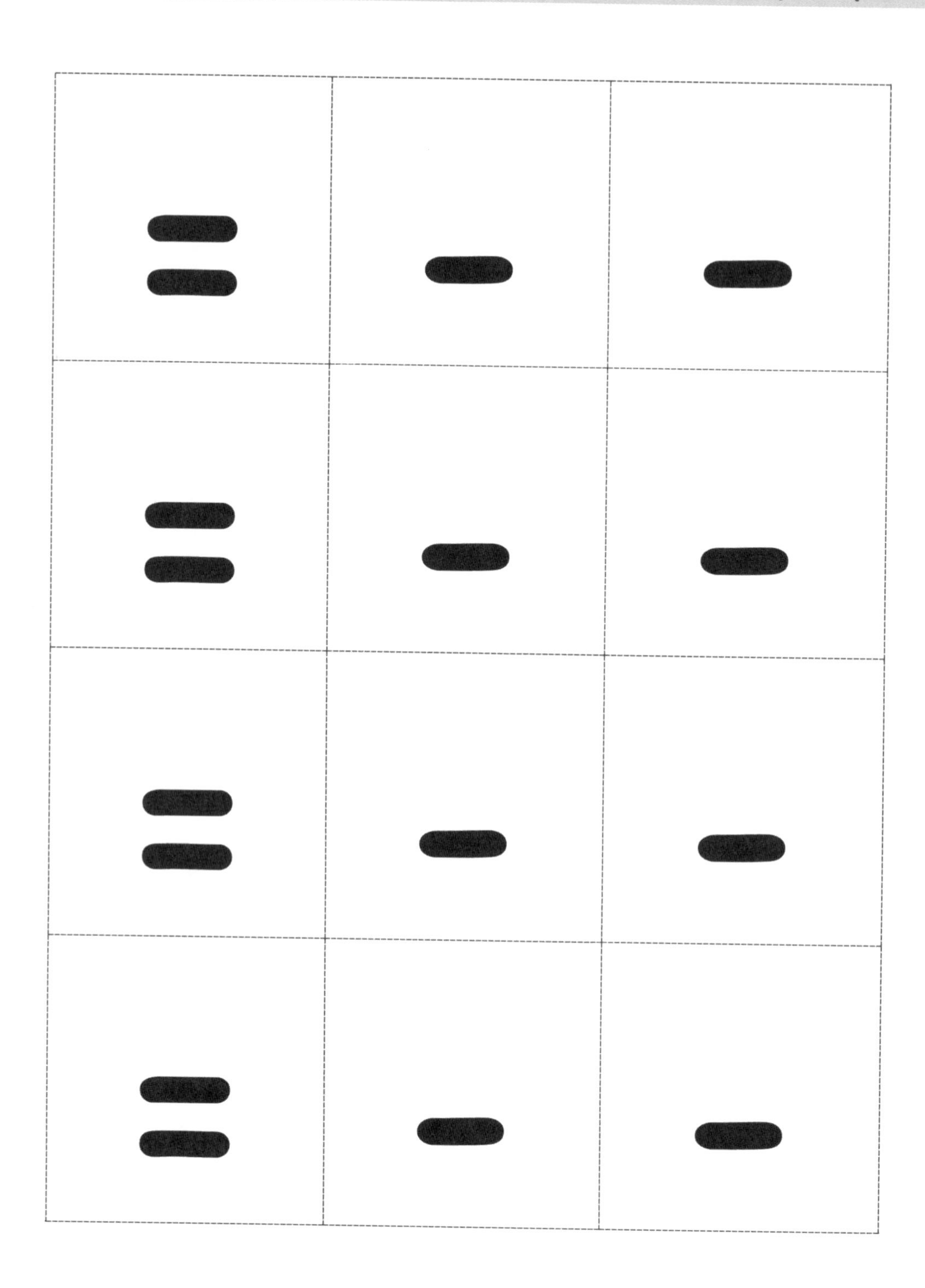

minus and equal symbol cards

Lesson 16

Objective: Relate counting on to making ten and taking from ten.

Suggested Lesson Structure

- Fluency Practice (14 minutes)
- Application Problem (5 minutes)
- Concept Development (31 minutes)
- Student Debrief (10 minutes)

Total Time (60 minutes)

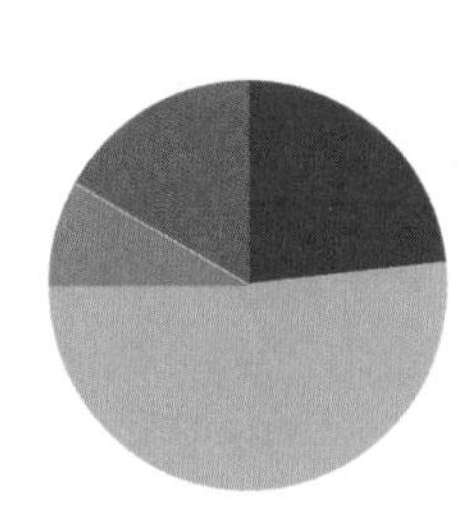

Fluency Practice (14 minutes)

- Subtract 9 **1.OA.6** (10 minutes)
- 5 and 4 Less **1.OA.6** (2 minutes)
- Happy Counting by Twos: Odd Numbers **1.OA.5** (2 minutes)

Subtract 9 (10 minutes)

Materials: (S) Personal white board, 5-group row insert (Lesson 12 Fluency Template 2)

Note: This fluency activity reviews the take from ten subtraction strategy. The goal is for students to be able to use this strategy as mental math. For the first two problems, have students cross off the circles to show their subtraction. Then, have students cover the circles and imagine subtracting them.

T: Look at your 5-group row insert. Draw more circles to the right of your 5-group to show a total of 12.

S: (Draw 2 more circles).

T: Say 12 as a number bond, with 10 as a part.

S: 10 and 2 make 12.

T: Turn your circles into a number bond.

S/T: (Draw lines to make a number bond with the numeral 12 on top.)

T: Show me 12 – 9. Think about whether you should subtract from the part with ten or the part with two.

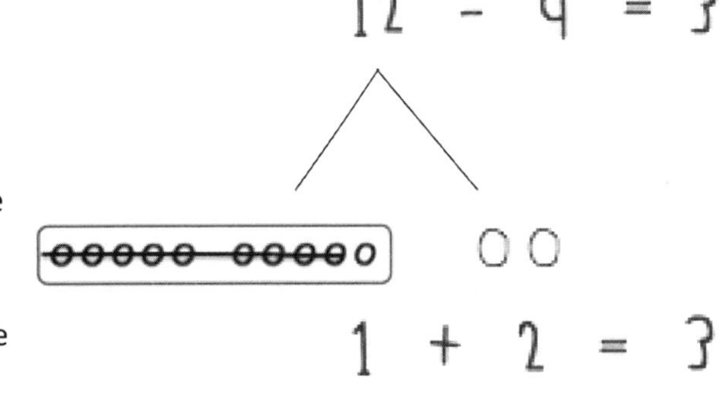

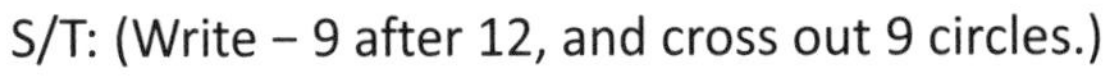

S/T: (Write – 9 after 12, and cross out 9 circles.)

T: Below your circles, write an addition sentence to show what is left.

S: (Write 1 + 2 = 3.)

T: What is 12 – 9?

S: 3.

Continue with other numbers between 11 and 20. As soon as possible, reduce the number of steps (e.g., show me 14 – 9).

5 and 4 Less (2 minutes)

Materials: (T) 5-group row cards (Lesson 12 Fluency Template 1)

Note: This activity supports Grade 1's core fluency standard of adding and subtracting within 10 and helps students to see 4 less as related to 5 less (take out the five except for 1). Lead struggling students to visualize 5 less by hiding a 5-group. Make the connection to seeing the number on their fingers and hiding one hand.

Flash a card for two to three seconds. Students say the number that is 5 less and then 4 less.

Happy Counting by Twos: Odd Numbers (2 minutes)

Note: A review of counting on allows students to maintain fluency with adding and subtracting 2.

Repeat the Happy Counting activity from Lesson 4, counting by twos from 1 to 19 and back. This range may be adjusted to meet the needs of students.

Application Problem (5 minutes)

There were 16 coats on the rack. Nine students took their coats to go outside. How many coats were still on the rack?

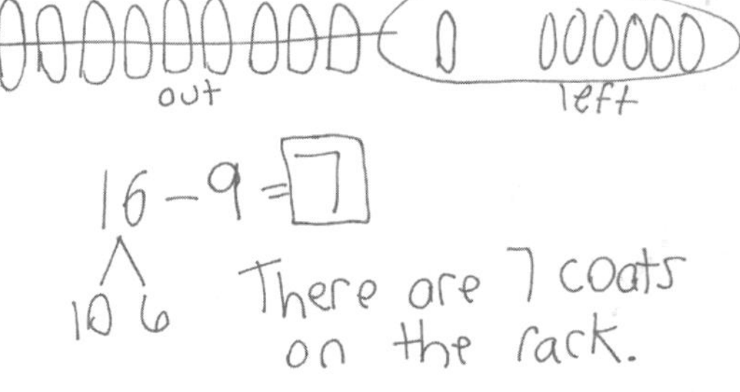

Extension: If 4 more students take their coats to go outside, how many coats will still be hanging?

Note: In this problem, students may use the take from ten strategy or count on strategy. While circulating, look for students who used these strategies, and ask them to share during the Student Debrief.

Concept Development (31 minutes)

Materials: (S) Personal white board

Have students sit in a semicircle in the meeting area with their personal white boards.

T: (Write 11 – 9 = ____.) Solve 11 – 9 on your personal white board.

S: (Solve on personal white board as the teacher circulates and selects two students: one who is using the count on strategy and another using the take from ten strategy.)

S: I started with 9 and counted on. Niiiine, 10, 11. Two fingers are up.

T: Let's all try counting on.

T/S: Niiiine, 10, 11. (Put up a finger for each count after 9.)

T: (Ask the second student.) How did you solve 11 – 9?

S: I took 9 from 10 and did 1 + 1 and got 2.

T: Let's all use the take from ten strategy to solve on our personal white boards.

S: (Show a number bond to break apart 11 to solve.)

T: What did you do?

S: 10 – 9 is 1; 1 + 1 is 2.

T: Everyone, let's use the take from ten strategy using our fingers to check! Start by showing 11 fingers.

S: We can't! We only have 10 fingers!

MP.4

T: Oh boy. We can't quite do that, can we? We'll just have to use our imaginations. First, put up your 10 fingers.

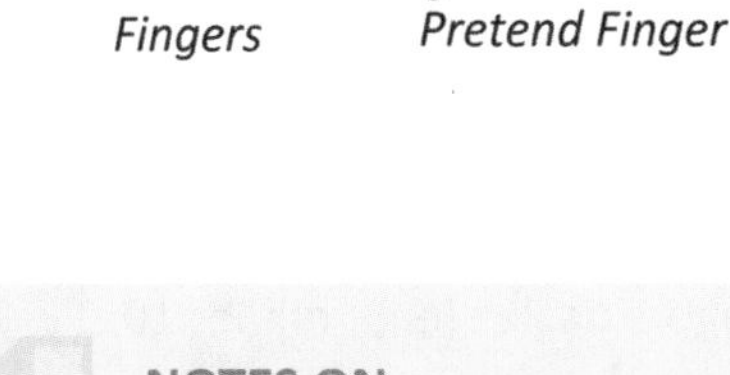

S: (Show 10 fingers.)

T: How many more fingers do we need to imagine?

S: 1.

T: Visualize, or picture in your mind, 1 more finger next to your 10. Now, take away 9, all at once.

S: (Hold 1 finger up.)

T: How many fingers do you have up?

S: 1.

T: How many pretend fingers are still up?

S: 1.

T: So, how many fingers are there altogether, including pretend fingers? Let's count. Nod your head when you count your pretend fingers so we are sure we counted them.

S/T: Ooone, 2. (Nod head while saying 2.)

T: What is 11 – 9?

S: 2.

NOTES ON MULTIPLE MEANS OF REPRESENTATION:

Sharing strategies is important for students to articulate the way they chose to solve a problem. Other students hear how their classmates are thinking, and this may guide them in understanding the strategies at a deeper level. The teacher can see who is using Level 1, Level 2, or Level 3 strategies in the classroom.

T: Which strategy was easier for you? Turn and talk to your partner.

S: (Discuss.)

T: I heard many students say that they were all easy. They took about the same amount of time. Let's try another problem to see if one strategy is a better shortcut than the other.

Invite all students to solve 17 – 9 using the two strategies (take from ten, modeled with a number bond and with pretend fingers, and counting on). This allows students to experience that the take from ten strategy is more efficient. Generate a discussion about the difficulty of trying to count 7 pretend fingers since they are hard to keep track of. Repeat the process subtracting 9 from 12 to 18 out of sequence so that students have a chance to practice the take from ten strategy. A suggested sequence is 13 – 9, 17 – 9, 15 – 9, 12 – 9. Discuss the increased efficiency of taking from ten as the minuend, or total, gets bigger when subtracting 9, gradually abandoning the counting on strategy and exclusively using the take from ten strategy.

For 14 – 9 and on, use the following paradigm to demonstrate a more efficient way to count on when using pretend fingers. Students find that trying to keep track of more than 3 pretend fingers through head nodding becomes difficult.

T: Let's try 14 – 9. Show 10 fingers, and imagine 4 more.

S: (Show 10 fingers.)

T: Now, take away 9, all at once. How many fingers do you have up?

S: 1.

T: How many pretend fingers are still up?

S: 4.

T: Instead of nodding our heads 4 times to count on, can you see how many fingers there are altogether?

S: Yes. We can just add 1 and 4. That's 5.

As the strategy becomes more familiar, invite students to visualize the entire process instead of using their fingers.

Note: Although using the take from ten strategy is more efficient than counting on one at a time, starting with 13 – 9, some students may find counting on by keeping track on their fingers easier (e.g., niiine, 10, 11, 12, 13, as they put up a finger for each number) because they have not yet mastered the take from ten strategy. It is not wrong for students to say counting on is easier, but with continued practice, they may embrace the Level 3 strategy of taking from ten.

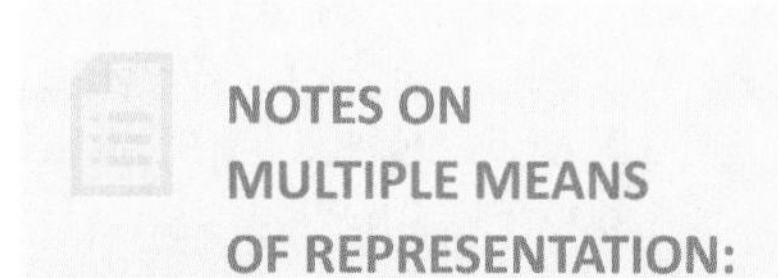

NOTES ON MULTIPLE MEANS OF REPRESENTATION:

When using word problems in class or sending them home as homework, be sure to provide help for nonreaders. Tell parents they can read the problems to their children since the focus is on students' problem-solving skills and not their reading ability.

Problem Set (10 minutes)

Students should do their personal best to complete the Problem Set within the allotted 10 minutes. For some classes, it may be appropriate to modify the assignment by specifying which problems they work on first. Some problems do not specify a method for solving. Students should solve these problems using the RDW approach used for Application Problems.

Name Maria Date ____

Solve the problem by counting on (a) and using a number bond to take from ten (b).

1. Lucy had 12 balloons at her birthday party. She gave 9 balloons to her friends. How many balloons did she have left?

 a. 12 - 9 = 3

 b. 12 - 9 = 3 (10 2)

 Lucy had 3 balloons left.

2. Justin had 15 blueberries on his plate. He ate 9 of them. How many does he have left to eat?

 a. 15 - 9 = 6

 b. 15 - 9 = 6 (10 5)

 Justin has 6 blueberries left to eat.

Complete the subtraction sentences by using the take from ten strategy and counting on. Tell which strategy you would prefer to use for Problems 3 and 4.

3. a. 11 - 9 = 2 (9 + 2 = 11) b. 11 - 9 = 2 (10 1) ☐ take from ten ☑ count on

4. a. 18 - 9 = 9 (9 + 9 = 18) b. 18 - 9 = 9 (10 8) ☑ take from ten ☐ count on

5. Think about how to solve the following subtraction problems:

16 - 9	12 - 9	18 - 9
11 - 9	15 - 9	14 - 9
13 - 9	19 - 9	17 - 9

Choose which problems you think are easier to count on from 9 and which are easier to use the take from ten strategy for. Write the problems in the boxes below.

Problems to use the *count on* strategy with:	Problems to use the *take from ten* strategy with:
11-9 12-9 13-9	14-9 16-9 18-9 15-9 17-9 19-9

Were there any problems that were just as easy using either method? Did you use a different method for any problems? 11-9 is easy either way. 18-9=9 I used my doubles.

Student Debrief (10 minutes)

Lesson Objective: Relate counting on to making ten and taking from ten.

The Student Debrief is intended to invite reflection and active processing of the total lesson experience.

Invite students to review their solutions for the Problem Set. They should check work by comparing answers with a partner before going over answers as a class. Look for misconceptions or misunderstandings that can be addressed in the Debrief. Guide students in a conversation to debrief the Problem Set and process the lesson.

Any combination of the questions below may be used to lead the discussion.

- In Problem 3, how is the take from ten strategy similar to counting on?
- We found that counting on from 9 took different amounts of time, depending on what number we were subtracting from. Is this also true when using the take from ten strategy? Does it take longer to take from ten when the starting number is larger? Explain your reasoning.
- We used our pretend fingers to show the take from ten strategy. (Model 12 – 9.) How is this like counting on? What did we do to make our count on strategy more efficient? Look at Problem 5. Which strategy did you choose for each problem? Explain your reasoning.
- Guide students to see that counting on one at a time becomes less efficient as the difference becomes larger.

- As time allows, expand the discussion to point out that the modifications to counting on (mentioned in the previous bullet) do make it more efficient and on par with the take from ten strategy.
- What new math strategy did we use today to solve subtraction problems more efficiently? (Taking from ten using fingers.)
- Look at your Application Problem. How did you choose to solve it? Explain your thinking. How could the strategies discussed today be used to solve this problem?

Exit Ticket (3 minutes)

After the Student Debrief, instruct students to complete the Exit Ticket. A review of their work will help with assessing students' understanding of the concepts that were presented in today's lesson and planning more effectively for future lessons. The questions may be read aloud to the students.

Name ______________________________ Date ______________

Solve the problem by counting on (a) and using a number bond to take from ten (b).

1. Lucy had 12 balloons at her birthday party. She gave 9 balloons to her friends. How many balloons did she have left?

   a. 12 - 9 = ______

   b. 12 - 9 = _____

   Lucy had ___ balloons left.

---

2. Justin had 15 blueberries on his plate. He ate 9 of them. How many does he have left to eat?

   a. 15 - 9 = ______

   b. 15 - 9 = _____

   Justin has ___ blueberries left to eat.

Complete the subtraction sentences by using the take from ten strategy and counting on. Tell which strategy you would prefer to use for Problems 3 and 4.

3. a. 11 - 9 = ____ b. 11 - 9 = ____ ☐ take from ten ☐ count on

4. a. 18 - 9 = ____ b. 18 - 9 = ____ ☐ take from ten ☐ count on

5. Think about how to solve the following subtraction problems:

16 - 9	12 - 9	18 - 9
11 - 9	15 - 9	14 - 9
13 - 9	19 - 9	17 - 9

Choose which problems you think are easier to count on from 9 and which are easier to use the take from ten strategy. Write the problems in the boxes below.

Problems to use the *count on* strategy with:	Problems to use the *take from ten* strategy with:

Were there any problems that were just as easy using either method? Did you use a different method for any problems?

Name ______________________________ Date ______________

Complete the subtraction sentences by using both the count on and take from ten strategies.

1. a. $13 - 9 =$ ____

   b. $13 - 9 =$ ____

2. a. $17 - 9 =$ ____

   b. $17 - 9 =$ ____

Name ______________________ Date ____________

Complete the subtraction sentences by using either the count on or take from ten strategy. Tell which strategy you used.

1. 17 - 9 = ____

☐ take from ten

☐ count on

2. 12 - 9 = ____

☐ take from ten

☐ count on

3. 16 - 9 = ____

☐ take from ten

☐ count on

4. 11 - 9 = ____

☐ take from ten

☐ count on

5. Nicholas collected 14 leaves. He pasted 9 into his notebook. How many of his leaves were not pasted into his notebook? Choose the count on or take from ten strategy to solve.

I chose this strategy:

☐ take from ten

☐ count on

6. Sheila had 17 oranges. She gave 9 oranges to her friends. How many oranges does Sheila have left? Choose the count on or take from ten strategy to solve.

I chose this strategy:

☐ take from ten

☐ count on

7. Paul has 12 marbles. Lisa has 18 marbles. They each rolled 9 marbles down a hill. How many marbles did each student have left? Tell which strategy you chose for each student.

Paul has _____ marbles left.

Lisa has _____ marbles left.

8. Just as you did today in class, think about how to solve the following problems, and talk to your parent or caregiver about your ideas.

15 - 9	13 - 9	17 - 9
18 - 9	19 - 9	12 - 9
11 - 9	14 - 9	16 - 9

Circle the problems you think are easier to solve by counting on from 9. Put a rectangle around those that are easier to solve using the take from ten strategy. Remember, some might be just as easy using either method.

Lesson 17

Objective: Model subtraction of 8 from teen numbers.

Suggested Lesson Structure

■ Fluency Practice	(14 minutes)
■ Application Problem	(5 minutes)
■ Concept Development	(31 minutes)
■ Student Debrief	(10 minutes)
Total Time	**(60 minutes)**

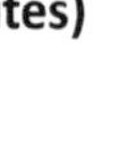

Fluency Practice (14 minutes)

- Subtract 9 **1.OA.6** (4 minutes)
- Sprint: Subtract 9 **1.OA.6** (10 minutes)

Subtract 9 (4 minutes)

Materials: (T) Subtract 9 flash cards (Fluency Template)

Note: This fluency activity reviews the take from ten subtraction strategy when the subtrahend is 9.

Show a subtract 9 flash card (e.g., 12 – 9).

T: Say 12 the Say Ten way.
S: Ten 2.
T: 10 – 9 is...?
S: 1.
T: 1 + 2 is...? (Point to the 2.)
S: 3.
T: 12 – 9 is...?
S: 3.

Sprint: Subtract 9 (10 minutes)

Materials: (S) Subtract 9 Sprint

Note: This Sprint reviews the take from ten subtraction strategy when the subtrahend is 9.

Application Problem (5 minutes)

Gisella had 13 markers in her bag. Eight markers fell out of the bag. How many markers does Gisella have now?

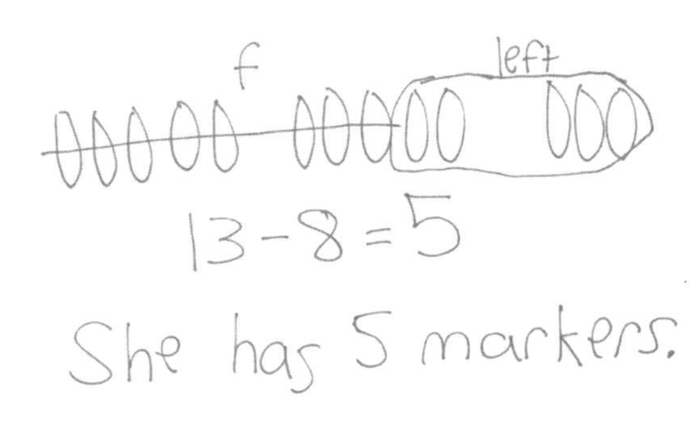

Note: While circulating, notice which students already recognize the application of the take from ten strategy, previously applied only to subtracting 9. Notice which students are crossing off one at a time instead of crossing off 8 quickly. Student strategy choices are discussed in the Student Debrief.

Concept Development (31 minutes)

Materials: (T) Linking cubes of different colors (S) Personal white board

Note: Using different color linking cubes helps students realize that not all objects need to be identical in a given set.

Have students sit in a semicircle in the meeting area with their personal white boards.

T: (Project and read aloud.) Ayan had 15 building blocks. He used 8 of them to make a car. How many blocks were left?

T: How would you solve this problem? Use your personal white board to show your work. (Circulate and observe student strategies as they solve.)

T: How did you solve?

S: I drew 15 squares. I crossed off 8, and I had 7 pieces left. → I counted on from 8 to 15. Eiiiight, 9, 10, 11, 12, 13, 14, 15. I have 7 fingers up, so 7 blocks. → I used the take from ten strategy. I saw that I can take apart 15 into 10 and 5. I took away 8 from 10 and did 2 + 5 = 7. Seven blocks.

T: No matter which strategies these students used, did they get the same answer?

S: Yes!

T: (Show a stick of 15 cubes, 10 in one color and 5 in another color.) Here is a stick of 15 linking cubes to show how many building blocks Ayan had in the beginning. To use the take from ten strategy, let's break this apart into...?

S: 10 and 5.

T: (Break off and separate into two sticks.) We need to take away...?

S: 8 pieces.

T: From 10 or 5?

S: 10.

T: (Take away 8 from 10.) 10 minus 8 is...?

S: 2.

NOTES ON MULTIPLE MEANS OF ACTION AND EXPRESSION:

It is important to guide students to evaluate their thinking, as well as their partners', during the turn and talks. This provides students an opportunity to evaluate their process and analyze errors.

T: 2 and 5 make...?

S: 7.

T: Let's check by using our fingers. Show me 15 fingers. How many pretend fingers are up?

S: (Show 10 fingers.) 5.

T: Take away 8, all at once.

S: (Show 2 fingers.)

T: How many fingers are up?

S: 2.

T: How many pretend fingers are there?

S: 5.

T: How many fingers, including pretend fingers, are there altogether?

S: 7.

T: What addition sentence helped you solve 15 – 8?

S: 2 + 5 = 7.

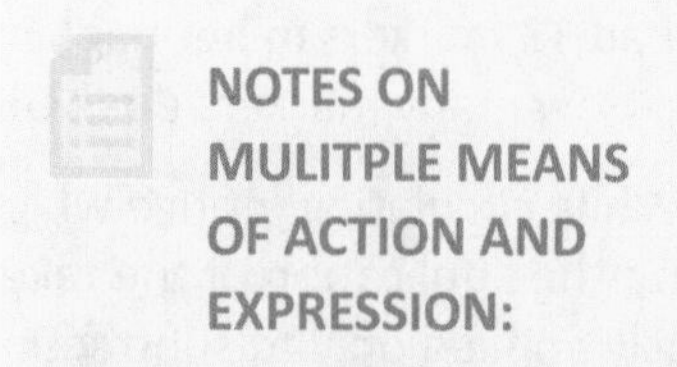

NOTES ON MULITPLE MEANS OF ACTION AND EXPRESSION:

Adapt what is expected of certain students depending on their level of understanding. Some students may be ready to move away from draw and circle 10 to just break apart the teen number with a number bond in their work.

Repeat the process following the suggested sequence: 11 – 8, 12 – 8, 14 – 8, 15 – 8, 17 – 8, 18 – 8 (take 8 from 8 rather than 10), and 19 – 8 (take 8 from 9). Linking cubes may be used to aid student understanding for the first few problems, but then move toward using fingers. At 18 – 8 and 19 – 8, reintroduce the linking cubes, as they provide a clearer visual representation for determining from where to quickly subtract 8. If time allows, have students work with a partner to practice subtracting 8 using the take from ten strategy with fingers and writing the addition sentence to help solve.

Problem Set (10 minutes)

Students should do their personal best to complete the Problem Set within the allotted 10 minutes. For some classes, it may be appropriate to modify the assignment by specifying which problems they work on first. Some problems do not specify a method for solving. Students should solve these problems using the RDW approach used for Application Problems.

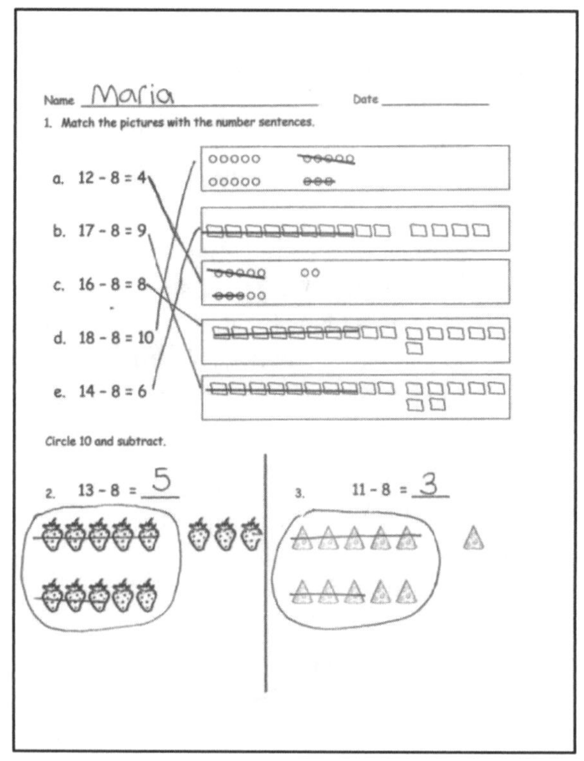

Name Maria Date

1. Match the pictures with the number sentences.

a. 12 - 8 = 4

b. 17 - 8 = 9

c. 16 - 8 = 8

d. 18 - 8 = 10

e. 14 - 8 = 6

Circle 10 and subtract.

2. 13 - 8 = 5

3. 11 - 8 = 3

Student Debrief (10 minutes)

Lesson Objective: Model subtraction of 8 from teen numbers.

The Student Debrief is intended to invite reflection and active processing of the total lesson experience.

Invite students to review their solutions for the Problem Set. They should check work by comparing answers with a partner before going over answers as a class. Look for misconceptions or misunderstandings that can be addressed in the Debrief. Guide students in a conversation to debrief the Problem Set and process the lesson.

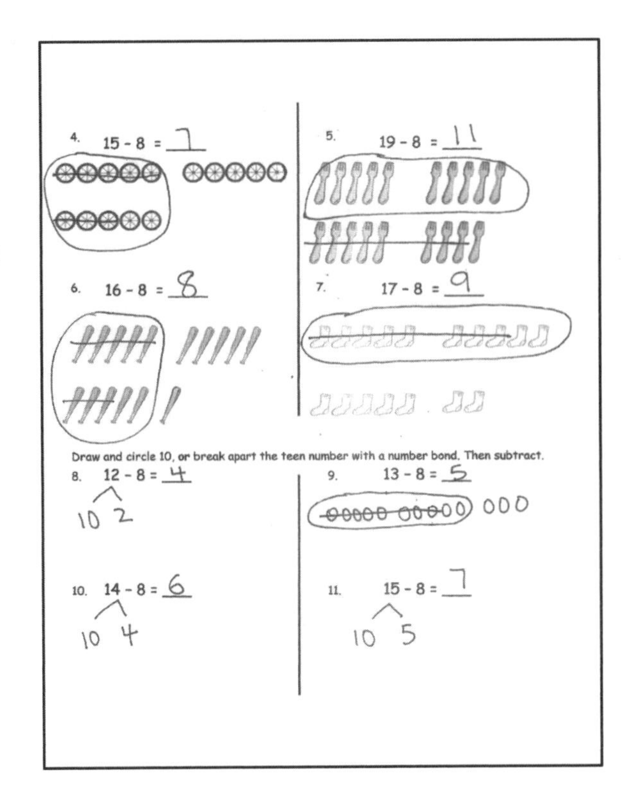

Any combination of the questions below may be used to lead the discussion.

- Look at Problem 5. Where did you take 8 from? Why is it wiser to take 8 from 9 than 10?
- Look at the way a student solved Problem 6. How is her solution similar to and different from yours?

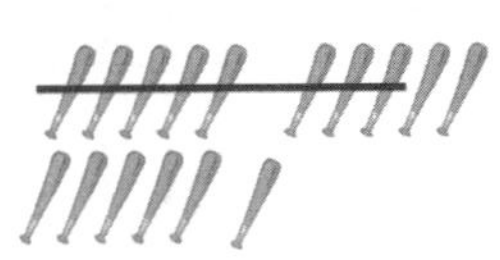

- How can knowing 15 – 9 = 6 help you solve 15 – 8? Explain your thinking.
  - When we take from ten to solve these two problems, what is different about how we get our solution? (In 15 – 9, we add 1 to 5. In 15 – 8, we add 2 to 5.)
  - How is 15 – 9 different from 15 – 8? How much less are we taking away? How would that change the answer? (We took away 1 less, so the answer will have 1 more.)
  - Following this pattern, how would you solve 15 – 7?
- Look at the Application Problem. How did you choose to solve it? Explain your thinking. How could the strategy discussed today be used to solve this problem?

Exit Ticket (3 minutes)

After the Student Debrief, instruct students to complete the Exit Ticket. A review of their work will help with assessing students' understanding of the concepts that were presented in today's lesson and planning more effectively for future lessons. The questions may be read aloud to the students.

A

Number Correct:

Name ____________________ Date __________

*Write the missing number. Pay attention to the addition or subtraction sign.

1.	10 - 9 = □		16.	10 - 9 = □	
2.	1 + 2 = □		17.	11 - 9 = □	
3.	10 - 9 = □		18.	12 - 9 = □	
4.	1 + 3 = □		19.	15 - 9 = □	
5.	10 - 9 = □		20.	14 - 9 = □	
6.	1 + 1 = □		21.	13 - 9 = □	
7.	10 - 9 = □		22.	17 - 9 = □	
8.	1 + 2 = □		23.	18 - 9 = □	
9.	12 - 9 = □		24.	9 + □ = 13	
10.	10 - 9 = □		25.	9 + □ = 14	
11.	1 + 3 = □		26.	9 + □ = 16	
12.	13 - 9 = □		27.	9 + □ = 15	
13.	10 - 9 = □		28.	9 + □ = 17	
14.	1 + 5 = □		29.	9 + □ = 18	
15.	15 - 9 = □		30.	9 + □ = 19	

B

Number Correct:

Name ______________________ Date ____________

*Write the missing number. Pay attention to the addition or subtraction sign.

1.	10 - 9 = □		16.	10 - 9 = □	
2.	1 + 1 = □		17.	11 - 9 = □	
3.	10 - 9 = □		18.	13 - 9 = □	
4.	1 + 2 = □		19.	14 - 9 = □	
5.	10 - 9 = □		20.	13 - 9 = □	
6.	1 + 3 = □		21.	12 - 9 = □	
7.	10 - 9 = □		22.	15 - 9 = □	
8.	1 + 4 = □		23.	16 - 9 = □	
9.	14 - 9 = □		24.	9 + □ = 12	
10.	10 - 9 = □		25.	9 + □ = 13	
11.	1 + 3 = □		26.	9 + □ = 15	
12.	13 - 9 = □		27.	9 + □ = 14	
13.	10 - 9 = □		28.	9 + □ = 15	
14.	1 + 2 = □		29.	9 + □ = 17	
15.	12 - 9 = □		30.	9 + □ = 16	

Name ______________________________ Date ______________

1. Match the pictures with the number sentences.

a. 12 - 8 = 4

b. 17 - 8 = 9

c. 16 - 8 = 8

d. 18 - 8 = 10

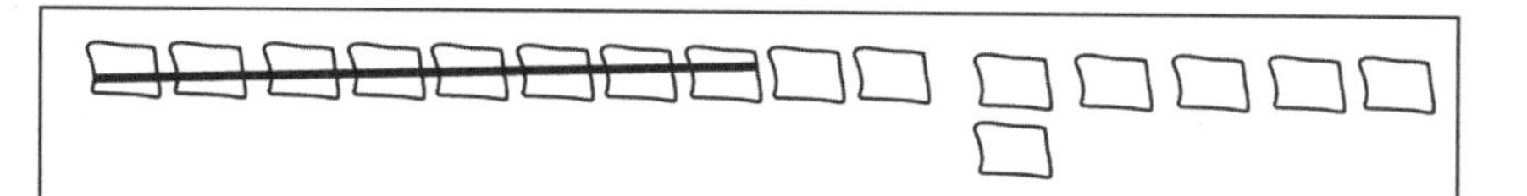

e. 14 - 8 = 6

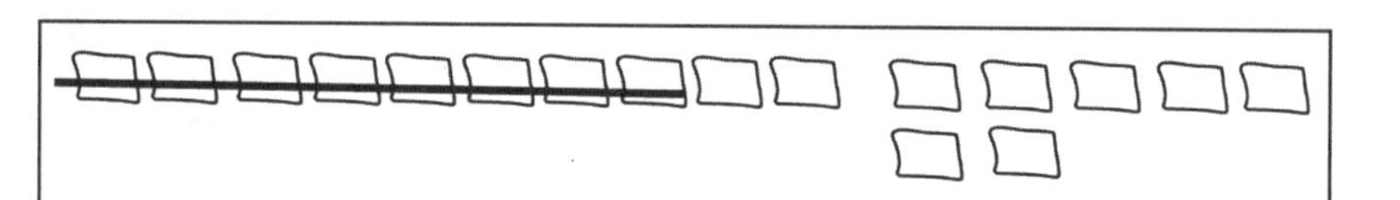

Circle 10 and subtract.

2. 13 - 8 = ______

3. 11 - 8 = ______

EUREKA MATH

4. 15 - 8 = _____

5. 19 - 8 = _____

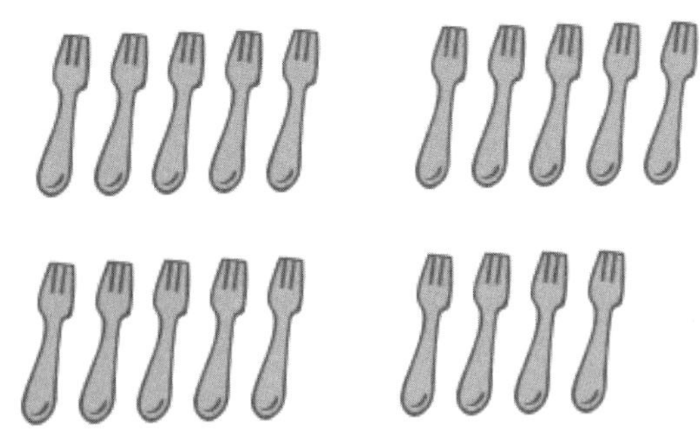

6. 16 - 8 = _____

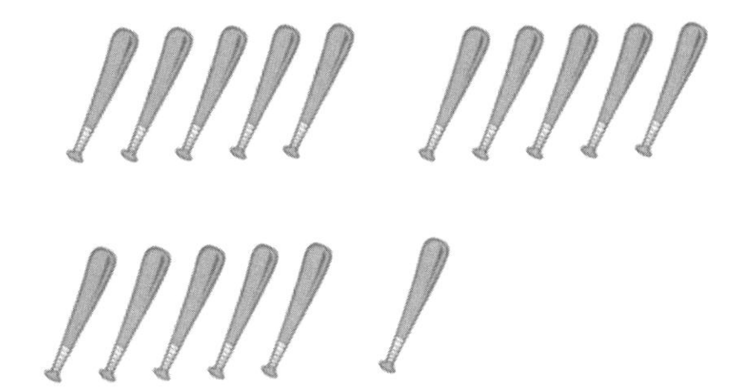

7. 17 - 8 = _____

Draw and circle 10, **or** break apart the teen number with a number bond. Then subtract.

8. 12 - 8 = ____

9. 13 - 8 = ____

10. 14 - 8 = ____

11. 15 - 8 = ____

Name ____________________ Date ____________

1. Draw and circle 10. Then subtract.

a. 12 - 8 = ____	b. 14 - 8 = ____

2. Use a number bond to break apart the teen number. Then subtract.

15 - 8 = ____

Name ____________________ Date ____________

1. Match the number sentence to the picture or to the number bond.

a. 13 - 7 = ____

b. 16 - 8 = ____

c. 11 - 8 = ____

d. 13 - 8 = ____

13 / 10 3	10 - 7 = 3 3 + 3 = 6

13 / 10 3	10 - 8 = 2 2 + 3 = 5

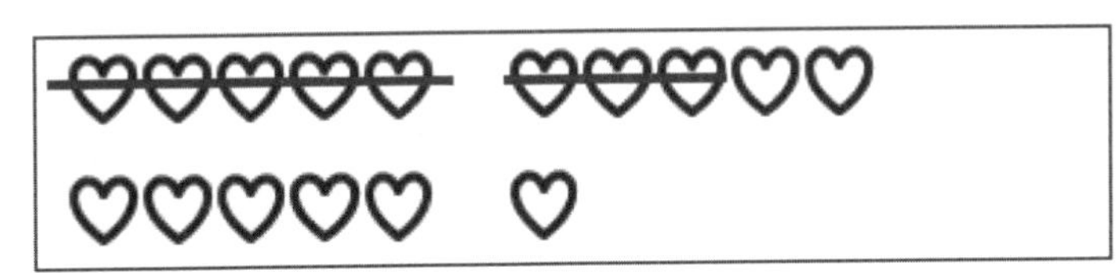

2. Show how you would solve 14 - 8, either with a number bond or a drawing.

Circle 10. Then subtract.

3. Milo has 17 rocks. He throws 8 of them into a pond. How many does he have left?

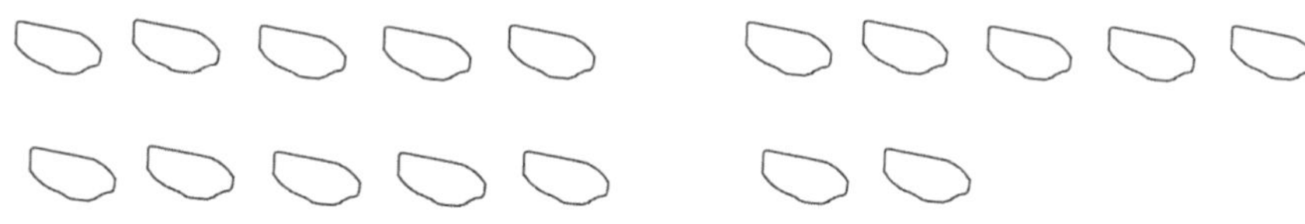

Milo has ____ rocks left.

Draw and circle 10. Then subtract.

4. Lucy has $12. She spends $8. How much money does she have now?

Lucy has $_____ now.

Draw and circle 10, or use a number bond to break apart the teen number and subtract.

5. Sean has 15 dinosaurs. He gives 8 to his sister. How many dinosaurs does he keep?

Sean keeps _____ dinosaurs.

---

6. Use the picture to fill in the math story. Show a number sentence.

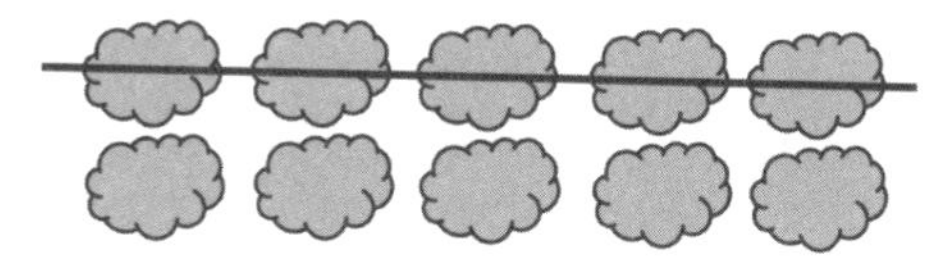

Olivia saw _____ clouds in the sky. _____ clouds went away. How many clouds are left?	Try it! Can you show how to solve this problem with a number bond?

10 - 9	11 - 9
12 - 9	13 - 9
14 - 9	15 - 9
16 - 9	17 - 9
18 - 9	19 - 9

subtract 9 flash cards

Lesson 18

Objective: Model subtraction of 8 from teen numbers.

Suggested Lesson Structure

■ Fluency Practice	(12 minutes)
■ Application Problem	(5 minutes)
■ Concept Development	(33 minutes)
■ Student Debrief	(10 minutes)
Total Time	**(60 minutes)**

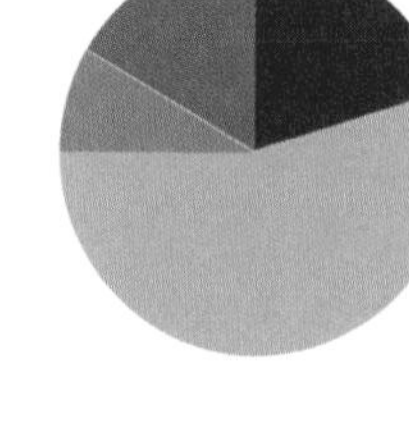

Fluency Practice (12 minutes)

- Cold Call: Subtract 9 **1.OA.6** (4 minutes)
- Hide Zero Number Sentences **1.NBT.2** (2 minutes)
- Number Path **1.OA.6** (6 minutes)

Cold Call: Subtract 9 (4 minutes)

Materials: (T) Subtract 9 flash cards (Lesson 17 Fluency Template)

Note: This fluency activity reviews the take from ten subtraction strategy when the subtrahend is 9.

Show a subtract 9 flash card (e.g., 12 – 9). Play Cold Call. Flash a card, and then call on a student or group of students to answer. Students do not raise their hands to be chosen. If students continue to need help subtracting 9, use the following vignette.

T: Say 12 the Say Ten way.
S: Ten 2.
T: 10 – 9 is...?
S: 1.
T: 1 + 2 is...? (Point to the 2.)
S: 3.
T: So, 12 – 9 is...?
S: 3.

Hide Zero Number Sentences (2 minutes)

Materials: (S) Hide Zero cards (Fluency Template 1)

Note: This fluency activity strengthens the understanding of place value and prepares students to understand ten as a unit by the module's end.

Show students numbers from 10 to 19 with Hide Zero cards (e.g., 15). Students say an addition sentence with 10 as an addend (e.g., 10 + 5 = 15). As students say the sentence, break apart the Hide Zero cards to model the equation. Students can also say the numbers the Say Ten way and the regular way.

Number Path (6 minutes)

Materials: (T/S) Personal white board, number path 1–20 (Fluency Template 2), counter

Note: Using a number path to get to and from 10 prepares students to relate counting on and taking from ten in Lesson 19.

T: Put your counter on 8.
S: (Place the counter on 8.)
T: How many spaces do you need to move to land on 10? (Pause to provide thinking time.)
S: 2.
T: Let's check. Move your counter to 10.
S: (Move the counter to 10.)
T: Were you right?
S: Yes!
T: Write an equation to show what you did.
S: (Write 8 + 2 = 10.)

Continue moving to and from 10 within 10. Next, start at 10, and move the counters to and from teen numbers. Ask questions about how students determined the number of spaces they moved. Did they count each space, or did they "just know"?

Application Problem (5 minutes)

Juliana rolls 8 cars down a ramp. If she started with 15 cars at the top of the ramp, how many cars does Juliana still have at the top of the ramp?

Note: This Application Problem provides another context for students to subtract 8 from a teen number. While it is still a *take from with result unknown* problem type, the problem is somewhat more complex based on the order of the sentences within the story. In this story, the quantity being subtracted is given first.

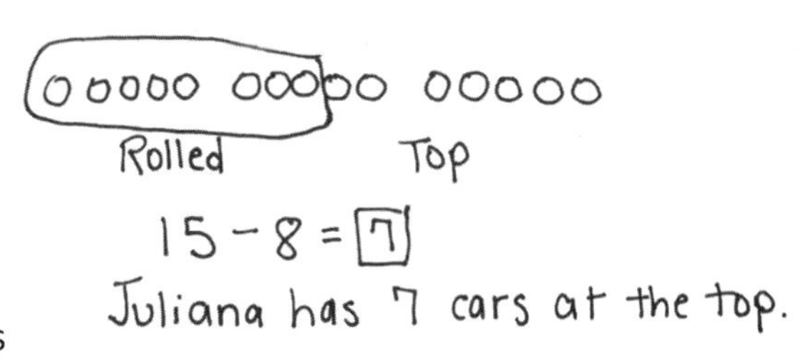

Concept Development (33 minutes)

Materials: (S) Personal white board

Have students gather in the meeting area with their personal white boards.

T: (Project 15 – 8 = ___.) Show me 15 fingers. How many pretend fingers are up?
S: (Show 10 fingers.) 5.
T: Take away 8 all at once. How many fingers, including pretend fingers, are there now?
S: 7.
T: What addition sentence helped you solve 15 – 8?
S: 2 + 5 = 7.
T: Let's use 5-group drawings to show how we used our fingers. How did we show 15 with our fingers?
S: We used 10 fingers and 5 pretend fingers.

MP.4

T: (Decompose 15 by drawing a 5-group row on the board. Leave extra space between the first 10 circles and the last 5 circles.)
T: (Draw a frame around the 10 circles.) This is so everyone can see 10 and 5 more easily, just like how we've framed 10 objects together in the past.
T: How did you take away 8 all at once using your fingers? How can we show that in our drawing?
S: We took down 8 fingers, so cross off 8 from the ten. → We can just hide 8 circles from the ten.
T: If we cross off or hide 8 circles from 10, how many circles would that leave us in the frame?
S: 2.
T: Great. (Hide 8 circles.) How many circles do you see now?
S: 7.
T: What addition sentence do you see in your picture?
S: 2 + 5 = 7.

Repeat the process above with the following sequence: 11 – 8, 16 – 8, 13 – 8, 17 – 8, 12 – 8, 14 – 8, 18 – 8, and 19 – 8. Invite students to draw 5-group rows on their personal white boards. After solving a few problems using both strategies as a whole class, have students work with their partners. Alternate having Student A solve the problem using fingers and pretend fingers while Student B shows her work with 5-group row drawings.

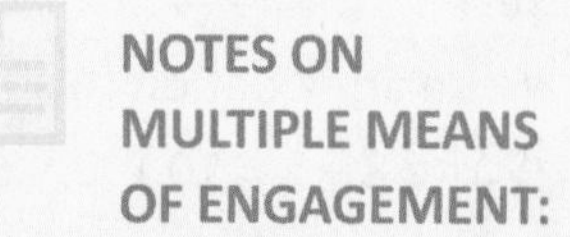

NOTES ON MULTIPLE MEANS OF ENGAGEMENT:

Having students work in partners frequently develops their cooperative learning skills. Some students have trouble working together with a partner while others shine through as leaders. Be sure to talk about how to work well in a team if any problems develop.

When it seems appropriate, ask students to close their eyes to see if they can visualize what is happening when they subtract 8, encouraging them to move away from using their fingers or drawings and work instead toward using mental math. Encourage students to share what they are picturing in their minds as they are solving.

Problem Set (10 minutes)

Students should do their personal best to complete the Problem Set within the allotted 10 minutes. For some classes, it may be appropriate to modify the assignment by specifying which problems they work on first. Some problems do not specify a method for solving. Students should solve these problems using the RDW approach used for Application Problems.

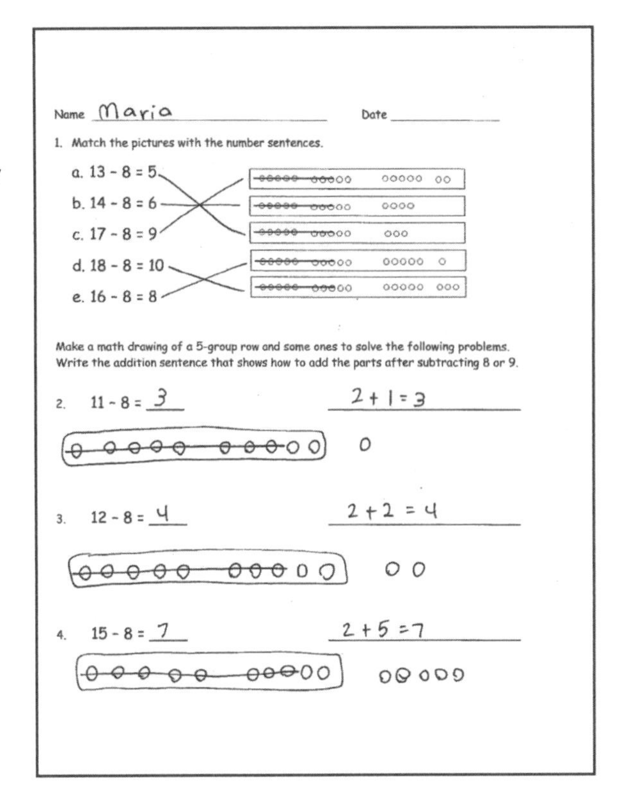
Name Maria Date

1. Match the pictures with the number sentences.

a. 13 - 8 = 5
b. 14 - 8 = 6
c. 17 - 8 = 9
d. 18 - 8 = 10
e. 16 - 8 = 8

Make a math drawing of a 5-group row and some ones to solve the following problems. Write the addition sentence that shows how to add the parts after subtracting 8 or 9.

2. 11 - 8 = 3 2 + 1 = 3
3. 12 - 8 = 4 2 + 2 = 4
4. 15 - 8 = 7 2 + 5 = 7

Student Debrief (10 minutes)

Lesson Objective: Model subtraction of 8 from teen numbers.

The Student Debrief is intended to invite reflection and active processing of the total lesson experience.

Invite students to review their solutions for the Problem Set. They should check work by comparing answers with a partner before going over answers as a class. Look for misconceptions or misunderstandings that can be addressed in the Debrief. Guide students in a conversation to debrief the Problem Set and process the lesson.

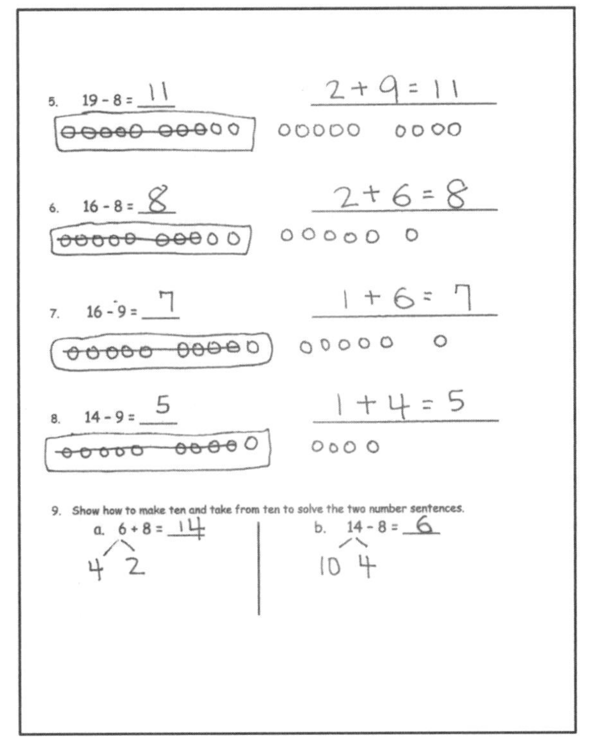
5. 19 - 8 = 11 2 + 9 = 11
6. 16 - 8 = 8 2 + 6 = 8
7. 16 - 9 = 7 1 + 6 = 7
8. 14 - 9 = 5 1 + 4 = 5

9. Show how to make ten and take from ten to solve the two number sentences.

a. 6 + 8 = 14
4 2

b. 14 - 8 = 6
10 4

Any combination of the questions below may be used to lead the discussion.

- What pattern did you notice every time we took away 8 from a teen number?
- How did you solve 18 – 8 and 19 – 8? How is solving these problems different from solving the other – 8 problems?
- How did solving Problem 7 help you solve Problem 8?
- Look at Problem 9. How are (a) and (b) related? Using these examples, explain how the make ten strategy is related to the take from ten strategy.
- How can we use what we learned about taking away 8 from a teen number to solve a – problem?
- What tools did we use today to help us subtract 8 from a teen number? (Our fingers and 5-group drawings.) How did they help us?
- How is the way you subtract 8 from a teen number different from the way you subtract 9?

- Look at the Application Problem. How did you choose to solve it? Explain your thinking. How could the strategies discussed today be used to solve this problem?

Exit Ticket (3 minutes)

After the Student Debrief, instruct students to complete the Exit Ticket. A review of their work will help you assess the students' understanding of the concepts that were presented in the lesson today and plan more effectively for future lessons. You may read the questions aloud to the students.

Name ______________________________ Date ______________

1. Match the pictures with the number sentences.

a. 13 - 8 = 5

b. 14 - 8 = 6

c. 17 - 8 = 9

d. 18 - 8 = 10

e. 16 - 8 = 8

Make a math drawing of a 5-group row and some ones to solve the following problems. Write the addition sentence that shows how to add the parts after subtracting 8 or 9.

2. 11 - 8 = _____ ______________________

3. 12 - 8 = _____ ______________________

4. 15 - 8 = _____ ______________________

5. 19 - 8 = ____ ______________________

6. 16 - 8 = ____ ______________________

7. 16 - 9 = ____ 

8. 14 - 9 = ____ ______________________

9. Show how to make ten and take from ten to solve the two number sentences.

a. 6 + 8 = ____

b. 14 - 8 = ____

Name ______________________________ Date ______________

Draw 5-group rows, and cross out to solve. Complete the number sentences. Write the 2+ addition sentence that helped you add the two parts.

1. 14 - 8 = _____

2 + _____ = _____

2. 17 - 8 = _____

2 + _____ = _____

Name ______________________________ Date ______________

Draw 5-group rows, and cross out to solve. Write the 2+ addition sentence that helped you add the two parts.

1. Annabelle had 13 goldfish. Eight goldfish ate fish food. How many goldfish did not eat fish food?

_____ goldfish did not eat fish food.

2. Sam collected 15 buckets of rain water. He used 8 buckets to water his plants. How many buckets of rain water does Sam have left?

Sam has _____ buckets of rain water left.

3. There were 19 turtles swimming in the pond. Some turtles climbed up onto the dry rocks, and now there are only 8 turtles swimming. How many turtles are on the dry rocks?

There are _____ turtles on the dry rocks.

Show making ten or taking from ten to solve the number sentences.

4. 7 + 8 = ______

5. 15 - 8 = ______

Find the missing number by drawing 5-group rows.

6. 11 - 9 = ______

7. 14 - 9 = ______

8. Draw 5-group rows to show the story. Cross out or use number bonds to solve. Write a number sentence to show how you solved the problem.

There were 14 people at home. Ten people were watching a football game. Four people were playing a board game. Eight people left. How many people stayed?

______ people stayed at home.

10		20	
0	1	2	3
4	5	6	7
8	9		

hide zero cards, numeral side (copy double-sided with next page)

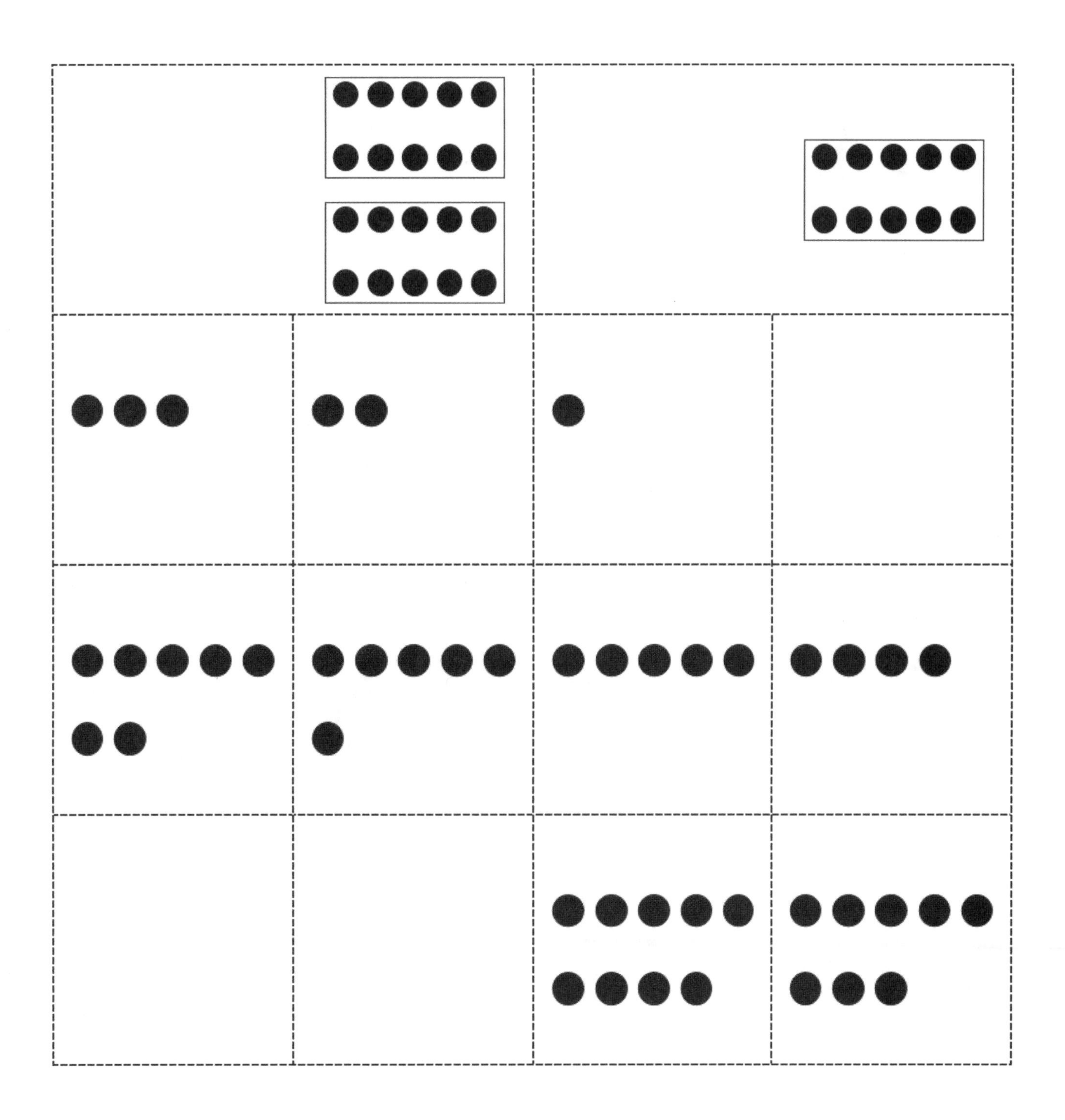

hide zero cards, dot side (copy double-sided with previous page)

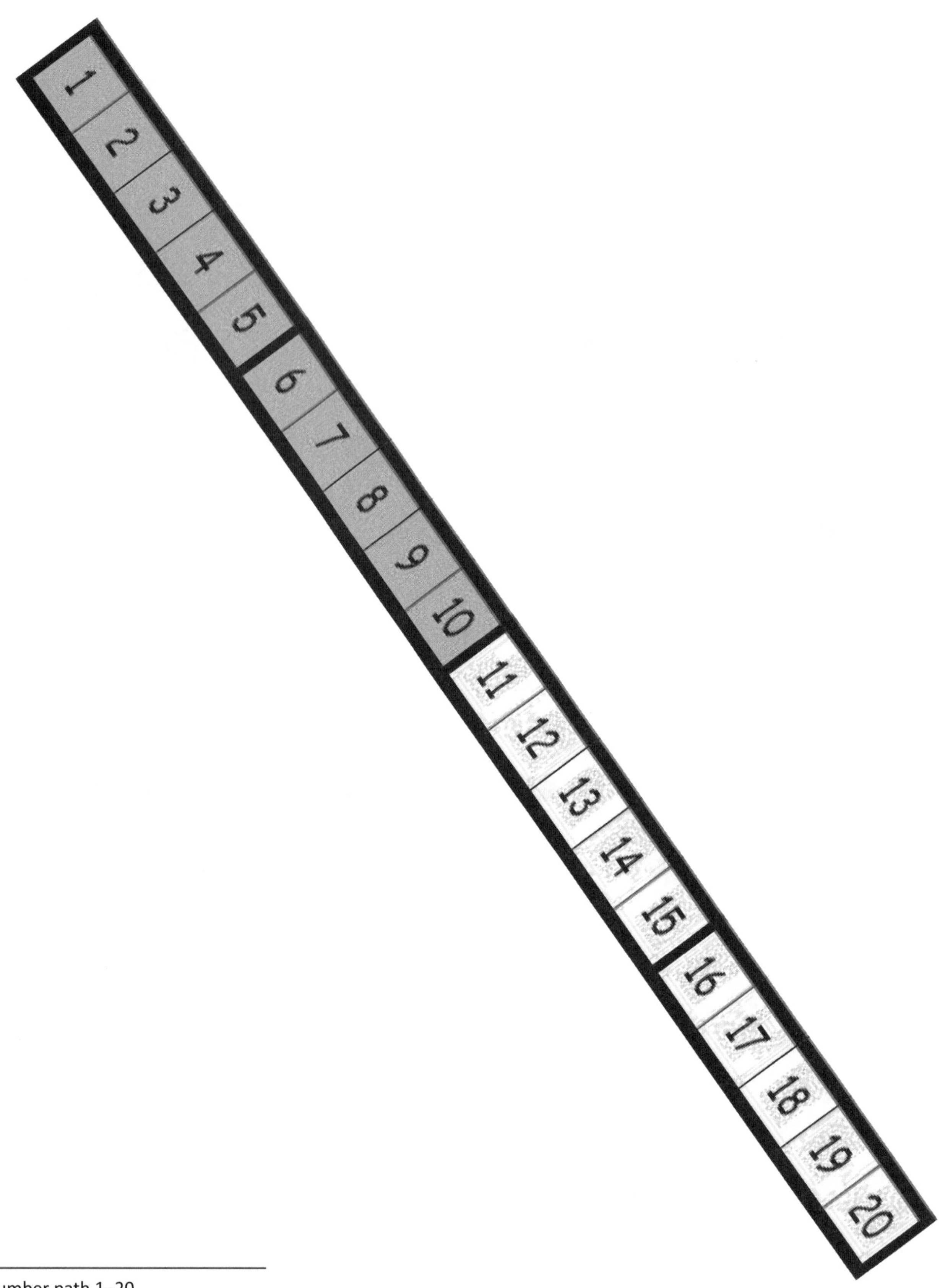

number path 1–20

Lesson 19

Objective: Compare efficiency of counting on and taking from ten.

Suggested Lesson Structure

Fluency Practice	(12 minutes)
Application Problem	(8 minutes)
Concept Development	(30 minutes)
Student Debrief	(10 minutes)
Total Time	**(60 minutes)**

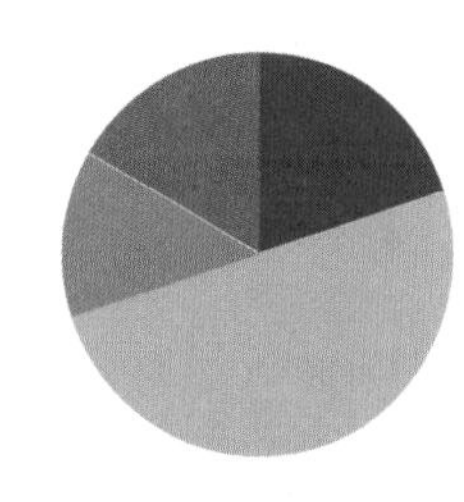

Fluency Practice (12 minutes)

- Subtract 9 and 8 and Relate to Addition **1.OA.6** (6 minutes)
- Say Ten Counting **1.NBT.5** (4 minutes)
- Get to 10 **1.OA.6** (2 minutes)

Subtract 9 and 8 and Relate to Addition (6 minutes)

Materials: (S) Personal white board, 5-group row insert (Lesson 12 Fluency Template 2)

Note: When reviewing the take from ten subtraction strategy, remember that the goal is for students to eventually be able to solve these problems mentally. Therefore, for the first two problems, have students cross off the circles. Then, challenge those who are ready to imagine subtracting the circles to solve with their eyes closed.

T: Draw more circles to show 12.
T: Say 12 as a number bond, with 10 as a part.
S: 10 and 2 make 12.
T: Turn your circles into a number bond.
S/T: (Draw lines to make a number bond with the numeral 12 on top.)
T: Show me 12 – 9. Think about whether you should subtract from the part with ten or the part with two.
S/T: (Write – 9 after 12, and cross out 9.)

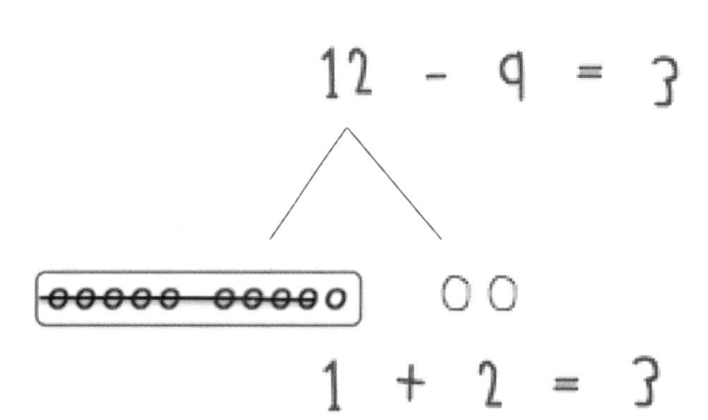

T: Below your circles, write an addition sentence to show what is left.
S: (Write 1 + 2 = 3.)

T: What is 12 – 9?

S: 3.

T: Say 12 – 9 = 3 as a related addition sentence. (Call on a student.)

S: 9 + 3 = 12.

Continue with other numbers between 11 and 20, alternating between subtracting 9 and subtracting 8. As soon as possible, reduce steps (e.g., show me 11 – 8).

Say Ten Counting (4 minutes)

Materials: (S) Personal white board

Note: Say Ten counting strengthens understanding of place value. It is used throughout Grade 1 Fluency Practice, beginning in Module 1, Lesson 4. A description of Say Ten counting, as shared with children in kindergarten, can be found in Grade Kindergarten, Module 5, Lesson 4.

Practice Say Ten counting from 0 to 40 and back. Count for two minutes. Then, have students see how many numbers they can write from 10 to 40 in two minutes.

Get to 10 (2 minutes)

Materials: (T) 20-bead Rekenrek

Note: Practice with getting to 10 from single-digit and teen numbers prepares students for today's lesson as they are encouraged to count on or back strategically, stopping at 10 and continuing to the desired number.

T: (Show 8 on the Rekenrek.) What number do you see?

S: 8.

T: How can I get to 10?

S: Add 2.

T: (Move 2 beads to make ten.) Good.

T: (Show 12.) What number do you see?

S: 12.

Continue with other numbers within 20.

Application Problem (8 minutes)

Carla, Jose, and Yannis each have 8 cherries. They all get more cherries to put in their bowls. Now, Carla has 12 cherries, Jose has 14 cherries, and Yannis has 16 cherries. How many more cherries did they each put in their bowls? Write a number sentence for each answer.

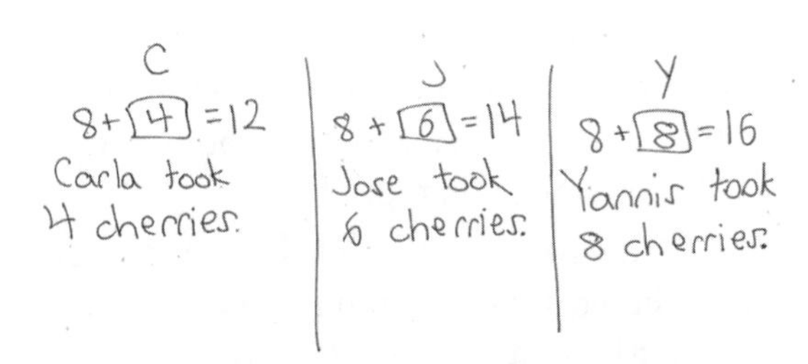

Note: This Application Problem enables students to consider three different missing addends all starting from 8. Consider adjusting the story to include only Carla or only Carla and Jose, depending on students' needs. During the Student Debrief, students connect their solutions to one child's quantity of cherries as a possible stepping-stone for solving the other children's quantities of cherries.

Concept Development (30 minutes)

Materials: (T) Number path 1–20 (Lesson 18 Fluency Template 2) (S) Personal white board, number path 1–20 (Lesson 18 Fluency Template 2)

Have students come to the meeting area and sit in a semicircle with their materials.

T: (Write 13 – 8 = ___.) Let's count on by tracking on our fingers to solve 13 – 8.

S: Eiiight, 9, 10, 11, 12, 13. (Put up a finger for each number starting with 9.)

T: What is 13 – 8?

S: 5.

T: Let's count on using a more efficient strategy. You are an expert at making ten, so let's count on from 8 to 13, this time by making ten. Show me 8 fingers.

S: (Extend 8 fingers.)

T: How many fingers do we need to pop up to make ten? Show me.

S: 2. (Extend the rest of the fingers.)

T: We need to now imagine more fingers popping up. How many more pretend fingers do we need to get to 13?

S: 3.

T: How many more fingers, including pretend fingers, did we need to get from 8 to 13?

S: 5.

T: Let's use the number path to show what we did with our fingers.

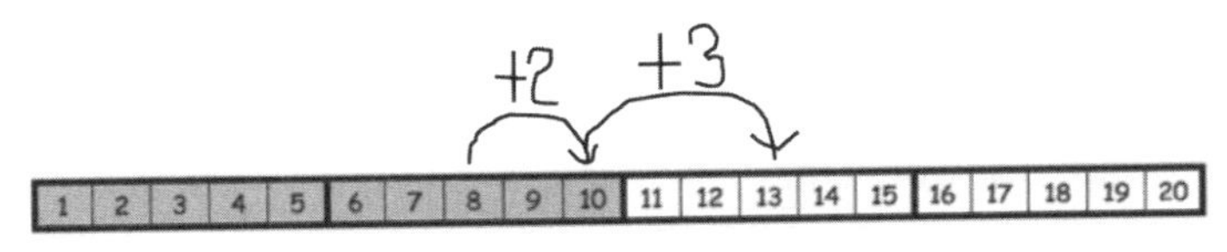

MP.4

T: (Project the number path.) Let's see what counting up by making ten looks like on the number path. How many do we need to get from 8 to 10?

S: 2.

T: I can just jump 2 squares to get to 10 from 8. (Draw a curved arrow from 8 to 10, and write + 2.)

T: I need to get to 13. What is 13 the Say Ten way?

S: Ten 3.

T: How many do we need to get from 10 to 13?

S: 3.

T: I don't need to count on tennnn, 11, 12, 13. I can just jump 3 squares to get to 13 from 10. (Draw a curved arrow from 10 to 13, and write + 3.)

T: How many squares did we jump in all from 8 to 13?
How many do we need to get from 8 to 13?

S: 5.

T: How did you know so quickly?
S: 2 and 3 is 5. → 2 + 3 = 5.
T: Great job counting on to make ten first.
T: Let's check this work using the take from ten strategy using our fingers and a number bond. Put up 13 fingers. How many of your fingers and pretend fingers are up?
S: 10 fingers and 3 pretend ones.
T: (Write the number bond for 13.) Subtract 8 fingers all at once.
S: (Show 2 fingers.)
T: Where did you take away the 8 from?
S: From the 10 fingers.
T: What is 10 – 8? (Point to 10 in the number bond and 8 in the expression.)
S: 2.
T: How many more pretend fingers do you have?
S: 3.
T: (Point to 3 in the number bond.) What is 2 and 3?
S: 5.
T: So, what is 13 – 8? Say the number sentence.
S: 13 – 8 = 5.

MP.4

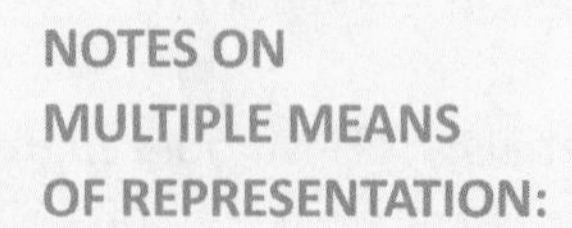

NOTES ON MULTIPLE MEANS OF REPRESENTATION:

Teachers feel a sense of pride as their students use strategies to make math easy. It is exciting when students are able to explain how they are thinking and relate counting on to make ten and take from ten. Use these students to show others who may want or need some extra help.

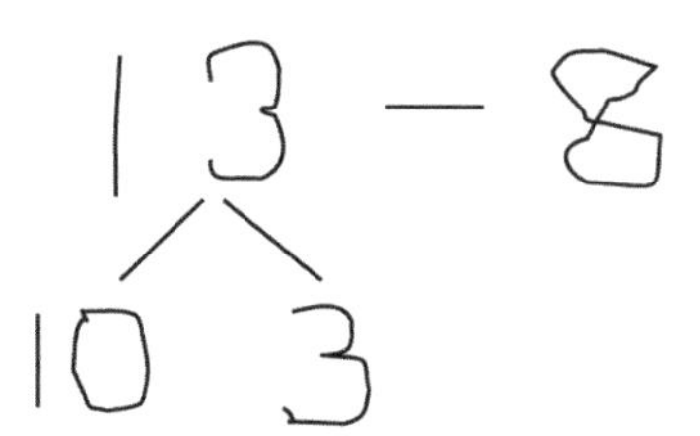

Repeat the process using the number path and the take from ten strategy following the suggested sequence: 11 – 8, 14 – 8, 15 – 8, 12 – 8, 17 – 8, and 16 – 8. When it seems appropriate, encourage students to imagine using their fingers and move toward using only the number bond to solve. This is an opportunity for partner work. After a few modeled problems, allow students to work in partnerships, with Partner A solving and Partner B checking, and then changing roles.

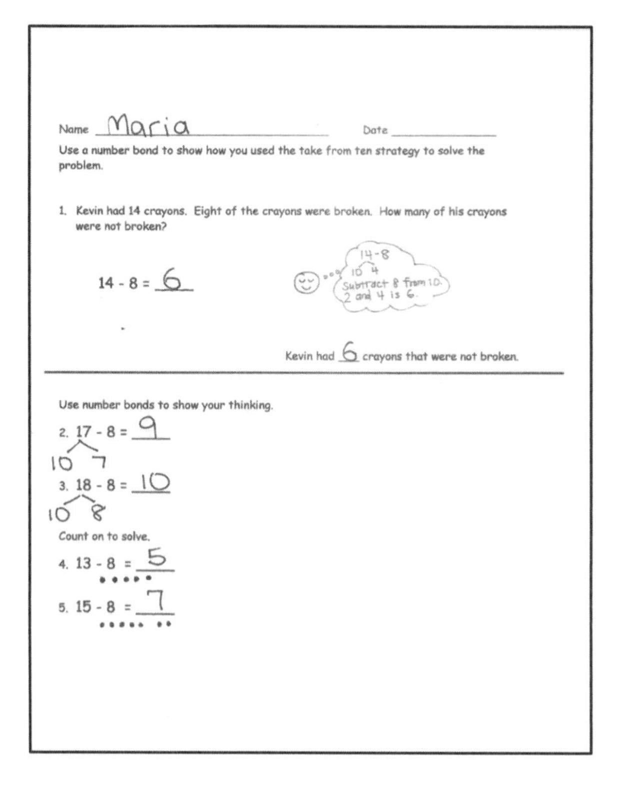

Name Maria Date

Use a number bond to show how you used the take from ten strategy to solve the problem.

1. Kevin had 14 crayons. Eight of the crayons were broken. How many of his crayons were not broken?

14 - 8 = 6

14-8
10 4
Subtract 8 from 10.
2 and 4 is 6.

Kevin had 6 crayons that were not broken.

Use number bonds to show your thinking.

2. 17 - 8 = 9
10 7

3. 18 - 8 = 10
10 8

Count on to solve.

4. 13 - 8 = 5

5. 15 - 8 = 7

Problem Set (10 minutes)

Note: If needed, allow students to use their personal white boards with the number path insert to help them complete the Problem Set.

Students should do their personal best to complete the Problem Set within the allotted 10 minutes. For some classes, it may be appropriate to modify the assignment by specifying which problems they work on first. Some problems do not specify a method for solving. Students should solve these problems using the RDW approach used for Application Problems.

Student Debrief (10 minutes)

Lesson Objective: Compare efficiency of counting on and taking from ten.

The Student Debrief is intended to invite reflection and active processing of the total lesson experience.

Invite students to review their solutions for the Problem Set. They should check work by comparing answers with a partner before going over answers as a class. Look for misconceptions or misunderstandings that can be addressed in the Debrief. Guide students in a conversation to debrief the Problem Set and process the lesson.

Any combination of the questions below may be used to lead the discussion.

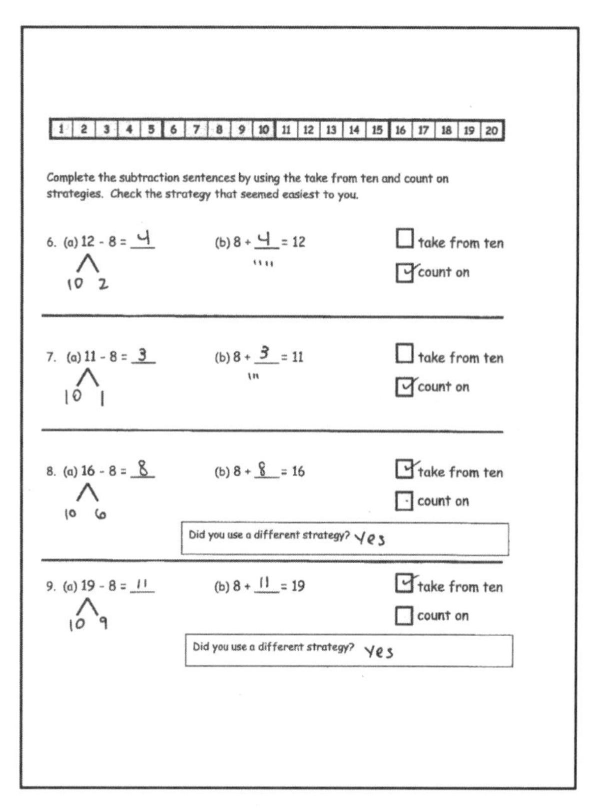

1 2 3 4 5 6 7 8 9 10 11 12 13 14 15 16 17 18 19 20

Complete the subtraction sentences by using the take from ten and count on strategies. Check the strategy that seemed easiest to you.

6. (a) 12 - 8 = 4 (b) 8 + 4 = 12 ☐ take from ten ☑ count on

7. (a) 11 - 8 = 3 (b) 8 + 3 = 11 ☐ take from ten ☑ count on

8. (a) 16 - 8 = 8 (b) 8 + 8 = 16 ☑ take from ten ☐ count on
Did you use a different strategy? Yes

9. (a) 19 - 8 = 11 (b) 8 + 11 = 19 ☑ take from ten ☐ count on
Did you use a different strategy? Yes

- Look at Problems 6 through 9. Which strategy do you prefer, counting on or the take from ten strategy? (It is important to emphasize that they are both good shortcuts rather than discussing which strategy is more efficient.) Why?
- How are these two strategies, counting on to make ten and take from ten, similar to each other? Use 15 – 8, and turn and talk to your partner. (For both of them, we do 2 + 5. For counting on, we are adding 2 to 8 to get to 10 and then adding 5 to get to 15. In the take from ten strategy, you take 8 from 10 and get 2. You add 2 to 5 that's still left and get 7.)
- Explain to your partner how counting on to make ten is related to taking from ten.
- What new math tool did we use today to show counting on to make ten? (Using the number path to count on by using 2 hops to get to 10 and then adding the hops used to get to the teen number.)
- Look at the Application Problem. How did you solve it? How could we use today's strategies to solve the problem? How could knowing how many cherries Carla took help you solve how many cherries the other children took?

Exit Ticket (3 minutes)

After the Student Debrief, instruct students to complete the Exit Ticket. A review of their work will help with assessing students' understanding of the concepts that were presented in today's lesson and planning more effectively for future lessons. The questions may be read aloud to the students.

Name ________________________ Date ____________

Use a number bond to show how you used the take from ten strategy to solve the problem.

1. Kevin had 14 crayons. Eight of the crayons were broken. How many of his crayons were not broken?

14 - 8 = ____

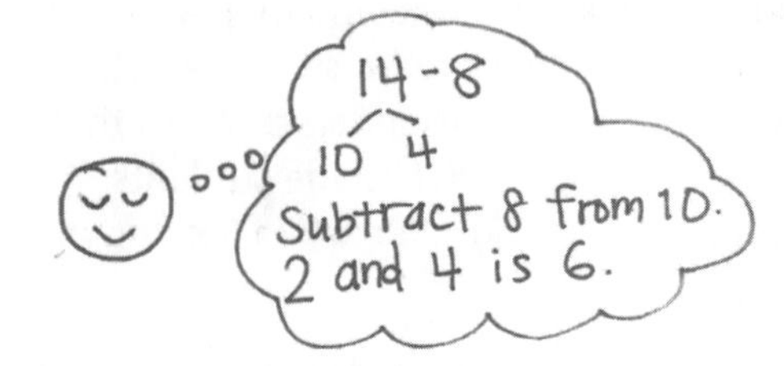

Kevin had ___ crayons that were not broken.

Use number bonds to show your thinking.

2. 17 - 8 = ____

3. 18 - 8 = ____

Count on to solve.

4. 13 - 8 = ____

5. 15 - 8 = ____

1	2	3	4	5	6	7	8	9	10	11	12	13	14	15	16	17	18	19	20

Complete the subtraction sentences by using the take from ten and count on strategies. Check the strategy that seemed easiest to you.

6. a. 12 - 8 = ____ b. 8 + ____ = 12 ☐ take from ten ☐ count on

---

7. a. 11 - 8 = ____ b. 8 + ____ = 11 ☐ take from ten ☐ count on

---

8. a. 16 - 8 = ____ b. 8 + ____ = 16 ☐ take from ten ☐ count on

Did you use a different strategy?

---

9. a. 19 - 8 = ____ b. 8 + ____ = 19 ☐ take from ten ☐ count on

Did you use a different strategy?

Name ______________________________ Date ______________

Complete the subtraction sentences by using the take from ten strategy and count on.

1. a. 11 - 8 = ____   b. 8 + ____ = 11

2. a. 15 - 8 = ____   b. 8 + ____ = 15

Name ______________________________ Date ______________

Complete the subtraction sentences by using the take from ten strategy and count on.

1. a. 12 - 8 = ____ b. 8 + ____ = 12

2. a. 15 - 8 = ____ b. 8 + ____ = 15

Choose the count on strategy or the take from ten strategy to solve.

3. 11 - 8 = ____

4. 17 - 8 = ____

Use a number bond to show how you solved using the take from ten strategy.

5. Elise counted 16 worms on the pavement. Eight worms crawled into the dirt. How many worms did Elise still see on the pavement?

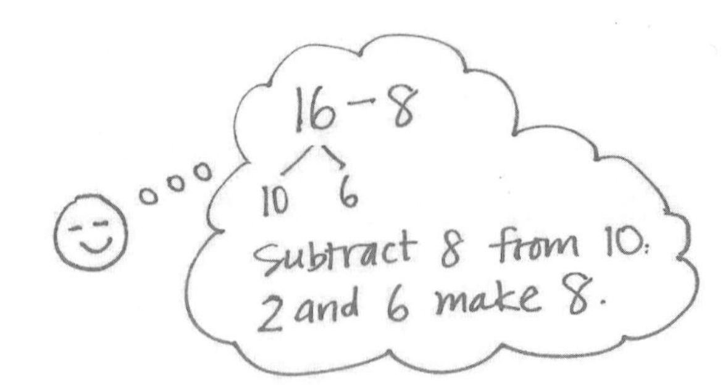

16 - 8 = _____

Elise still saw _____ worms on the pavement.

---

6. John ate 8 orange slices. If he started with 13, how many orange slices does he have left?

John has _____ orange slices left.

---

7. Match the addition number sentence to the subtraction number sentence. Fill in the missing numbers.

a. 12 - 8 = ______

8 + ______ = 11

b. 15 - 8 = ______

8 + ______ = 18

c. 18 - 8 = ______

8 + ______ = 12

d. 11 - 8 = ______

8 + ______ = 15

Lesson 20

Objective: Subtract 7, 8, and 9 from teen numbers.

Suggested Lesson Structure

■ Fluency Practice	(18 minutes)
■ Application Problem	(5 minutes)
■ Concept Development	(27 minutes)
■ Student Debrief	(10 minutes)
Total Time	**(60 minutes)**

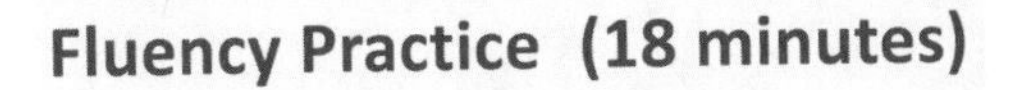

Fluency Practice (18 minutes)

- Number Path: Get to 10 **1.OA.6** (8 minutes)
- Sprint: Subtract 8 **1.OA.6** (10 minutes)

Number Path: Get to 10 (8 minutes)

Materials: (T) Subtract 9 flash cards (Lesson 17 Fluency Template), subtract 8 flash cards (Fluency Template)
(S) Personal white board, number path 1–20 (Lesson 18 Fluency Template 2)

Note: Using a number path to get to and from 10 reviews Lesson 19, when students were encouraged to relate taking from ten to counting on.

T: (Show the flash card 15 – 8.)
T: Write 15 – 8 as an addition sentence. Use a box for the number we don't know.
S: (Write 8 + ☐ = 15.)
T: How many spaces do you need to move to land on 10?
S: 2.
T: Hop from 8 to 10. Use your finger if you need help. Were you right?
S: Yes!
T: Now, hop to 15. How many spaces did you move?
S: 5.
T: 2 + 5 = __?
S: 7.
T: So, what is the missing number in your addition sentence?
S: 7.
T: Say the subtraction sentence.
S: 15 – 8 = 7.

8 + ☐ = 15

1	2	3	4	5	6	7	8	9	10	11	12	13	14	15	16	17	18	19	20

Repeat the sequence with the other flash cards.

Sprint: Subtract 8 (10 minutes)

Materials: (S) Subtract 8 Sprint

Note: This Sprint reviews the take from ten subtraction strategy when the subtrahend is 8.

Application Problem (5 minutes)

Imran has 8 crayons in his pencil box and 7 crayons in his desk. How many crayons does Imran have in total?

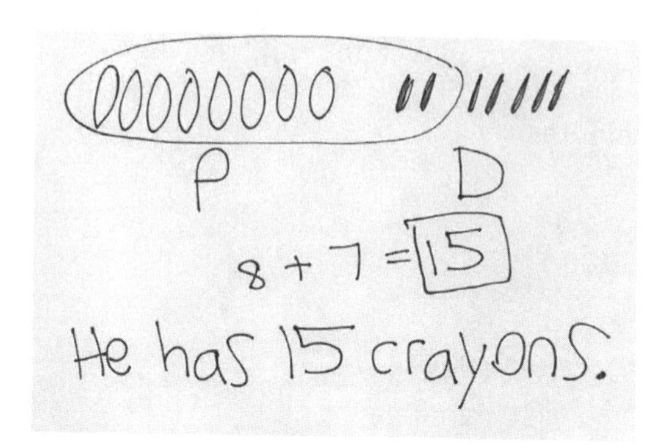

Note: Because students have been focusing on subtraction, some students may try to subtract 7 from 8 to solve. Look for such misunderstandings that can be addressed through discussion during the Student Debrief or individual support.

Concept Development (27 minutes)

Materials: (S) Personal white board, number path 1–20 (Lesson 18 Fluency Template 2), numeral cards 7–19 and subtraction symbol (Template)

Have students come to the meeting area and sit in a semicircle with their personal white boards.

T: (Write 13 – 9 = ____.) Solve and share with your partner what you did to get your answer.

S: (Discuss solution and strategies.)

T: Explain what you did to get your answer.

S: We made a 5-group drawing. → We used the take from ten strategy using fingers. → We made a picture in our minds. We just took away 9 from 10 and did 1 + 3. That's 4.

T: Everyone, use the number path to show how you can count on to make ten first. Don't forget to use two arrows to show your thinking.

S: (Solve by starting from 9. The arrows land on 10 and 13.)

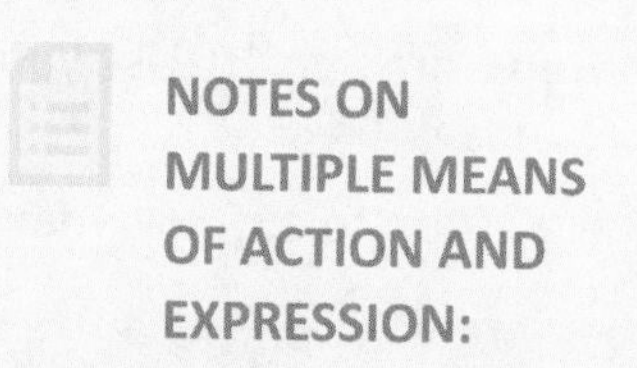

NOTES ON MULTIPLE MEANS OF ACTION AND EXPRESSION:

To support students who need extra pictorial support, draw a number bond (e.g., 13 decomposed to 10 and 3 circles in 5-group rows) along with the number bond.

T: What addition number sentence helped you to solve 13 – 9?

S: 1 + 3 = 4.

T: How is counting on the number path similar to using our fingers and pretend fingers?

S: After we drop 9 fingers, we have 1 more finger left from 10 fingers. We then add 1 to 3 pretend fingers. This is just like hopping 1 square to get to 10 and 3 more to get to 13. We had to add 1 and 3 both times.

Continue by following the suggested sequence: 13 – 7, 13 – 8, 15 – 9, and 15 – 7. Have Partner A and Partner B alternate between using the number path and their fingers to show their work.

T: (Write 12 – 7 = ____.) Let's use a number bond to solve 12 – 7. Visualize 5-group rows showing 12. What two parts do you see?

S: 10 and 2.

T: (Make a number bond for 12. Point to – 7.) Where would you take 7 away from?

S: Take 7 away from 10.

T: (Point to 10 and then 7 on the board.) Take 7 away in your mind. What is 10 – 7?

S: 3.

T: How many circles are there altogether? What two parts can you picture?

S: There are 5 circles. 2 and 3 make 5.

Continue the process, and invite students to solve using a number bond by following the suggested sequence: 11 – 7, 11 – 8, 13 – 9, 12 – 8, 17 – 8, 16 – 7, 19 – 7, and 19 – 8.

T: Now, we are going to play Simple Strategies! (Assign partners based on readiness levels. Instruct each pair to combine their numeral cards and make two piles: digits 11–19 and digits 7–9.) Here's how you play:

1. Partner A picks a card from the teen numbers pile.
2. Partners use the 9 card and the subtraction sign to make a subtraction fact. (Put the 8 and 7 cards aside for later use.)
3. Partner A solves by using any of the strategies from today's lesson.
4. Partner B writes down the addition fact that helped to solve the problem (e.g., for 13 – 9, write 1 + 3).
5. Switch roles. Keep the 9 card up each time the partners begin a new expression using a new teen number card.

As students play, the teacher circulates and moves students to working with – 8, then – 7, as appropriate.

Problem Set (10 minutes)

Students should do their personal best to complete the Problem Set within the allotted 10 minutes. For some classes, it may be appropriate to modify the assignment by specifying which problems they work on first. Some problems do not specify a method for solving. Students should solve these problems using the RDW approach used for Application Problems.

Note: Students may use drawings that reflect the strategies they learned from the past few days. For example, they may use 5-group drawings, arrows on a number path, or number bonds.

Student Debrief (10 minutes)

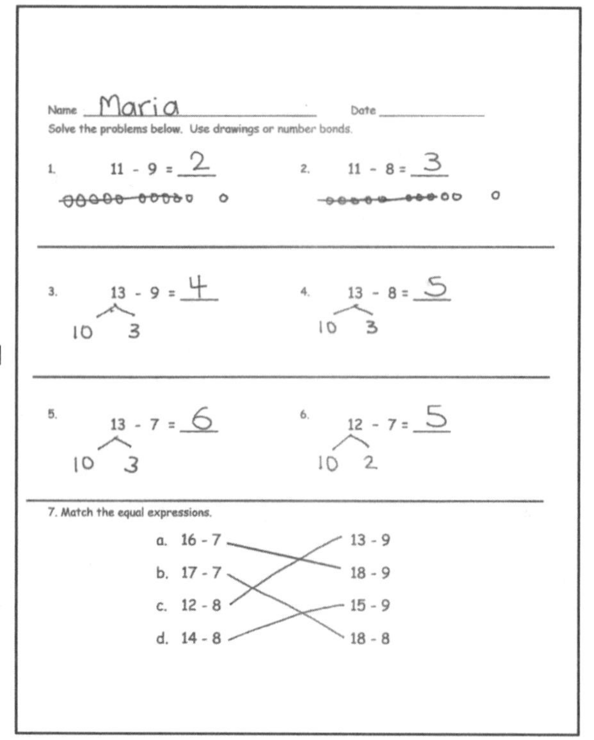

Name Maria Date

Solve the problems below. Use drawings or number bonds.

1. 11 - 9 = 2
2. 11 - 8 = 3
3. 13 - 9 = 4
4. 13 - 8 = 5
5. 13 - 7 = 6
6. 12 - 7 = 5

7. Match the equal expressions.

a. 16 - 7 — 13 - 9
b. 17 - 7 — 18 - 9
c. 12 - 8 — 15 - 9
d. 14 - 8 — 18 - 8

Lesson Objective: Subtract 7, 8, and 9 from teen numbers.

The Student Debrief is intended to invite reflection and active processing of the total lesson experience.

Invite students to review their solutions for the Problem Set. They should check work by comparing answers with a partner before going over answers as a class. Look for misconceptions or misunderstandings that can be addressed in the Debrief. Guide students in a conversation to debrief the Problem Set and process the lesson.

Any combination of the questions below may be used to lead the discussion.

- Look at your work from Simple Strategies! What did you notice about the addition facts for – 9 problems? – 8 problems? – 7 problems?
- Look at Problem 8 on your Problem Set. What is happening to the solution as you move from Part (a) to Part (c)? Explain why this is happening.
- Look at Problems 8 and 9. What do you notice? Explain how Problem 8 (a) and (b) relate to Problem 9 (a) and (b).
- Look at Problems 9 and 10. What do you notice? Explain how the rows are related. If there was a column (d) here, what might the number sentences be?
- Look at Problem 12. What did you do to solve these? Explain your thinking.
- How could knowing Problem 11(a) help you solve Problem 11(b)?
- Share your Application Problem with a partner. How did you solve it?

Complete the subtraction sentences to make them true.

	a.	b.	c.
8.	12 - 9 = 3	13 - 9 = 4	14 - 9 = 5
9.	12 - 8 = 4	13 - 8 = 5	14 - 8 = 6
10.	11 - 7 = 4	12 - 7 = 5	13 - 7 = 6
11.	16 - 9 = 7	18 - 9 = 9	17 - 9 = 8
12.	16 - 7 = 9	15 - 6 = 9	15 - 8 = 7
13.	15 - 9 = 6	11 - 8 = 3	16 - 9 = 7

Exit Ticket (3 minutes)

After the Student Debrief, instruct students to complete the Exit Ticket. A review of their work will help with assessing students' understanding of the concepts that were presented in today's lesson and planning more effectively for future lessons. The questions may be read aloud to the students.

A

Number Correct:

Name ____________________ Date ____________

*Write the missing number. Pay attention to the addition or subtraction sign.

1.	10 - 8 = □		16.	10 - 8 = □	
2.	2 + 2 = □		17.	11 - 8 = □	
3.	10 - 8 = □		18.	12 - 8 = □	
4.	2 + 3 = □		19.	15 - 8 = □	
5.	10 - 8 = □		20.	14 - 8 = □	
6.	2 + 4 = □		21.	13 - 8 = □	
7.	10 - 8 = □		22.	17 - 8 = □	
8.	2 + 1 = □		23.	18 - 8 = □	
9.	11 - 8 = □		24.	8 + □ = 11	
10.	10 - 8 = □		25.	8 + □ = 12	
11.	2 + 2 = □		26.	8 + □ = 15	
12.	12 - 8 = □		27.	8 + □ = 14	
13.	10 - 8 = □		28.	8 + □ = 16	
14.	2 + 5 = □		29.	8 + □ = 17	
15.	15 - 8 = □		30.	8 + □ = 18	

B

Name ________________________ Number Correct: Date ____________

*Write the missing number. Pay attention to the addition or subtraction sign.

1.	10 - 8 = □		16.	10 - 8 = □	
2.	2 + 1 = □		17.	11 - 8 = □	
3.	10 - 8 = □		18.	13 - 8 = □	
4.	2 + 2 = □		19.	14 - 8 = □	
5.	10 - 8 = □		20.	13 - 8 = □	
6.	2 + 3 = □		21.	12 - 8 = □	
7.	10 - 8 = □		22.	15 - 8 = □	
8.	2 + 2 = □		23.	16 - 8 = □	
9.	12 - 8 = □		24.	8 + □ = 10	
10.	10 - 8 = □		25.	8 + □ = 11	
11.	2 + 3 = □		26.	8 + □ = 13	
12.	13 - 8 = □		27.	8 + □ = 12	
13.	10 - 8 = □		28.	8 + □ = 13	
14.	2 + 2 = □		29.	8 + □ = 15	
15.	12 - 8 = □		30.	8 + □ = 16	

Name ______________________________ Date ______________

Solve the problems below. Use drawings or number bonds.

1. 11 - 9 = _____

2. 11 - 8 = _____

3. 13 - 9 = _____

4. 13 - 8 = _____

5. 13 - 7 = _____

6. 12 - 7 = _____

7. Match the equal expressions.

a. 16 - 7	13 - 9
b. 17 - 7	18 - 9
c. 12 - 8	15 - 9
d. 14 - 8	18 - 8

Complete the subtraction sentences to make them true.

	a.	b.	c.
8.	12 - 9 = ____	13 - 9 = ____	14 - 9 = ____
9.	12 - 8 = ____	13 - 8 = ____	14 - 8 = ____
10.	11 - 7 = ____	12 - 7 = ____	13 - 7 = ____
11.	16 - 9 = ____	18 - 9 = ____	17 - 9 = ____
12.	16 - ____ = 9	15 - ____ = 9	15 - ____ = 7
13.	15 - ____ = 6	11 - ____ = 3	16 - ____ = 7

EUREKA MATH

Name ____________________ Date ________

Solve the problems below. Use drawings or number bonds.

a. 14 - 9 = _____

b. 14 - 7 = _____

c. 14 - 8 = _____

---

d. 16 - 7 = _____

e. 16 - 9 = _____

f. 16 - 8 = _____

Name ______________________ Date ____________

Complete the number sentences to make them true.

1. 15 - 9 = _____
2. 15 - 8 = _____
3. 15 - 7 = _____

4. 17 - 9 = _____
5. 17 - 8 = _____
6. 17 - 7 = _____

7. 16 - 9 = _____
8. 16 - 8 = _____
9. 16 - 7 = _____

10. 19 - 9 = _____
11. 19 - 8 = _____
12. 19 - 7 = _____

13. Match equal expressions.

a. 19 - 9 12 - 7

b. 13 - 8 18 - 8

14. Read the math story. Use a drawing or a number bond to show how you know who is right.

a. Elsie says that the expressions 17 - 8 and 18 - 9 are equal. John says they are not equal. Who is right?

b. John says that the expressions 11 - 8 and 12 - 8 are not equal. Elsie says they are. Who is right?

c. Elsie says that to solve 17 - 9, she can take one from 17 and give it to 9 to make 10. So, 17 - 9 is equal to 16 - 10. John thinks Elsie made a mistake. Who is correct?

d. John and Elsie are trying to find several subtraction number sentences that start with numbers larger than 10 and have an answer of 7. Help them figure out number sentences. They started the first one.

16 - 9 = _____

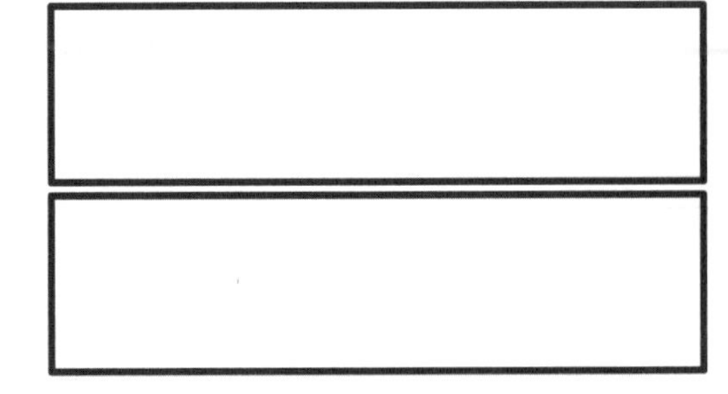

10 - 8 =	11 - 8 =
12 - 8 =	13 - 8 =
14 - 8 =	15 - 8 =
16 - 8 =	17 - 8 =
18 - 8 =	

subtract 8 flash cards

7	8	9	10
11	12	13	14
15	16	17	18
19	–		

numeral cards 7–19 and subtraction symbol

Lesson 21

Objective: Share and critique peer solution strategies for *take from with result unknown* and *take apart with addend unknown* word problems from the teens.

Suggested Lesson Structure

■ Fluency Practice	(13 minutes)
■ Application Problem	(5 minutes)
■ Concept Development	(32 minutes)
■ Student Debrief	(10 minutes)
Total Time	**(60 minutes)**

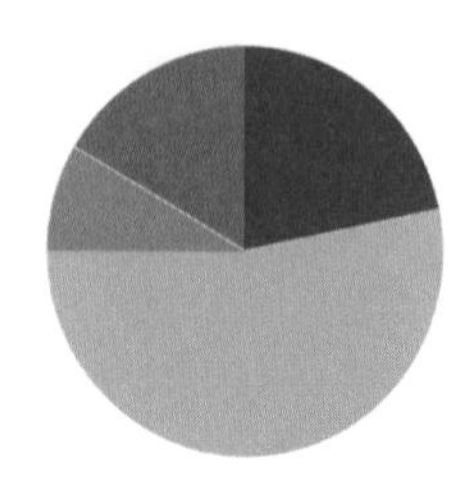

Fluency Practice (13 minutes)

- Subtraction with Hide Zero Cards **1.OA.6** (3 minutes)
- Sprint: Subtract 7, 8, 9 **1.OA.6** (10 minutes)

Subtraction with Hide Zero Cards (3 minutes)

Materials: (T) Hide Zero cards (Lesson 18 Fluency Template 1)

Note: This fluency activity reviews subtracting 7, 8, and 9 using the Hide Zero cards, which helps prepare students to understand ten as a unit by the module's end.

T: (Show 15.) Say 15 the Say Ten way.
S: Ten 5.
T: (Break apart the cards to show 10 and 5. Hold up 10.) 10 – 9 = __?
S: 1.
T: (Hold up 5.) 1 + 5 = __?
S: 6.
T: (Put the cards back together to show 15.) So, 15 – 9 = __?
S: 6.

Continue subtracting 9, 8, and then 7 from teen numbers.

Sprint: Subtract 7, 8, 9 (10 minutes)

Materials: (S) Subtract 7, 8, 9 Sprint

Note: Subtracting 7, 8, and 9 from teen numbers allows students to practice the take from ten subtraction strategy.

Application Problem (5 minutes)

There are 16 reading mats in the classroom. If 9 reading mats are being used, how many reading mats are still available?

Note: While the Application Problem provides the opportunity to continue exploring subtracting 9 from a teen number, it also directly connects with students' work during today's Problem Set. By using the same quantities as the upcoming Problem Set, students have a context for comparing and analyzing other student samples.

16 − 9 = 7
10 6
There are 7 more mats to use.

Concept Development (32 minutes)

Materials: (T) Student work samples—take from ten strategies (Template) (S) Personal white board

Have students come to the meeting area and sit in a semicircle.

T: (Project and read.) Colby is reading a book that is 14 pages long. She has already read 8 pages. How many more pages does Colby need to read to finish the book? Turn and talk to your partner about how you would solve this problem.

T: (Project Student A's sample.) How did Student A solve this problem? Explain to your partner what this student was thinking. What strategy did Student A use?

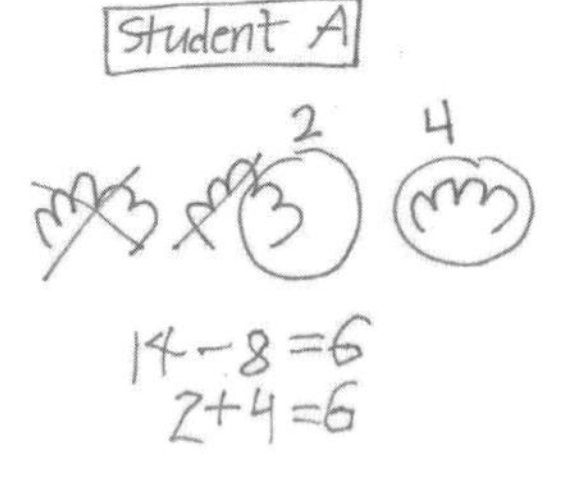

S: She drew 14 fingers as 10 and 4. She took away 8 fingers from 10 and got 2. She then added 2 and 4 to get 6. She used the take from ten strategy! That's the right answer!

T: (Label Student A's work sample *Take from Ten Strategy.*)

T: Can you think of another good way to make a math drawing?

S: Use a 5-group row drawing. That's another easy way to see the take from ten strategy.

T: (Project Student B's sample.) How did Student B solve the problem?

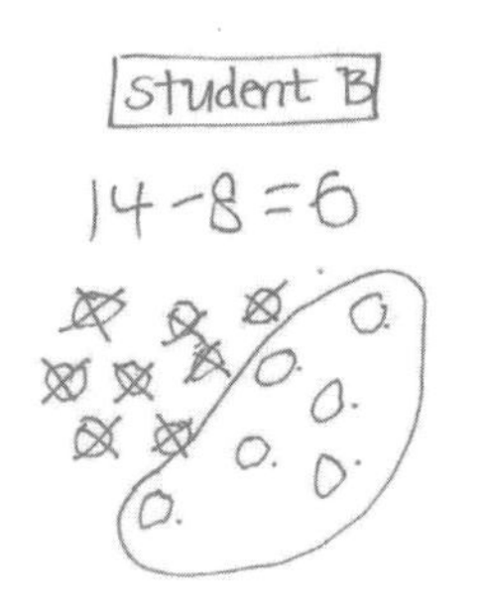

S: He drew a picture, but it's a little hard to see because the shapes are not organized. He drew 14 circles and took away 8 and circled the leftovers. He counted the leftovers: 1, 2, 3, 4, 5, 6.

T: (Label Student B's sample *Draw a Picture.*)

T: (Project Student C's sample.) Take a look at Student C's work. Her answer is 14. Is that correct? Did she do her work correctly? Turn and talk to your partner.

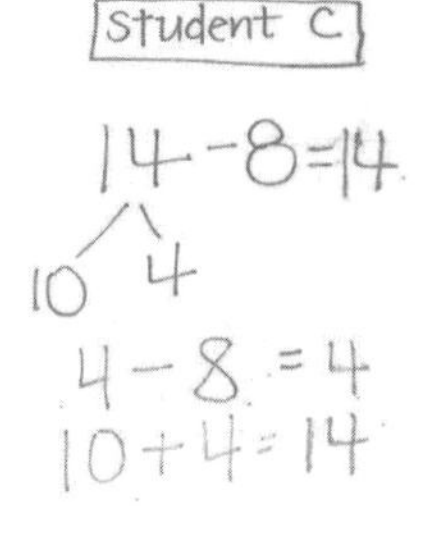

S: No.

T: What do you mean? What did she do wrong here? Well, did she do anything right?

S: She broke apart 14 into 10 and 4. That's correct. But look at her number sentence. She says 4 − 8 = 4. This is not correct.

T: (Use fingers or the number path to show students her mistake. If she were to take 8 from 4, the answer is less than 0.)

S: Her answer is 14. That doesn't make sense. We started with 14 and took away 8. Her answer has to be 8 less than 14.

T: I love the way you looked at her work so carefully. How can you help her get the correct answer? How would you teach her? What strategy did she try to use? Turn and talk to your partner.

S: I would tell her that you should always check what number you are taking away. In this problem, you have to take away 8. You need to subtract 8 from 10.

T: (Label Student C's sample work *Take from 10.*)

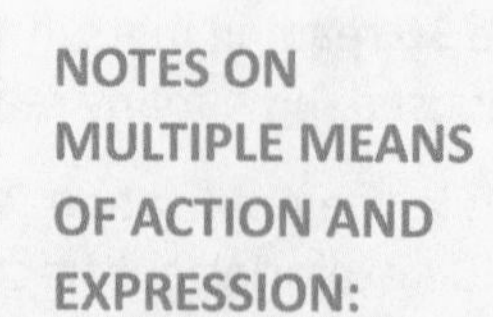

NOTES ON MULTIPLE MEANS OF ACTION AND EXPRESSION:

Direct students to analyze errors so they understand why they made a mistake. Being able to articulate the mistake helps develop their math comprehension at a deeper level.

Repeat the process, and analyze the work samples for Students D and E. Ask students to compare the strategies in these last two samples. Be sure to label the strategy used for each student's sample work.

Student D

8 + [6] = 14

8 9 10 11 12 13 14

(6)

Student E

+2 +4

8 → 10 → 14

2 + 4 = 6

T: Except for Student C's work, do these all show ways to solve the problem correctly? Which way seems like it's a better shortcut? Turn and talk to your partner.

S: (Discuss while the teacher circulates.)

T: (Project and read aloud.) Antalya collected 15 leaves. Nine are yellow. The rest are red. How many leaves are red? Solve this problem by showing your work clearly on your personal white board.

MP.2

Have student swap boards with their partners, and discuss the following:

- Study what strategy your partner used.
- Did you get the same answer?
- Take turns to explain your partner's strategy.
- Are your strategies similar? How? Are they different? How?
- What did your partner do well?
- Was one strategy a better shortcut than the other? Explain.

NOTES ON MULTIPLE MEANS OF ENGAGEMENT:

Make sure to validate different accurate and efficient strategies students are using or attempting to use. Be aware that students think in different ways. Encourage and cultivate strategic competence in the classroom by allowing students to explain their thinking. Help them understand their missteps.

If time allows, repeat the partner work following the suggested sequence: 12 – 7, 18 – 7 (What did you take 7 away from?), and 15 – 9.

Problem Set (10 minutes)

Students should do their personal best to complete the Problem Set within the allotted 10 minutes. For some classes, it may be appropriate to modify the assignment by specifying which problems they work on first. Some problems do not specify a method for solving. Students should solve these problems using the RDW approach used for Application Problems.

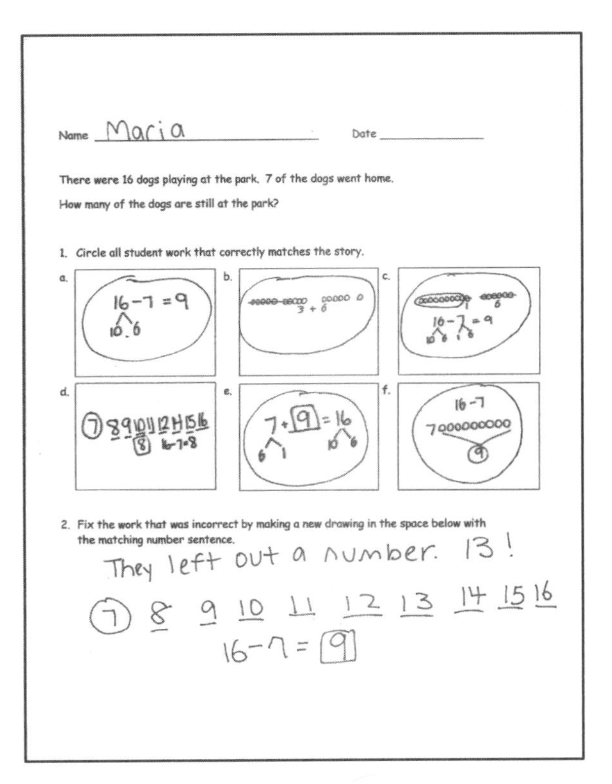
Name Maria Date

There were 16 dogs playing at the park. 7 of the dogs went home.
How many of the dogs are still at the park?

1. Circle all student work that correctly matches the story.

a. 16 − 7 = 9
b. 3 + 6
c. 16 − 7 = 9
d. 7 8 9 10 11 12 14 15 16 8 16 − 7 = 8
e. 7 + 9 = 16
f. 16 − 7 9

2. Fix the work that was incorrect by making a new drawing in the space below with the matching number sentence.

They left out a number. 13!
7 8 9 10 11 12 13 14 15 16
16 − 7 = 9

Student Debrief (10 minutes)

Lesson Objective: Share and critique peer solution strategies for *take from with result unknown* and *take apart with addend unknown* word problems from the teens.

The Student Debrief is intended to invite reflection and active processing of the total lesson experience.

Invite students to review their solutions for the Problem Set. They should check work by comparing answers with a partner before going over answers as a class. Look for misconceptions or misunderstandings that can be addressed in the Debrief. Guide students in a conversation to debrief the Problem Set and process the lesson.

Any combination of the questions below may be used to lead the discussion.

- Compare your solution to Problems 2 and 3 with your partner. How is your work similar or different from your partner's?
- Explain how your partner solved Problem 3.
- Study the ways 16 − 7 was solved. Which solutions seem to be the longest way to solve the problem? Which seem to be the best shortcut?
- (Project Sample C and Problem 1(d).) What have you learned from studying the mistakes from these students' work?
- Look at your Application Problem with a partner. Did you solve it the same way or a different way? Is your strategy or your partner's strategy similar to one of the samples in our Problem Set? If so, explain how it is similar. Is your strategy or your partner's strategy different from all of the samples in the Problem Set? If so, explain your strategy.

Solve on your own. Show your thinking by drawing or writing.
Write a statement to answer the question.

3. There were 12 sugar cookies in the box. My friend and I ate 5 of them. How many cookies are left in the box?

5 2
A L
12 − 5 = 7
There are 7 cookies left.

4. Megan checked out 17 books from the library. She read 9 of them. How many does she have left to read?

1 7
R L
17 − 9 = 8
10 7
Megan has 8 books left to read.

When you are done, share your solutions with a partner. How did your partner solve each problem? Be ready to share how your partner solved the problem.

Exit Ticket (3 minutes)

After the Student Debrief, instruct students to complete the Exit Ticket. A review of their work will help with assessing students' understanding of the concepts that were presented in today's lesson and planning more effectively for future lessons. The questions may be read aloud to the students.

A

Number Correct:

Name ______________________ Date ______________

*Write the missing number.

1.	10 - 9 = □		16.	12 - 7 = □	
2.	11 - 9 = □		17.	13 - 7 = □	
3.	13 - 9 = □		18.	14 - 7 = □	
4.	10 - 8 = □		19.	15 - 9 = □	
5.	11 - 8 = □		20.	15 - 8 = □	
6.	13 - 8 = □		21.	15 - 7 = □	
7.	10 - 7 = □		22.	17 - 7 = □	
8.	11 - 7 = □		23.	16 - 7 = □	
9.	13 - 7 = □		24.	17 - 7 = □	
10.	12 - 9 = □		25.	16 - □ = 9	
11.	13 - 9 = □		26.	16 - □ = 8	
12.	14 - 9 = □		27.	17 - □ = 8	
13.	12 - 8 = □		28.	17 - □ = 9	
14.	13 - 8 = □		29.	17 - □ = 16 - 8	
15.	14 - 8 = □		30.	□ - 7 = 17 - 8	

B

Name ______________________ Date ____________

Number Correct: 

*Write the missing number.

1.	10 - 9 = □		16.	11 - 7 = □	
2.	11 - 9 = □		17.	12 - 7 = □	
3.	12 - 9 = □		18.	15 - 7 = □	
4.	10 - 8 = □		19.	15 - 9 = □	
5.	11 - 8 = □		20.	15 - 8 = □	
6.	12 - 8 = □		21.	15 - 7 = □	
7.	10 - 7 = □		22.	15 - 8 = □	
8.	11 - 7 = □		23.	16 - 8 = □	
9.	12 - 7 = □		24.	16 - 7 = □	
10.	11 - 9 = □		25.	16 - □ = 9	
11.	12 - 9 = □		26.	16 - □ = 8	
12.	15 - 9 = □		27.	16 - □ = 7	
13.	11 - 8 = □		28.	16 - □ = 9	
14.	12 - 8 = □		29.	16 - □ = 15 - 8	
15.	15 - 8 = □		30.	□ - 8 = 15 - 7	

Name ______________________________ Date ______________

There were 16 dogs playing at the park. Seven of the dogs went home. How many of the dogs are still at the park?

1. Circle all the student work that correctly matches the story.

a. 16 − 7 = 9 (10, 6)

b. 3 + 6

c. 16 − 7 = 9 (10, 6, 1, 6)

d. 7 8 9 10 11 12 H 15 16 — 8 — 16 − 7 = 8

e. 7 + 9 = 16 (6, 1; 10, 6)

f. 16 − 7 — 9

2. Fix the work that was incorrect by making a new drawing in the space below with the matching number sentence.

Solve on your own. Show your thinking by drawing or writing.
Write a statement to answer the question.

3. There were 12 sugar cookies in the box. My friend and I ate 5 of them. How many cookies are left in the box?

4. Megan checked out 17 books from the library. She read 9 of them. How many does she have left to read?

When you are done, share your solutions with a partner. How did your partner solve each problem? Be ready to share how your partner solved the problem.

Name ______________________________ Date ______________

Meg thinks using the take from ten strategy is the best way to solve the following word problem. Bill thinks that solving the problem using the count on strategy is a better way. Solve both ways, and explain which strategy you think is best.

Strategies:

- Take from 10
- Make 10
- Count on
- I just knew

Mike and Sally have 6 cats. They have 14 pets in all. How many pets do they have that are *not* cats?

Meg's strategy	Bill's strategy

I think ______________ strategy is best because ______________

____

____.

Name ______________________________ Date ________________

Olivia and Jake both solved the word problems.
Write the strategy used under their work.
Check their work. If incorrect, solve correctly.
If solved correctly, solve using a different strategy.

Strategies:

- Take from 10
- Make 10
- Count on
- I just knew

1. A fruit bowl had 13 apples. Mike ate 6 apples from the fruit bowl. How many apples were left?

Olivia's work

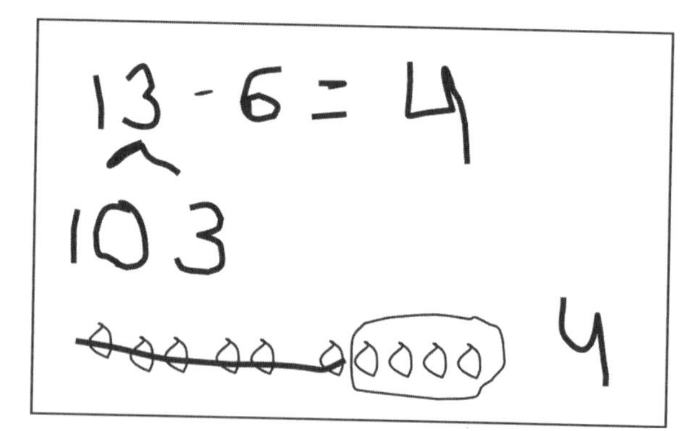

Jake's work

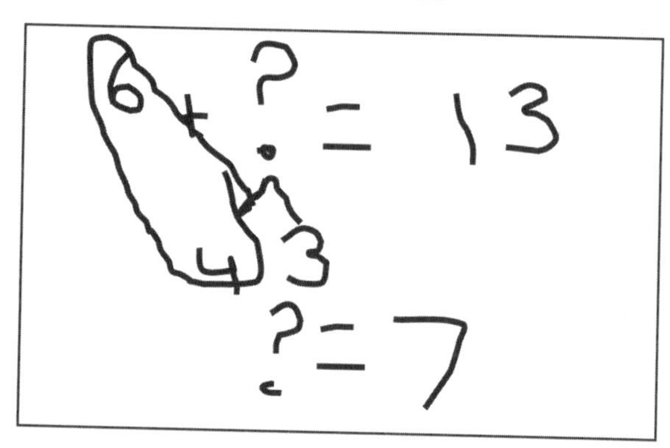

a. Strategy: ____________________

b. Strategy: ____________________

c. Explain your strategy choice below.

2. Drew has 17 baseball cards in a box. He has 8 cards with Red Sox players, and the rest are Yankees players. How many Yankees player cards does Drew have in his box?

Olivia's work

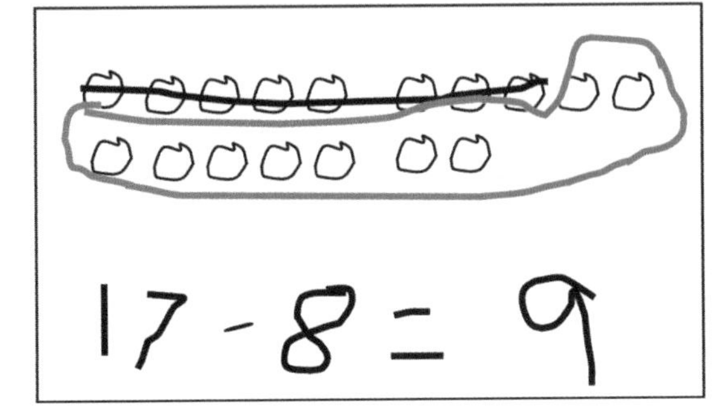

Jake's work

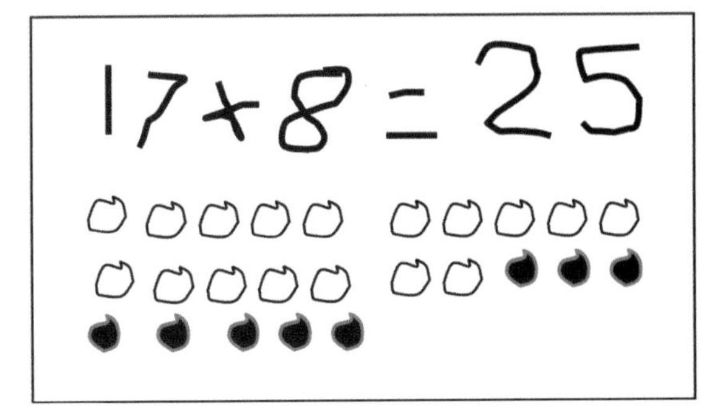

a. Strategy: ____________________

b. Strategy: ____________________

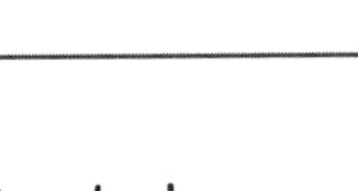

c. Explain your strategy choice below.

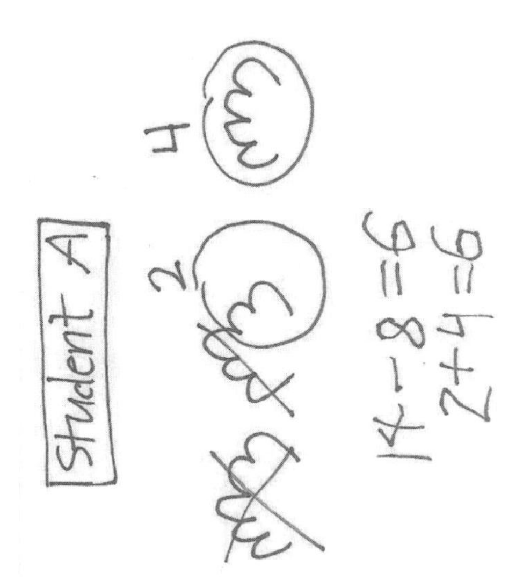

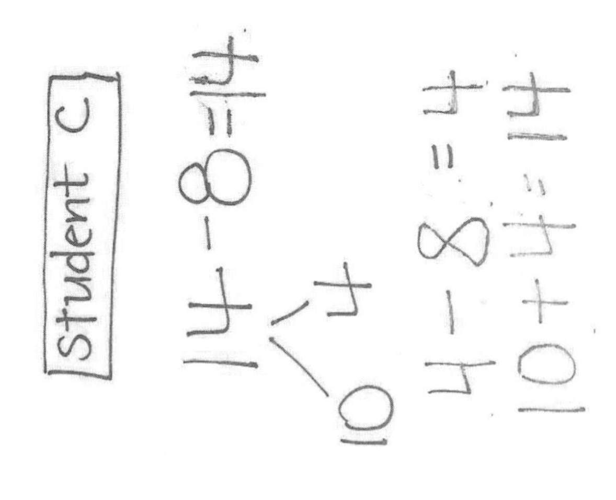

Student D

8 + 6 = 14

8 9 10 11 12 13 14

6

Student E

8 +2 10 +4 14

2 + 4 = 6

student work samples—take from ten strategies

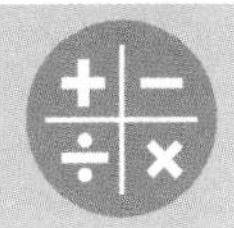

Topic C

Strategies for Solving *Change* or *Addend Unknown* Problems

1.OA.1, 1.OA.4, 1.OA.6, 1.OA.5, 1.OA.7, 1.OA.8

Focus Standards:		1.OA.1	Use addition and subtraction within 20 to solve word problems involving situations of adding to, taking from, putting together, taking apart, and comparing, with unknowns in all positions, e.g., by using objects, drawings, and equations with a symbol for the unknown number to represent the problem.
		1.OA.4	Understand subtraction as an unknown-addend problem. *For example, subtract 10 – 8 by finding the number that makes 10 when added to 8.*
		1.OA.6	Add and subtract within 20, demonstrating fluency for addition and subtraction within 10. Use mental strategies such as counting on; making ten (e.g., 8 + 6 = 8 + 2 + 4 = 10 + 4 = 14); decomposing a number leading to a ten (e.g., 13 – 4 = 13 – 3 – 1 = 10 – 1 = 9); using the relationship between addition and subtraction (e.g., knowing that 8 + 4 = 12, one knows 12 – 8 = 4); and creating equivalent but easier or known sums (e.g., adding 6 + 7 by creating the known equivalent 6 + 6 + 1 = 12 + 1 = 13).
Instructional Days:		4	
Coherence	**-Links from:**	GK–M4	Number Pairs, Addition and Subtraction to 10
	-Links to:	G2–M3	Place Value, Counting, and Comparison of Numbers to 1,000
		G2–M5	Addition and Subtraction Within 1,000 with Word Problems to 100

Topic C provides students with practice solving *add to with change unknown, take from with change unknown, put together with addend unknown,* and *take apart with addend unknown* word problems (**1.OA.1**). Drawing on the momentum gained from Topic B, Lesson 22 allows students to attack *put together/take apart with addend unknown* word problems such as, "Maria has 15 baseballs. Eight of them are old, and some of them are brand new. How many brand new baseballs does Maria have?" Students solve these problems using both the Level 2 counting on strategy and Level 3 subtraction strategies (**1.OA.4**).

Lesson 23 allows students to use counting on as it relates to subtraction, take from ten strategies, or the get to ten Level 3 strategy, as they solve *add to with change unknown* problems (**1.OA.6**). The get to ten strategy has students solving 12 – 3 as 12 – 2 – 1, understanding that decomposing the subtrahend to easily get to the ten yields a simpler, more manageable subtraction problem. It is the way a student can make ten when there is an unknown addend. It is a step away from counting on, where, rather than counting on by ones, students consider how much it takes to get to ten and then add on the rest to get to the teen number. For many students, the language of get to ten helps them bridge from counting on to a more efficient strategy. Up to this point, make ten for the students has shown both addends, and they are strategic about which number to break apart so that they can bond two numbers to make ten. This is a different, though related, process.

Lesson 24 presents students with *take from with change unknown* problems where they continue to select various strategies for solving. Students again relate various addition strategies to their recently acquired subtraction strategies, but in this new word problem type, the strategies they select and discuss help them better make sense of these problems. Students begin to recognize that although stories may be *take from with change unknown* problems, they can apply many strategies such as counting on, counting back, taking from ten, or getting to ten to accurately solve this challenging problem type.

Topic C closes with Lesson 25, where students move away from the context of story problems to find matching expressions to create true number sentences. They work solely with equations to show and talk about how they would re-represent a given addition or subtraction problem using a Level 2 or Level 3 strategy. For example, when given 9 + 6, students decompose the 6 into 1 and 5 and then can add using their new number sentence, 10 + 5 (i.e., 9 + 6 = 10 + 5) (**1.OA.7**), using pictures and words.

A Teaching Sequence Toward Mastery of Strategies for Solving *Change* or *Addend Unknown* Problems	
Objective 1:	**Solve *put together/take apart with addend unknown* word problems, and relate counting on to the take from ten strategy.** **(Lesson 22)**
Objective 2:	**Solve *add to with change unknown* problems, relating varied addition and subtraction strategies.** **(Lesson 23)**
Objective 3:	**Strategize to solve *take from with change unknown* problems.** **(Lesson 24)**
Objective 4:	**Strategize and apply understanding of the equal sign to solve equivalent expressions.** **(Lesson 25)**

Lesson 22

Objective: Solve *put together/take apart with addend unknown* word problems, and relate counting on to the take from ten strategy.

Suggested Lesson Structure

■ Fluency Practice	(15 minutes)
■ Concept Development	(35 minutes)
■ Student Debrief	(10 minutes)
Total Time	**(60 minutes)**

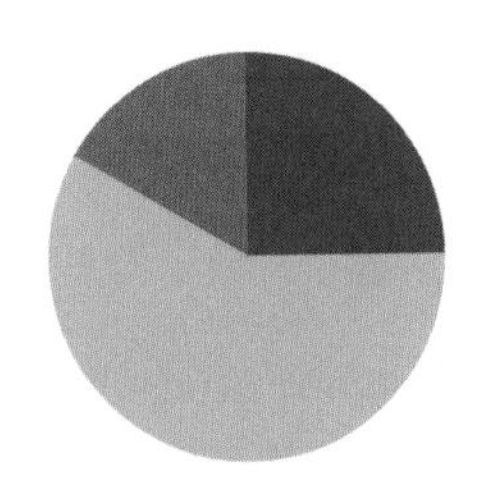

Fluency Practice (15 minutes)

- Subtraction with Hide Zero Cards **1.OA.6** (3 minutes)
- Count by Fives **1.OA.5** (2 minutes)
- Sprint: Missing Addend Within 10 **1.OA.6** (10 minutes)

Subtraction with Hide Zero Cards (3 minutes)

Materials: (T) Hide Zero cards (Lesson 18 Fluency Template 1)

Note: This fluency activity reviews subtracting 7, 8, and 9 using the Hide Zero cards, which helps prepare students to understand ten as a unit by the module's end. Since this is the second time students are doing this activity, have volunteers describe the steps necessary to apply the take from ten strategy.

T: (Show 15.) What do I need to do if I want to subtract 9?
S: Take apart 15.
T: (Break apart the cards to show 10 and 5.) Now what?
S: Take 9 from 10.
T: 10 – 9 = __?
S: 1.
T: What should I do next?
S: Add 1 to the 5.
T: 1 + 5 = __?
S: 6.
T: (Put the cards back together to show 15.) So, 15 – 9 = __?
S: 6.

Continue subtracting 9, 8, and then 7 from teen numbers.

Count by Fives (2 minutes)

Materials: (T) 100-bead Rekenrek

Note: Counting by fives promotes fluency with adding and subtracting 5.

Use the Rekenrek to count up and down by fives within 40. Students say the numbers as you move the beads. This time, count both forward and backward on your way up to 40 (e.g., 5, 10, 5, 10, 15, 20, 15, 20). Alternate between counting the Say Ten and regular way.

Sprint: Missing Addend Within 10 (10 minutes)

Materials: (S) Missing Addend Within 10 Sprint

Note: This review activity is intended to strengthen students' ability to fluently add and subtract within 10 while preparing students for the problem types that are presented in today's lesson.

Concept Development (35 minutes)

Materials: (S) Personal white board

Note: The Application Problem is embedded within the Concept Development since it directly pertains to the objective of today's lesson.

Students may sit with a partner in the meeting area (or at their seats) with their materials.

T: (Project the following problem: Mark has 14 crayons. Eight of the crayons are on the table, and some more crayons are in the box. How many crayons are in the box?)

S: (Solve and then share work.)

T: (Circulate, noticing students' accuracy with creating a drawing that matches the story and taking note of the varying ways students solved the problem.)

T: Explain your drawing to your partner, and discuss how you solved the problem.

S: (Share work.)

T: (Continue to circulate and take note of the language students are using to explain their thinking.)

T: (Point to today's problem.) Step 1: When we want to solve a problem, we read or listen to the problem. Let's read it together again. (Write on the board: 1. Read.)

S/T: Mark has 14 crayons. Eight of the crayons are on the table, and some more crayons are in the box. How many crayons are in the box?

T: Step 2: Draw as much of the math story as you can. You made some great drawings to match this story. What did you draw? (Write on the board: 2. Draw.)

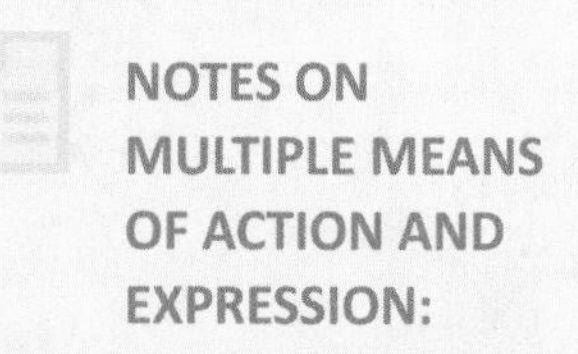

When students are required to draw, remind them to complete math drawings so they do not spend time making beautiful pictures. The use of lines or dots with labels is very efficient.

T: (Project work with a document camera or redraw on the board as students explain their drawings.)

S: I drew 14 lines in a row, like the 14 crayons in the problem. I circled 8 of them and labeled those with a T to show they were the ones on the table. → I started by drawing 8 circles for the 8 crayons on the table. Then, I started drawing dark circles until I got to 14.

T: Everyone look at your work. As I read the story, find the part of your math drawing that matches the sentence.

T: (Read from projection.) *Mark has 14 crayons*. Does your drawing show Mark has 14 crayons? Point to where your drawing shows the 14 crayons. Circle it with your finger.

MP.2

T: *Eight of the crayons are on the table*. Where does your picture show the 8 crayons that are on the table?

T: Are these 8 *more* crayons, or are they a *part* of Mark's 14 crayons?

S: They are a part of Mark's crayons!

T: How can we tell these crayons from the other crayons in the story?

S: We can make those crayons circles and the other ones dots. → We can label these crayons with a *T* since they are on the table. → We can circle them.

T: If you didn't already label them with a *T* or with the word *table*, add a label. Let's put a box around them too, so we can see them together clearly. (Write on the board: *and label* after *2. Draw*.)

T: ... *and some more crayons are in the box*. Can you find these crayons in your drawing? Point to them and circle them with your finger.

T: What could we label this set of crayons?

S: *B* for box.

T: If you didn't label these crayons, add *B* or the word *Box* to show these are the crayons in the box.

T: Now, we come to the question part of the word problem. *How many crayons are in the box?* Can we find the answer to this question in our drawing?

S: Yes, 6 crayons!

T: What number sentence would match this story? (Write on the board: *3. Write a number sentence*.)

S: 8 + 6 = 14.

T: Which number in the number sentence is the answer, or solution, to the question?

S: 6.

T: We have to make sure we put a rectangle around this number so we know it is the solution. If you didn't add a box, do that now.

T: What is the answer to our question? (Write: *Write or tell a statement of the solution.*)

S: There are 6 crayons in the box.

T: When we read the problem and draw the parts of the story, it can help us understand the problem and help us write the number sentence and the answer, or solution, sentence. Let's try to **read, draw, and write (RDW)** (point to list of steps now listed on board) to solve more problems.

Use the steps listed on the board as you repeat the process above with three more *put together/take apart with addend unknown* problems using the suggested sequence of story problems:

- There are 12 milk bottles in the crate. Nine are plain milk bottles, and the rest are chocolate milk bottles. How many are chocolate milk bottles?
- Ani puts some pink barrettes in her hair. She already had 7 blue barrettes in her hair. If Ani now has 11 barrettes in her hair, how many pink barrettes did she use?
- Laurie reads 5 books about frogs and then reads some books about toads. Laurie counts and realizes she has just read 13 books! How many books about toads did Laurie read?

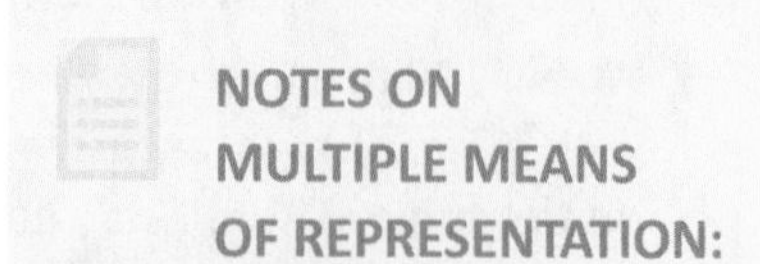

NOTES ON MULTIPLE MEANS OF REPRESENTATION:

Choose numbers and tasks differentiated for your learners. If students are having difficulty visualizing the story problems, use smaller numbers. For classes of students where they are very successful, use larger numbers within 20.

Have students project or draw their work on the board as the class shares and discusses each part of the RDW process. Ask students to find in the drawing where they can count on to determine the solution as well as where they can take away, or cover, a part as a method to finding the solution.

Problem Set (10 minutes)

Students should do their personal best to complete the Problem Set within the allotted 10 minutes. For some classes, it may be appropriate to modify the assignment by specifying which problems they work on first. Some problems do not specify a method for solving. Students should solve these problems using the RDW approach used for Application Problems.

Name Maria Date

Read the word problem.
Draw and label.
Write a number sentence and a statement that matches the story.

1. This week, Maria ate 5 yellow plums and some red plums. If she ate 11 plums in all, how many red plums did Maria eat?

y r

5+6=11

Maria ate 6 red plums.

2. Tatyana counted 14 frogs. She counted 8 swimming in the pond and the rest sitting on lily pads. How many frogs did she count sitting on lily pads?

S L

8 6

8+6=14

6 frogs were sitting on lily pads.

Student Debrief (10 minutes)

Lesson Objective: Solve *put together/take apart with addend unknown* word problems, and relate counting on to the take from ten strategy.

The Student Debrief is intended to invite reflection and active processing of the total lesson experience.

Invite students to review their solutions for the Problem Set. They should check work by comparing answers with a partner before going over answers as a class. Look for misconceptions or misunderstandings that can be addressed in the Debrief. Guide students in a conversation to debrief the Problem Set and process the lesson.

Any combination of the questions below may be used to lead the discussion.

- Look at Problems 1 and 2. Did you use the same or different strategy to solve? Explain why you chose to use the strategy (or strategies) you did.
- How did drawing the parts of the story help you solve the story problems?
- What new math strategy did we use today to communicate precisely how we solved the story problem? (**RDW.**) Explain what it is and how we used it.

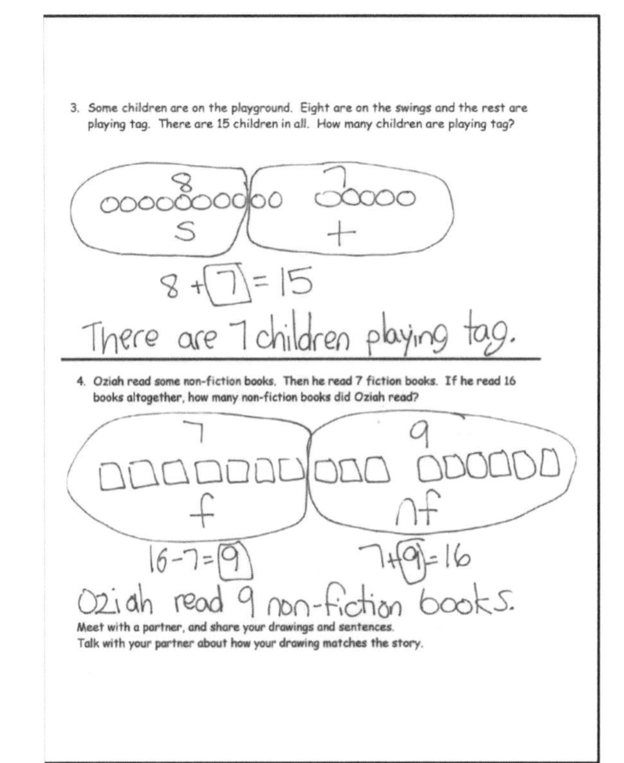
3. Some children are on the playground. Eight are on the swings and the rest are playing tag. There are 15 children in all. How many children are playing tag?

8 S 7 t

8 + [7] = 15

There are 7 children playing tag.

4. Oziah read some non-fiction books. Then he read 7 fiction books. If he read 16 books altogether, how many non-fiction books did Oziah read?

7 f 9 nf

16 − 7 = [9] 7 + [9] = 16

Oziah read 9 non-fiction books.

Meet with a partner, and share your drawings and sentences.
Talk with your partner about how your drawing matches the story.

Exit Ticket (3 minutes)

After the Student Debrief, instruct students to complete the Exit Ticket. A review of their work will help with assessing students' understanding of the concepts that were presented in today's lesson and planning more effectively for future lessons. The questions may be read aloud to the students.

A

Number Correct: 

Name ______________________ Date ____________

*Write the missing number.

1.	$2 + \square = 3$		16.	$2 + \square = 8$	
2.	$1 + \square = 3$		17.	$4 + \square = 8$	
3.	$\square + 1 = 3$		18.	$8 = \square + 6$	
4.	$\square + 2 = 4$		19.	$8 = 3 + \square$	
5.	$3 + \square = 4$		20.	$\square + 3 = 9$	
6.	$1 + \square = 4$		21.	$2 + \square = 9$	
7.	$1 + \square = 5$		22.	$9 = \square + 1$	
8.	$4 + \square = 5$		23.	$9 = 4 + \square$	
9.	$3 + \square = 5$		24.	$2 + 2 + \square = 9$	
10.	$3 + \square = 6$		25.	$2 + 2 + \square = 8$	
11.	$\square + 2 = 6$		26.	$3 + \square + 3 = 9$	
12.	$0 + \square = 6$		27.	$3 + \square + 2 = 9$	
13.	$1 + \square = 7$		28.	$5 + 3 = \square + 4$	
14.	$\square + 5 = 7$		29.	$\square + 4 = 1 + 5$	
15.	$\square + 4 = 7$		30.	$3 + \square = 2 + 6$	

B

Name ______________________ Number Correct: Date ____________

*Write the missing number.

1.	$1 + \square = 3$		16.	$3 + \square = 8$	
2.	$0 + \square = 3$		17.	$2 + \square = 8$	
3.	$\square + 3 = 3$		18.	$8 = \square + 1$	
4.	$\square + 2 = 4$		19.	$8 = 4 + \square$	
5.	$3 + \square = 4$		20.	$\square + 2 = 9$	
6.	$4 + \square = 4$		21.	$4 + \square = 9$	
7.	$4 + \square = 5$		22.	$9 = \square + 5$	
8.	$1 + \square = 5$		23.	$9 = 6 + \square$	
9.	$2 + \square = 5$		24.	$1 + 5 + \square = 9$	
10.	$4 + \square = 6$		25.	$3 + 2 + \square = 8$	
11.	$\square + 2 = 6$		26.	$2 + \square + 6 = 9$	
12.	$3 + \square = 6$		27.	$3 + \square + 4 = 9$	
13.	$3 + \square = 7$		28.	$5 + 4 = \square + 6$	
14.	$\square + 4 = 7$		29.	$\square + 3 = 6 + 2$	
15.	$\square + 5 = 7$		30.	$4 + \square = 2 + 7$	

Name ______________________________ Date ______________

Read the word problem.
Draw and label.
Write a number sentence and a statement that matches the story.

1. This week, Maria ate 5 yellow plums and some red plums. If she ate 11 plums in all, how many red plums did Maria eat?

---

2. Tatyana counted 14 frogs. She counted 8 swimming in the pond and the rest sitting on lily pads. How many frogs did she count sitting on lily pads?

3. Some children are on the playground. Eight are on the swings, and the rest are playing tag. There are 15 children in all. How many children are playing tag?

---

4. Oziah read some non-fiction books. Then, he read 7 fiction books. If he read 16 books altogether, how many non-fiction books did Oziah read?

Meet with a partner, and share your drawings and sentences.
Talk with your partner about how your drawing matches the story.

Name ______________________________ Date ______________

Read the word problem.
Draw and label.
Write a number sentence and a statement that matches the story.

Remember to draw a box around your solution in the number sentence.

1. Some students in Mrs. See's class are walkers. There are 17 students in her class in all. If 8 students ride the bus, how many students are walkers?

2. I baked 13 loaves of bread for a party. Some were burnt, so I threw them away. I brought the remaining 8 loaves to the party. How many loaves of bread were burnt?

Name ______________________________ Date ______________

Read the word problem.
Draw and label.
Write a number sentence and a statement that matches the story.

Remember to draw a box around your solution in the number sentence.

Strategies:

- Take from 10
- Make 10
- Count on
- I just knew

1. Michael and Anastasia pick 14 flowers for their mom. Michael picks 6 flowers. How many flowers does Anastasia pick?

2. Daquan bought 6 toy cars. He also bought some magazines. He bought 15 items in all. How many magazines did Daquan buy?

3. Henry and Millie baked 18 cookies. Nine of the cookies were chocolate chip. The rest were oatmeal. How many were oatmeal?

4. Felix made 8 birthday invitations with hearts. He made the rest with stars. He made 17 invitations in all. How many invitations had stars?

5. Ben and Miguel are having a bowling contest. Ben wins 9 times. They play 17 games in all. There are no tied games. How many times does Miguel win?

6. Kenzie went to soccer practice 16 days this month. Only 9 of her practices were on a school day. How many times did she practice on a weekend?

Lesson 23

Objective: Solve *add to with change unknown* problems, relating varied addition and subtraction strategies.

Suggested Lesson Structure

Fluency Practice	(15 minutes)
Application Problem	(5 minutes)
Concept Development	(30 minutes)
Student Debrief	(10 minutes)
Total Time	**(60 minutes)**

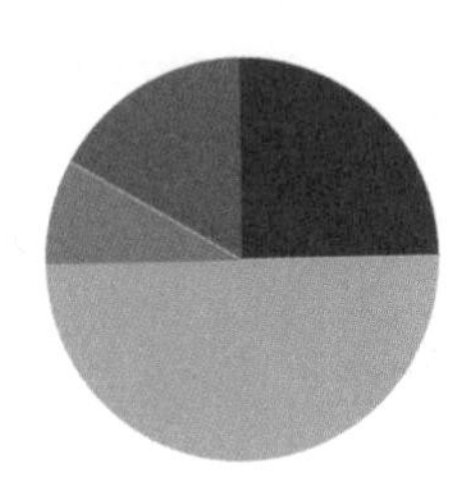

Fluency Practice (15 minutes)

- Subtraction with Partners **1.OA.6** (5 minutes)
- Sprint: Missing Addend Within 10 **1.OA.6** (10 minutes)

Subtraction with Partners (5 minutes)

Materials: (S) Personal white board

Note: This fluency activity reviews subtracting 7, 8, and 9 from teen numbers. Allow students who still require pictorial representations to draw 5-groups to solve.

Assign partners of equal ability. Partners assign each other a number from 11 to 17 (e.g., 12). On their personal white boards, they write number sentences with 9, 8, and 7 as the subtrahend and solve them (e.g., 12 – 9 = 3, 12 – 8 = 4, 12 – 7 = 5). Partners then exchange boards and check each other's work.

Sprint: Missing Addend Within 10 (10 minutes)

Materials: (S) Missing Addend Within 10 Sprint

Note: This fluency activity is intended to strengthen students' ability to fluently add and subtract within 10 while preparing students for the problem types that are presented in today's lesson.

Application Problem (5 minutes)

In the morning, there were 8 leaves on the floor under the ficus tree. During the day, more leaves fell on the floor. Now, there are 13 leaves on the floor. How many leaves fell during the day?

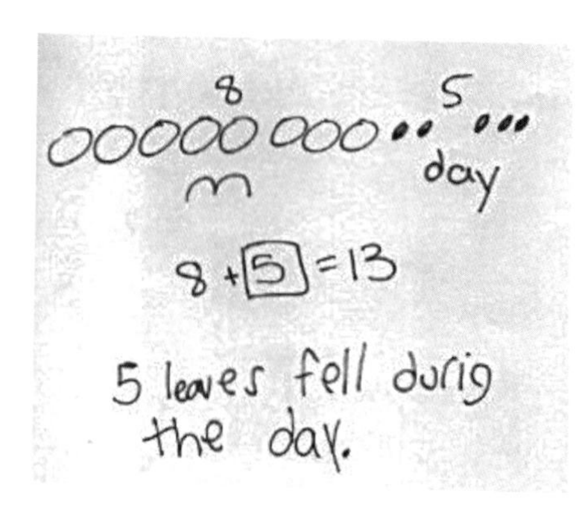

Note: In today's lesson, students grapple with an *add to with change unknown* problem. By giving students time to try this problem type independently, teachers have the opportunity to see how students are applying the RDW strategy without direct instruction on a specific method to solve.

Concept Development (30 minutes)

Materials: (S) Personal white board, work from the Application Problem

Students may sit next to their partners in the meeting area or at their seats with their materials.

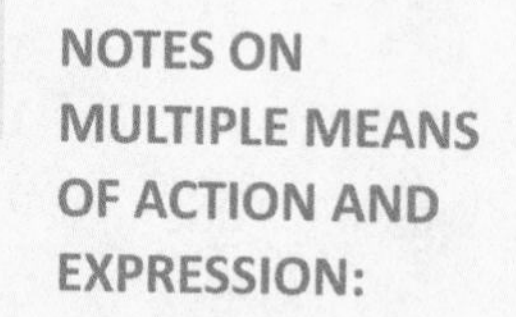

NOTES ON MULTIPLE MEANS OF ACTION AND EXPRESSION:

As students are using the RDW process, some may need help with organizing their work. Not all students use the blank white space well. Drawing lines or a grid for these students to work within helps them be more spatially organized.

T: (Project today's Application Problem.) Before we share our Application Problem with a partner, let's walk through the Read and Draw parts of the Read, Draw, Write strategy. We call this RDW. What does RDW stand for?

S: Read, draw, and write.

T: As I read the problem, find the part of your drawing that matches the story.

T: *In the morning, there were 8 leaves on the floor*. Point to where your drawing shows these 8 leaves on the floor. Do these leaves have a label so you can find them easily?

T: *During the day, more leaves fell on the floor*. Touch the part of your drawing that shows these leaves. Label this part if you haven't yet.

T: *Now, there are 13 leaves on the floor*. Can you find these leaves in your drawing? Is this a part of your leaves or is this the total number of leaves?

S: It's the total number of leaves. (Touch their drawings to show.)

T: *How many leaves fell during the day?*

S: 5 leaves!

T: Talk with your partner. How does your drawing help you see the story situation?

S: (Discuss.)

T: (Circulate and take note of students' language use, including terms such as *part, whole,* and *total.*)

T: What was missing, a part or the total number of leaves?

S: A part of the leaves.

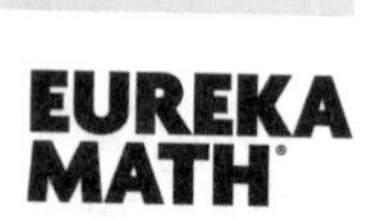

T: Now that we know we are missing a part, how could we solve this problem?

S: We can start at 8 and then count on until we get to 13. That would be 5. → We can draw all 13 and then cover, or take away, 8. We would have 5 left. → We can draw 13 as 10 and 3, so we can quickly cover the 8 without having to recount them. Then, it's easy to see the 2 and 3 that are left. That's 5.

T: I saw many of you draw your 8 leaves first and use counting on. How can we use our friendly number 10 to count on in bigger amounts, instead of counting by ones?

S: We can think about how many more we need to get to 10, and then add the rest all at once.

T: Let's try that here. We have 8 leaves, so how many would we draw to get to 10 leaves?

S: 2 more leaves.

T: From 10 leaves, how many more to get to the total, 13 leaves?

S: 3 leaves!

T: So, how many more leaves did we draw altogether?

S: 5 leaves!

T: Now that it's later in our Grade 1 year, we can go a little faster by jumping from 8 to 10 and then jumping to the total. Counting on in this way is a little faster.

T: Once we solve the problem, we have to write our number sentence and our statement. Which number sentence best matches what happened in the story? Talk with your partner.

S: (Discuss.)

S: 8 + 5 = 13 matches the story best because there were 8 leaves in the morning, and then 5 leaves joined the pile. There were 13 leaves at the end of the story. The part we did not know was the 5.

T: Which number needs a rectangle around it to show it is our answer, or our solution?

S: 5.

T: What is our statement, or our word sentence, that tells the answer to the question?

S: 5 leaves fell during the day.

T: Let's try some more. Remember to think about which number sentence best matches the story.

NOTES ON MULTIPLE MEANS OF REPRESENTATION:

If students have trouble coming up with the statement, have them re-read the question in the story problem. They can use the question to help write their answer sentence.

Present three more *add to with change unknown* story problems such as those below:

- Eight children were playing on the playground. More children came out to join the 8 children. Now, there are 14 children on the playground. How many children came out to join them on the playground?
- Some new baby ducks hatched at the farm. There were 5 ducks on the farm, and now there are 12 ducks. How many new baby ducks were hatched?
- Thirteen cars are in the parking lot. Six were already there in the morning. The rest of the cars arrived after lunch. How many cars arrived after lunch?

For each story, project the problem and read it aloud. Ask students, "Can you draw something? Listen again and ask yourself, 'What can I draw?'" Read the problem again to allow students to think about what they can draw from the problem. Encourage students to use the drawing to help them consider a strategy for solving by asking themselves, "What does my drawing show me?" Remind them to write a number sentence that matches the story and a word sentence, or solution statement, to answer the question. Give students approximately three minutes to work. Share one or two students' work by drawing it on the board or using a document camera. Have students explain their drawing, share their choices of labels, and explain how their number sentence matches the story.

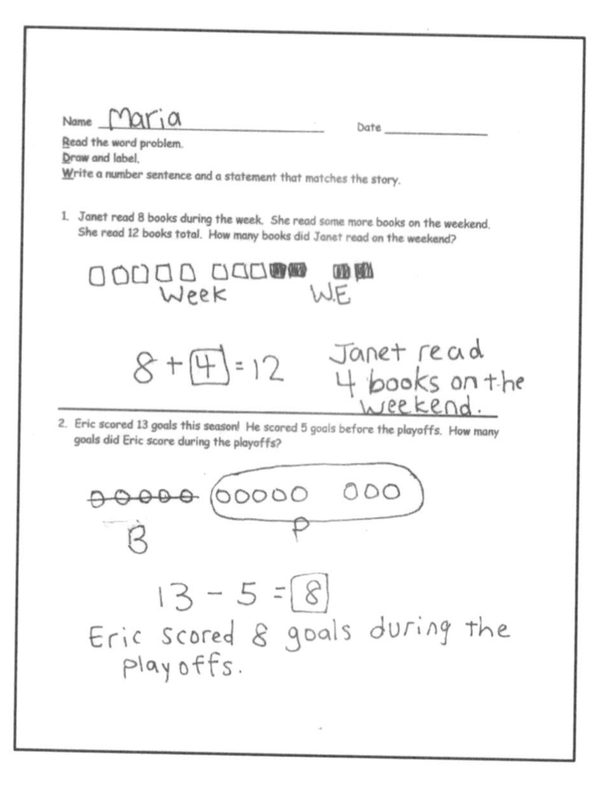
Name Maria Date

Read the word problem.
Draw and label.
Write a number sentence and a statement that matches the story.

1. Janet read 8 books during the week. She read some more books on the weekend. She read 12 books total. How many books did Janet read on the weekend?

Week WE

8 + 4 = 12

Janet read 4 books on the weekend.

2. Eric scored 13 goals this season! He scored 5 goals before the playoffs. How many goals did Eric score during the playoffs?

B P

13 − 5 = 8

Eric scored 8 goals during the playoffs.

Problem Set (10 minutes)

Students should do their personal best to complete the Problem Set within the allotted 10 minutes. For some classes, it may be appropriate to modify the assignment by specifying which problems they work on first. Some problems do not specify a method for solving. Students should solve these problems using the RDW approach used for Application Problems.

3. There were 8 ladybugs on a branch. Some more came. Then there were 15 ladybugs on the branch. How many ladybugs came?

8 7

L more

8 + 7 = 15

7 more ladybugs came.

4. Marco's friend gave him some baseball cards at school. If he was already given 9 baseball cards by his family, and he now has 19 cards in all, how many baseball cards did he get in school?

f S

19 − 9 = 10

10 9

Marco got 10 baseball cards in school.

Meet with a partner and share your drawings and sentences. Talk with your partner about how your drawing matches the story.

Student Debrief (10 minutes)

Lesson Objective: Solve *add to with change unknown* problems, relating varied addition and subtraction strategies.

The Student Debrief is intended to invite reflection and active processing of the total lesson experience.

Invite students to review their solutions for the Problem Set. They should check work by comparing answers with a partner before going over answers as a class. Look for misconceptions or misunderstandings that can be addressed in the Debrief. Guide students in a conversation to debrief the Problem Set and process the lesson.

Any combination of the questions below may be used to lead the discussion.

- Compare the way you solved Problems 1 and 2. How are the strategies you used the same or different?

- What do all of the story problems in the Problem Set have in common? (We always know the total and one of the parts. We had to look for the missing part.)
- Look at Problem 3. How did you use counting on? What did you do? How did that help you solve?

Exit Ticket (3 minutes)

After the Student Debrief, instruct students to complete the Exit Ticket. A review of their work will help with assessing students' understanding of the concepts that were presented in today's lesson and planning more effectively for future lessons. The questions may be read aloud to the students.

A

Name ______________________________ Number Correct: 

Date ____________

*Write the missing number.

1.	$2 + \square = 3$		16.	$2 + \square = 8$	
2.	$1 + \square = 3$		17.	$4 + \square = 8$	
3.	$\square + 1 = 3$		18.	$8 = \square + 6$	
4.	$\square + 2 = 4$		19.	$8 = 3 + \square$	
5.	$3 + \square = 4$		20.	$\square + 3 = 9$	
6.	$1 + \square = 4$		21.	$2 + \square = 9$	
7.	$1 + \square = 5$		22.	$9 = \square + 1$	
8.	$4 + \square = 5$		23.	$9 = 4 + \square$	
9.	$3 + \square = 5$		24.	$2 + 2 + \square = 9$	
10.	$3 + \square = 6$		25.	$2 + 2 + \square = 8$	
11.	$\square + 2 = 6$		26.	$3 + \square + 3 = 9$	
12.	$0 + \square = 6$		27.	$3 + \square + 2 = 9$	
13.	$1 + \square = 7$		28.	$5 + 3 = \square + 4$	
14.	$\square + 5 = 7$		29.	$\square + 4 = 1 + 5$	
15.	$\square + 4 = 7$		30.	$3 + \square = 2 + 6$	

B

Name ____________________ Number Correct: ______

Date ______

*Write the missing number.

1.	1 + □ = 3		16.	3 + □ = 8	
2.	0 + □ = 3		17.	2 + □ = 8	
3.	□ + 3 = 3		18.	8 = □ + 1	
4.	□ + 2 = 4		19.	8 = 4 + □	
5.	3 + □ = 4		20.	□ + 2 = 9	
6.	4 + □ = 4		21.	4 + □ = 9	
7.	4 + □ = 5		22.	9 = □ + 5	
8.	1 + □ = 5		23.	9 = 6 + □	
9.	2 + □ = 5		24.	1 + 5 + □ = 9	
10.	4 + □ = 6		25.	3 + 2 + □ = 8	
11.	□ + 2 = 6		26.	2 + □ + 6 = 9	
12.	3 + □ = 6		27.	3 + □ + 4 = 9	
13.	3 + □ = 7		28.	5 + 4 = □ + 6	
14.	□ + 4 = 7		29.	□ + 3 = 6 + 2	
15.	□ + 5 = 7		30.	4 + □ = 2 + 7	

Name ______________________________ Date ______________

Read the word problem.
Draw and label.
Write a number sentence and a statement that matches the story.

1. Janet read 8 books during the week. She read some more books on the weekend. She read 12 books total. How many books did Janet read on the weekend?

---

2. Eric scored 13 goals this season! He scored 5 goals before the playoffs. How many goals did Eric score during the playoffs?

3. There were 8 ladybugs on a branch. Some more came. Then, there were 15 ladybugs on the branch. How many ladybugs came?

---

4. Marco's friend gave him some baseball cards at school. If he was already given 9 baseball cards by his family, and he now has 19 cards in all, how many baseball cards did he get in school?

Meet with a partner and share your drawings and sentences. Talk with your partner about how your drawing matches the story.

Name ______________________________ Date ______________

Read the word problem.
Draw and label.
Write a number sentence and a statement that matches the story.

Shanika ate 7 mini-pretzels in the morning. She ate the rest of her mini-pretzels in the afternoon. She ate 13 mini-pretzels altogether that day. How many mini-pretzels did Shanika eat in the afternoon?

Name ______________________________ Date ______________

Read the word problem.
Draw and label.
Write a number sentence and a statement that matches the story.

1. Micah collected 9 pinecones on Friday and some more on Saturday. Micah collected a total of 14 pinecones. How many pinecones did Micah collect on Saturday?

2. Giana bought 8 star stickers to add to her collection. Now, she has 17 stickers in all. How many stickers did Giana have at first?

3. Samil counted 5 pigeons on the street. Some more pigeons came. There were 13 pigeons in all. How many pigeons came?

---

4. Claire had some eggs in the fridge. She bought 12 more eggs. Now, she has 18 eggs in all. How many eggs did Claire have in the fridge at first?

Lesson 24

Objective: Strategize to solve *take from with change unknown* problems.

Suggested Lesson Structure

- Fluency Practice (15 minutes)
- Application Problem (5 minutes)
- Concept Development (30 minutes)
- Student Debrief (10 minutes)

Total Time (60 minutes)

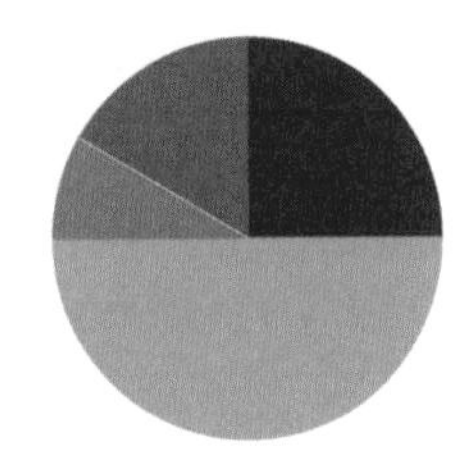

Fluency Practice (15 minutes)

- Count by Fives **1.OA.5** (5 minutes)
- Sprint: Missing Subtrahends Within 10 **1.OA.6** (10 minutes)

Count by Fives (5 minutes)

Materials: (T) Rekenrek

Note: Counting by fives promotes fluency with adding and subtracting 5.

Use the Rekenrek to count by fives to 40 and back. Students say the numbers as you move the beads. First, have students count the Say Ten way. Then, do it again, but have students count the regular way.

Sprint: Missing Subtrahends Within 10 (10 minutes)

Materials: (S) Missing Subtrahends Within 10 Sprint

Note: This review activity is intended to strengthen students' ability to fluently add and subtract within 10 while preparing students for the problem types that are presented in today's lesson.

Application Problem (5 minutes)

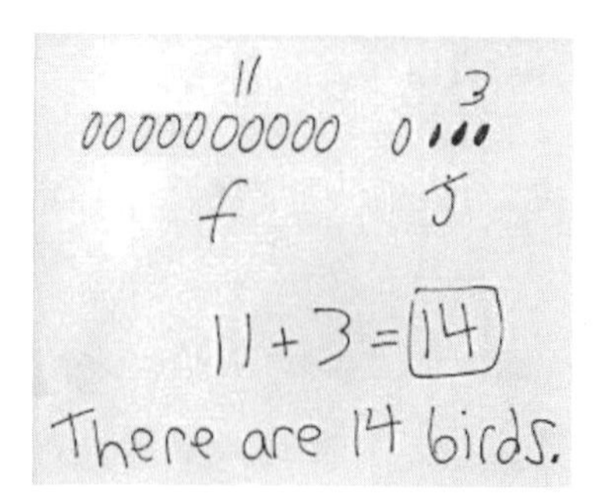

Yesterday, I saw 11 birds on a branch. Three birds joined them on the branch. How many birds were on the branch then?

Note: This problem is intentionally an *add to with result unknown* problem. Having spent two days on *change* or *addend unknown* situation types, students may be identifying a pattern in solving the problem type presented.

Misconceptions may arise through this Application Problem if students are overgeneralizing. Students use the context of this problem to transition into today's lesson where they are working with *take from with change unknown* problems. While students are completing the Application Problem, circulate and select a student's work in which the drawings accurately represent the story situation and are simple, labeled, and aligned in a single row. Use this work as the sample for sharing during the lesson.

Concept Development (30 minutes)

Materials: (S) Personal white board, work from Application Problem

Students may sit with their partners in the meeting area or at their seats with their materials.

T: (Project today's Application Problem.) We have been using the RDW process to solve problems. Before we share our Application Problem with our partners, what does RDW stand for again?

S: Read, draw, and write.

T: With your partner, share your solution, or answer. Be sure to discuss your drawings as you explain your idea. If you realize you forgot something or have to change something, you may do so.

T: (Project or redraw chosen student work.) This student's work uses simple shapes drawn in an organized line, which helps me see what we have. (See Application Problem image as an example.)

T: I need one volunteer to read the problem again for us and another volunteer to explain how the picture shows each part. (Choose students other than the one whose solution is being shared.)

S1: Yesterday, I saw birds in a tree.

S2: Here are the birds. (Points to the full line of shapes.)

S1: There were 11 birds on a branch, and then 3 birds joined them.

S2: These 11 birds are the ones on the branch first. I think that's why she wrote *f* under it. (Points to the first 11 birds.) Here are the 3 birds that joined in. That's why she wrote *j* under it. (Points to 3 birds at the end.)

S1: How many birds were in the tree?

S2: There are 14 birds. She wrote 11 + 3 = 14, and "You saw 14 birds," because that matches the story and the question. There were 11. Then 3 joined in, and now there are 14. (Points to the number sentence while explaining.)

T: You all did a great job reading, drawing, and writing to solve this problem. Let's try another problem.

G1-M2-TE-BK2-1.3.1-01.2016

T: (Project the following problem.) Today, I was passing the same tree. There were 11 birds in the tree when I first looked at it. I looked away. When I looked back, there were 5 birds. How many birds flew away?

S: (Begin to solve the problem.)

T: (Reread the question two more times to support struggling readers as students work.)

T: (Remind students to think about these questions: *Can you draw something? What can you draw? What does your drawing show you?* Give students approximately three minutes to RDW. Invite two or more students to solve on the board or on chart paper in pairs.)

T: (Project student work.) Let's look at the work these students did. They drew to show the 11 birds in the tree. Oh, and look at this. They drew a circle around 5 birds and wrote an *s* to show that these 5 birds that stayed were a part of all 11 birds that were in the tree. Let's draw another circle around these birds, the ones labeled *f*. These are the birds that flew away. (If neither group has a circle around them, draw a circle around each group.)

T: I'm going to use our lines from our number bonds to show that these two parts together make the total of 11 birds.

T: $11 - 6 = 5$. How many birds flew away? Let's put a rectangle around the solution.

S: Six birds flew away.

T: What strategies could you use to solve this?

S: I knew there were two parts, so I took away the 5 to find the other part. → I looked at the picture and counted them all. → I drew 11 in 5-group rows. It made it really easy to see 5 and then to see the other part. → I thought of my doubles plus one fact. $5 + 5$ is 10, so I needed $5 + 6$ to make 11.

NOTES ON MULTIPLE MEANS OF ACTION AND EXPRESSION:

Direct students to analyze the work their classmates have completed. As their teacher, you should also analyze their ability to solve the problem type presented. Look for common misconceptions and the way in which students explain how they got their answers to tell how students are progressing.

Repeat the process above for three more *take from with change unknown* story problems such as those listed below:

- Mina had 13 ants in her ant farm. Some ants escaped. Now, there are 9 ants in the ant farm. How many ants escaped?
- Jamal had 14 trains, but he only found 8 of his trains. How many of his trains are missing?
- June's baby brother hid some of her blocks. She has 7 blocks now. She used to have 15 blocks. How many blocks should June be looking for?

When sharing solutions and strategies, debrief quickly and move to the next problem. The goal is for students to love solving problems and to begin making connections between reading, drawing, and writing as a road to success as a problem solver!

Problem Set (10 minutes)

Students should do their personal best to complete the Problem Set within the allotted 10 minutes. For some classes, it may be appropriate to modify the assignment by specifying which problems they work on first. Some problems do not specify a method for solving. Students should solve these problems using the RDW approach used for Application Problems.

Student Debrief (10 minutes)

Lesson Objective: Strategize to solve *take from with change unknown* problems.

The Student Debrief is intended to invite reflection and active processing of the total lesson experience.

Invite students to review their solutions for the Problem Set. They should check work by comparing answers with a partner before going over answers as a class. Look for misconceptions or misunderstandings that can be addressed in the Debrief. Guide students in a conversation to debrief the Problem Set and process the lesson.

Any combination of the questions below may be used to lead the discussion.

- Look at Problems 1 and 2. How are your drawings similar? How are they different?
- How was your drawing similar to or different from your partner's drawing?
- What did today's problems have in common? How were they the same or different from yesterday's problems? What was unknown in the problem, a part or the total? Which strategies were easier for you to use when a part is missing instead of the total? (Note: Students might find Problem 2 tricky since the first number given is the part that is left and the whole is given later in the problem.)
- Which problem was tricky for you? What did you draw? How can we add to the drawing with more information from the problem? What does the drawing show you?
- How did your drawings help you with the problems? Use a specific problem to explain your thinking.

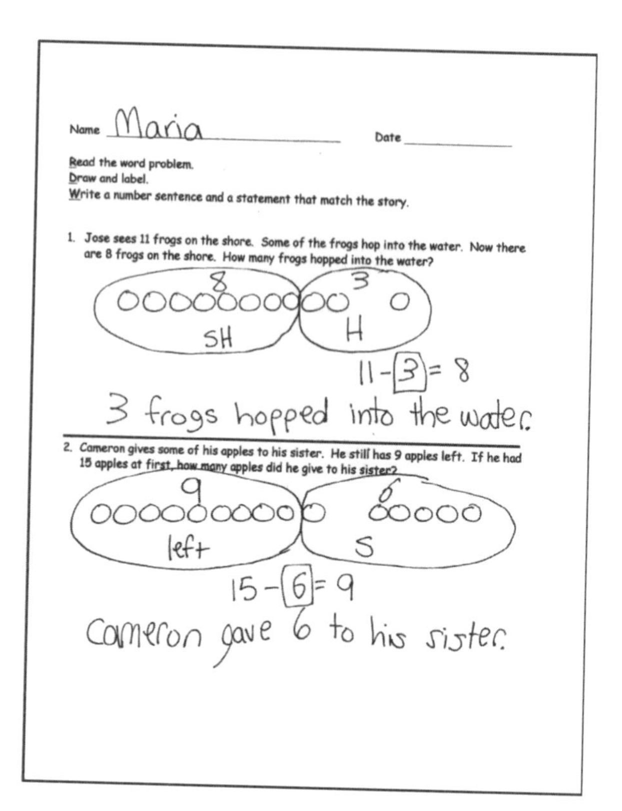

Name Maria Date

Read the word problem.
Draw and label.
Write a number sentence and a statement that match the story.

1. Jose sees 11 frogs on the shore. Some of the frogs hop into the water. Now there are 8 frogs on the shore. How many frogs hopped into the water?

11 - [3] = 8

3 frogs hopped into the water.

2. Cameron gives some of his apples to his sister. He still has 9 apples left. If he had 15 apples at first, how many apples did he give to his sister?

15 - [6] = 9

Cameron gave 6 to his sister.

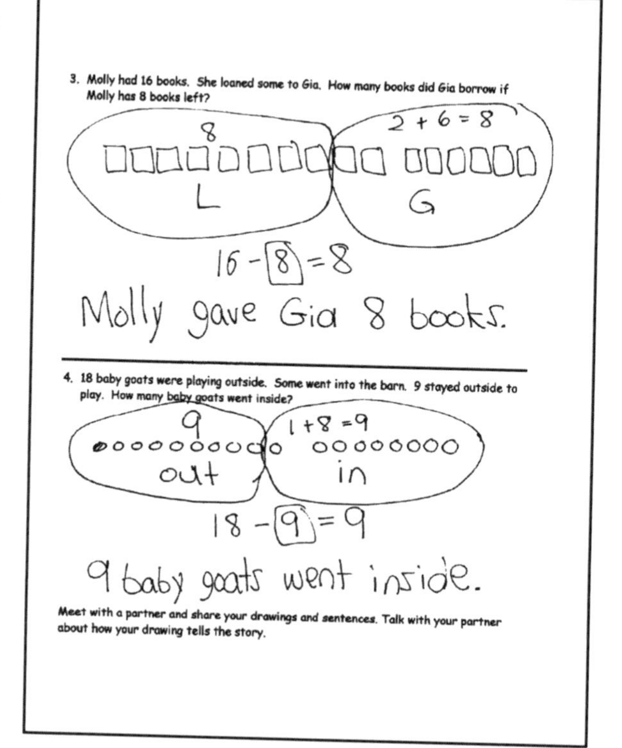

3. Molly had 16 books. She loaned some to Gia. How many books did Gia borrow if Molly has 8 books left?

16 - [8] = 8

Molly gave Gia 8 books.

4. 18 baby goats were playing outside. Some went into the barn. 9 stayed outside to play. How many baby goats went inside?

18 - [9] = 9

9 baby goats went inside.

Meet with a partner and share your drawings and sentences. Talk with your partner about how your drawing tells the story.

Exit Ticket (3 minutes)

After the Student Debrief, instruct students to complete the Exit Ticket. A review of their work will help with assessing students' understanding of the concepts that were presented in today's lesson and planning more effectively for future lessons. The questions may be read aloud to the students.

EUREKA MATH

A

Number Correct:

Name ____________________ Date ____________

*Write the missing number.

1.	2 - □ = 1		16.	6 - □ = 2	
2.	2 - □ = 2		17.	6 - □ = 3	
3.	2 - □ = 0		18.	6 - □ = 4	
4.	3 - □ = 2		19.	7 - □ = 3	
5.	3 - □ = 1		20.	7 - □ = 2	
6.	3 - □ = 0		21.	7 - □ = 1	
7.	3 - □ = 3		22.	8 - □ = 2	
8.	4 - □ = 4		23.	8 - □ = 3	
9.	4 - □ = 3		24.	4 = 8 - □	
10.	4 - □ = 2		25.	2 = 9 - □	
11.	4 - □ = 1		26.	3 = 9 - □	
12.	5 - □ = 0		27.	4 = 9 - □	
13.	5 - □ = 1		28.	10 - 3 = 9 - □	
14.	5 - □ = 2		29.	9 - □ = 10 - 5	
15.	5 - □ = 3		30.	9 - □ = 10 - 6	

B

Number Correct:

Name ______________________ Date ____________

*Write the missing number.

1.	2 - □ = 2		16.	6 - □ = 3	
2.	2 - □ = 1		17.	6 - □ = 4	
3.	2 - □ = 0		18.	6 - □ = 5	
4.	3 - □ = 3		19.	7 - □ = 4	
5.	3 - □ = 2		20.	7 - □ = 3	
6.	3 - □ = 1		21.	7 - □ = 2	
7.	3 - □ = 0		22.	8 - □ = 3	
8.	4 - □ = 4		23.	8 - □ = 4	
9.	4 - □ = 3		24.	5 = 8 - □	
10.	4 - □ = 2		25.	3 = 9 - □	
11.	4 - □ = 1		26.	4 = 9 - □	
12.	5 - □ = 5		27.	5 = 9 - □	
13.	5 - □ = 4		28.	10 - 4 = 9 - □	
14.	5 - □ = 3		29.	9 - □ = 10 - 6	
15.	5 - □ = 2		30.	9 - □ = 10 - 5	

Name ______________________________ Date ______________

Read the word problem.
Draw and label.
Write a number sentence and a statement that match the story.

1. Jose sees 11 frogs on the shore. Some of the frogs hop into the water. Now, there are 8 frogs on the shore. How many frogs hopped into the water?

---

2. Cameron gives some of his apples to his sister. He still has 9 apples left. If he had 15 apples at first, how many apples did he give to his sister?

3. Molly had 16 books. She loaned some to Gia. How many books did Gia borrow if Molly has 8 books left?

---

4. Eighteen baby goats were playing outside. Some went into the barn. Nine stayed outside to play. How many baby goats went inside?

Meet with a partner and share your drawings and sentences. Talk with your partner about how your drawing tells the story.

Name ______________________________ Date ______________

Read the word problem.
Draw and label.
Write a number sentence and a statement that matches the story.

There were 18 dogs splashing in a puddle. Some dogs left. There are 9 dogs still splashing in the puddle. How many dogs are left?

Name ______________________ Date ____________

Read the word problem.
Draw and label.
Write a number sentence and a statement that matches the story.

1. Toby dropped 12 crayons on the classroom floor. Toby picked up 9 crayons. Marnie picked up the rest. How many crayons did Marnie pick up?

---

2. There were 11 students on the playground. Some students went back into the classroom. If 7 students stayed outside, how many students went inside?

3. At the play, 8 students from Mr. Frank's room got a seat. If there were 17 children from Room 24, how many children did not get a seat?

---

4. Simone had 12 bagels. She shared some with friends. Now, she has 9 bagels left. How many did she share with friends?

Lesson 25

Objective: Strategize and apply understanding of the equal sign to solve equivalent expressions.

Suggested Lesson Structure

■ Fluency Practice	(15 minutes)
■ Application Problem	(7 minutes)
■ Concept Development	(28 minutes)
■ Student Debrief	(10 minutes)
Total Time	**(60 minutes)**

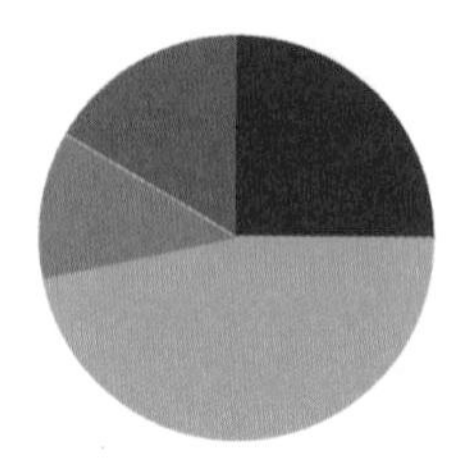

Fluency Practice (15 minutes)

- Make It Equal: Addition Expressions **1.OA.6** (5 minutes)
- Sprint: Make It Equal **1.OA.6** (10 minutes)

Make It Equal: Addition Expressions (5 minutes)

Materials: (S) Personal white board, counters

Note: This activity builds fluency with the make ten addition strategy and reinforces the meaning of the equal sign, which prepares students for today's lesson.

Write or project 9 + ☐ = 8 + ☐. Students find different numbers that make the equation true and check their answers with a partner. If necessary, students can use counters in addition to drawings that they can make on their personal white boards. During the last minute, ask for volunteers to share the equations they found. Write them on the board and ask if anyone notices a pattern (that the numbers are always consecutive).

Sprint: Make It Equal (10 minutes)

Materials: (S) Make It Equal Sprint

Note: This Sprint uses review addition facts to strengthen students' understanding of the equal sign.

Application Problem (7 minutes)

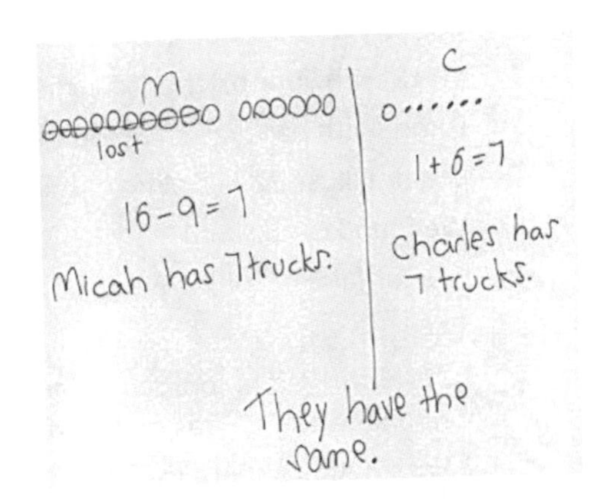

Micah had 16 trucks and lost 9 of them. Charles had 1 truck and received 6 more trucks from his mother. Who has more trucks, Micah or Charles?

Note: Students apply their prior understanding of *take from with result unknown* and *add to with result unknown* problems as they solve this two-part problem. This provides a context for exploring today's objective of further understanding the meaning of the equal sign by pairing equivalent expressions and constructing true number sentences.

Concept Development (28 minutes)

Materials: (T) Expression cards (Template) for use in small groups during Problem Set (S) Personal white board, work from Application Problem, linking cubes

Students may sit in the meeting area or at their seats, next to a partner, with all materials.

T: Who has more trucks, Micah or Charles? (Write Micah on the left side of the board and Charles on the right side of the board.)

S: Neither. They have the same number of trucks!

T: Talk with a partner. Use your drawings to help you prove to your partner that Micah and Charles have the same number of trucks.

S: (Discuss, using their drawings to explain.)

T: (Circulate and listen to ensure that all students see that Micah and Charles have the same number of trucks.) What number sentence did you write to match Micah's part of the story?

S: 16 – 9 = 7. (Write 16 – 9 = 7 below Micah.)

T: What number sentence did you write to match Charles' part of the story?

S: 1 + 6 = 7. (Write 1 + 6 = 7 below Charles.)

T: So, Micah and Charles have an equal number of trucks?

S: Yes!

T: (Write 16 – 9 under the Micah section and 1 + 6 under the Charles section.) We can say, then, that 16 – 9 is equal to 1 + 6. (Draw equal sign in between expressions.)

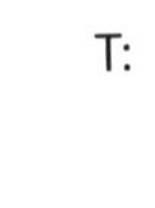

T: How does our story help us see that 16 – 9 = 1 + 6? (Point to each part while reading the number sentence.) Talk with your partners. (Listen as students explain their thinking to their partners.)

S: Since 16 – 9 is 7 and 1 + 6 is 7, they are equal. 16 – 9 equals 1 + 6. → Once I took the 9 from 10, Micah and Charles both show 1 and 6. They both have 7.

NOTES ON MULTIPLE MEANS OF ENGAGEMENT:

For those students who are able to quickly repeat the process, cultivate excitement by connecting on-level math to higher math, presenting numbers to 40.

T: Let's try to make some more cool number sentences like this.

T: Work with your partner to make at least two expressions that equal 12.

S: (Work with partners.) We found 10 + 2 and 11 + 1.

T: Great. I'll use 10 + 2. Who has another?

S: We found 6 + 6.

T: True or false? 10 + 2 = 6 + 6.

S: True!

T: Let's all write this cool number sentence on our personal white boards and read it together.

S: (Write number sentence.) 10 + 2 = 6 + 6.

After having generated several similar number sentences, start erasing some addends.

T: If I erase this 6 (erase the 6 after the equal sign), what number goes here to make this equation true?

S: 6. You would need to have two sixes to equal 12.

T: (Distribute an expression card to each student. Odd numbered classes need to pair two students together.) Solve the expression. You may use linking cubes or another strategy. If you're using linking cubes, you may need to borrow extras from a neighbor. After you solve the expression, make a linking cube stick to show your final amount.

T: There is someone in the room who has the same answer. Find that person, and create a number sentence together to show that your two expressions make equal amounts.

T: What true number sentences did we make?

NOTES ON MULTIPLE MEANS OF ENGAGEMENT:

Remember to challenge advanced learners. An extension activity can be given where number sentences are false, and students have to make them true. If given a false statement such as 3 + 5 = 7 + 2, they could change the 5 into 6 or the 2 into 1.

Problem Set (10 minutes)

Students should do their personal best to complete the Problem Set within the allotted 10 minutes. For some classes, it may be appropriate to modify the assignment by specifying which problems they work on first. Some problems do not specify a method for solving. Students should solve these problems using the RDW approach used for Application Problems.

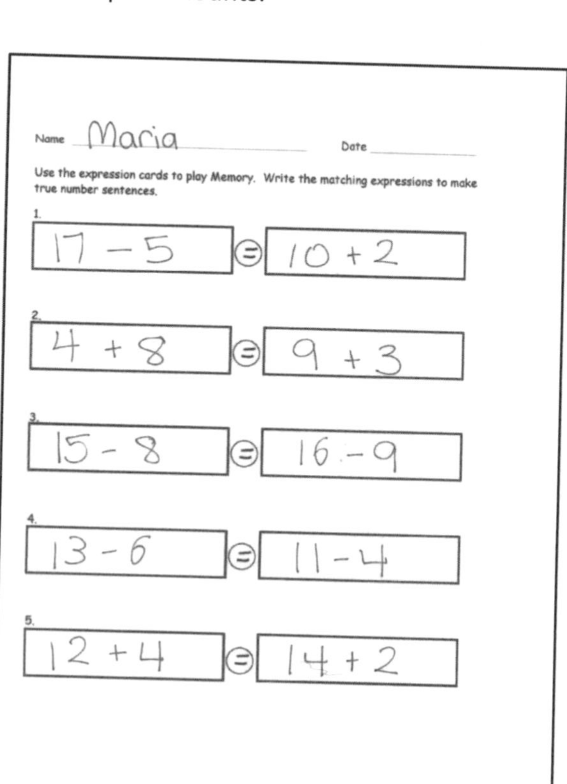

Name Maria Date

Use the expression cards to play Memory. Write the matching expressions to make true number sentences.

1. 17 − 5 = 10 + 2
2. 4 + 8 = 9 + 3
3. 15 − 8 = 16 − 9
4. 13 − 6 = 11 − 4
5. 12 + 4 = 14 + 2

Student Debrief (10 minutes)

Lesson Objective: Strategize and apply understanding of the equal sign to solve equivalent expressions.

The Student Debrief is intended to invite reflection and active processing of the total lesson experience.

Invite students to review their solutions for the Problem Set. They should check work by comparing answers with a partner before going over answers as a class. Look for misconceptions or misunderstandings that can be addressed in the Debrief. Guide students in a conversation to debrief the Problem Set and process the lesson.

6. Write a true number sentence using the expressions that you have left over. Use pictures and words to show how you know two of the expressions have the same unknown numbers. $9 + 7 = 10 + 6$ Both sides equal 16. $9 + 7 = 16$ $10 + 6 = 16$
7. Use other facts you know to write at least two true number sentences similar to the type above. $9 + 2 = 10 + 1$ $4 + 5 = 8 + 1$
8. The following addition number sentences are FALSE. Change one number in each problem to make a TRUE number sentence, and rewrite the number sentence.
   a. $8 + 5 = 10 + 2$ $8 + 4 = 10 + 2$
   b. $9 + 3 = 8 + 5$ $10 + 3 = 8 + 5$
   c. $10 + 3 = 7 + 5$ $10 + 2 = 7 + 5$
9. The following subtraction number sentences are FALSE. Change one number in each problem to make a TRUE number sentence, and rewrite the number sentence.
   a. $12 - 8 = 1 + 2$ $12 - 8 = 2 + 2$
   b. $13 - 9 = 1 + 4$ $14 - 9 = 1 + 4$
   c. $1 + 3 = 14 - 9$ $1 + 3 = 14 - 10$

Any combination of the questions below may be used to lead the discussion.

- (Write 14 + 2 = 10 + 6 on the board.) Show how you know these are equal expressions. What do you notice about the numbers when you break apart 14?
- (Point to the number sentence written on the board.) Which of the parts of the number sentence are the expression? What does it mean to use "=" between the two expressions? Explain the meaning of equal.
- Look at your Problem Set. Which expressions can you solve in your head? How can they help you solve other expressions that might be harder for you?
- Look at the true number sentences we made during today's partner activity. What did you notice about the expressions that made these number sentences true?
- Which expressions were missing a part? Which expressions were missing the whole, or total?
- How did the Application Problem connect to today's lesson?

Exit Ticket (3 minutes)

After the Student Debrief, instruct students to complete the Exit Ticket. A review of their work will help with assessing students' understanding of the concepts that were presented in today's lesson and planning more effectively for future lessons. The questions may be read aloud to the students.

A

Name ______________________ Number Correct: Date ____________

*Write the missing number.

1.	$\square = 4 + 1$		16.	$7 + 3 = 4 + \square$	
2.	$\square = 4 + 2$		17.	$6 + 4 = 5 + \square$	
3.	$\square = 4 + 3$		18.	$5 + 5 = 6 + \square$	
4.	$\square = 5 + 1$		19.	$5 + 3 = \square + 1$	
5.	$\square = 5 + 2$		20.	$5 + 4 = \square + 5$	
6.	$\square = 5 + 3$		21.	$4 + 5 = \square + 5$	
7.	$\square = 6 + 1$		22.	$2 + \square = 6 + 2$	
8.	$8 = 7 + \square$		23.	$4 + \square = 5 + 3$	
9.	$9 = 8 + \square$		24.	$\square + 4 = 5 + 2$	
10.	$9 = \square + 1$		25.	$\square + 6 = 4 + 3$	
11.	$9 = \square + 9$		26.	$4 + 2 = 1 + \square$	
12.	$8 = \square + 1$		27.	$3 + 4 = \square + 2$	
13.	$\square = 7 + 1$		28.	$4 + 4 = 2 + \square$	
14.	$10 = 8 + \square$		29.	$3 + \square = 2 + 7$	
15.	$10 = \square + 8$		30.	$\square + 2 = 2 + 6$	

B

Name ______________________ Number Correct: ☆

Date __________

*Write the missing number.

1.	□ = 3 + 1		16.	5 + 5 = 4 + □	
2.	□ = 3 + 2		17.	6 + 4 = 7 + □	
3.	□ = 3 + 3		18.	3 + 7 = 8 + □	
4.	□ = 4 + 1		19.	5 + 2 = □ + 1	
5.	□ = 4 + 2		20.	5 + 3 = □ + 5	
6.	□ = 4 + 3		21.	4 + 4 = □ + 4	
7.	□ = 5 + 1		22.	3 + □ = 6 + 3	
8.	8 = 1 + □		23.	4 + □ = 5 + 4	
9.	9 = 1 + □		24.	□ + 4 = 2 + 5	
10.	8 = □ + 7		25.	□ + 6 = 3 + 4	
11.	8 = □ + 8		26.	4 + 3 = 1 + □	
12.	7 = □ + 1		27.	4 + 4 = □ + 2	
13.	□ = 6 + 1		28.	4 + 5 = 2 + □	
14.	10 = 9 + □		29.	3 + □ = 2 + 6	
15.	10 = □ + 9		30.	□ + 2 = 2 + 7	

Name ______________________________ Date ______________

Use the expression cards to play Memory. Write the matching expressions to make true number sentences.

1. [ ] = [ ]

2. [ ] = [ ]

3. [ ] = [ ]

4. [ ] = [ ]

5. [ ] = [ ]

6. Write a true number sentence using the expressions that you have left over. Use pictures and words to show how you know two of the expressions have the same unknown numbers.

7. Use other facts you know to write at least two true number sentences similar to the type above.

8. The following addition number sentences are FALSE. Change one number in each problem to make a TRUE number sentence, and rewrite the number sentence.

   a. $8 + 5 = 10 + 2$ ____________________________

   b. $9 + 3 = 8 + 5$ ____________________________

   c. $10 + 3 = 7 + 5$ ____________________________

9. The following subtraction number sentences are FALSE. Change one number in each problem to make a TRUE number sentence, and rewrite the number sentence.

   a. $12 - 8 = 1 + 2$ ____________________________

   b. $13 - 9 = 1 + 4$ ____________________________

   c. $1 + 3 = 14 - 9$ ____________________________

Name ______________________________ Date ______________

You are given these new expression cards. Write matching expressions to make true number sentences.

8 + 9	12 - 7	19 - 2	2 + 15
3 + 2	10 + 7	14 - 9	1 + 4

[ ] = [ ]

[ ] = [ ]

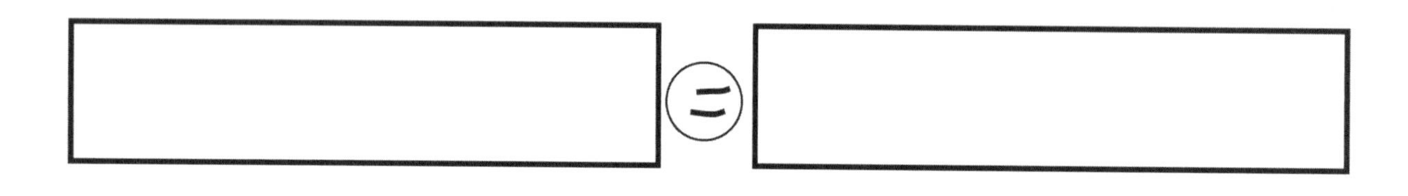

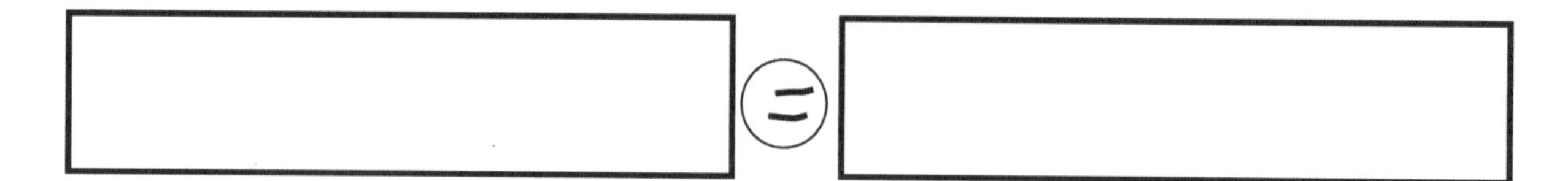

EUREKA MATH

Name ______________________________ Date ______________

1. Circle "true" or "false."

Equation	True or False?
a. 2 + 3 = 5 + 1	True / False
b. 7 + 9 = 6 + 10	True / False
c. 11 - 8 = 12 - 9	True / False
d. 15 - 4 = 14 - 5	True / False
e. 18 - 6 = 2 + 10	True / False
f. 15 - 8 = 2 + 5	True / False

2. Lola and Charlie are using expression cards to make true number sentences. Use pictures and words to show who is right.

   a. Lola picked 4 + 8, and Charlie picked 9 + 3. Lola says these expressions are equal, but Charlie disagrees. Who is right? Explain your thinking.

b. Charlie picked 11 - 4, and Lola picked 6 + 1. Charlie says these expressions are not equal, but Lola disagrees. Who is right? Use a picture to explain your thinking.

c. Lola picked 9 + 7, and Charlie picked 15 - 8. Lola says these expressions are equal but Charlie disagrees. Who is right? Use a picture to explain your thinking.

3. The following addition number sentences are FALSE. Change one number in each problem to make a TRUE number sentence, and rewrite the number sentence.

a. 10 + 5 = 9 + 5 ____________________

b. 10 + 3 = 8 + 4 ____________________

c. 9 + 3 = 8 + 5 ____________________

12 - 7	3 + 2
7 + 8	10 + 5

15 - 9	1 + 5
6 + 8	10 + 4

15 - 8	2 + 5
17 - 9	1 + 7

expression cards

11 - 7	3 + 1
6 + 7	10 + 3

17 - 8	2 + 7
4 + 8	10 + 2

7 + 9	10 + 6
11 - 8	2 + 1

expression cards

8 + 9	10 + 7
9 + 9	10 + 8

4 + 8	10 + 2
17 - 5	9 + 3

15 - 8	13 - 6
11 - 9	1 + 1

expression cards

12 + 4	10 + 6
14 + 2	9 + 7

expression cards

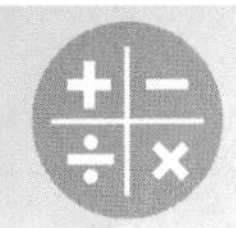

Topic D

Varied Problems with Decompositions of Teen Numbers as 1 Ten and Some Ones

1.OA.1, 1.OA.6, 1.NBT.2a, 1.NBT.2b, 1.NBT.5

Focus Standards:	1.OA.1	Use addition and subtraction within 20 to solve word problems involving situations of adding to, taking from, putting together, taking apart, and comparing, with unknowns in all positions, e.g., by using objects, drawings, and equations with a symbol for the unknown number to represent the problem.
	1.OA.6	Add and subtract within 20, demonstrating fluency for addition and subtraction within 10. Use mental strategies such as counting on; making ten (e.g., 8 + 6 = 8 + 2 + 4 = 10 + 4 = 14); decomposing a number leading to a ten (e.g., 13 – 4 = 13 – 3 – 1 = 10 – 1 = 9); using the relationship between addition and subtraction (e.g., knowing that 8 + 4 = 12, one knows 12 – 8 = 4); and creating equivalent but easier or known sums (e.g., adding 6 + 7 by creating the known equivalent 6 + 6 + 1 = 12 + 1 = 13).
	1.NBT.2ab	Understand that the two digits of a two-digit number represent amounts of tens and ones. Understand the following as special cases: a. 10 can be thought of as a bundle of ten ones—called a "ten." b. The numbers from 11 to 19 are composed of a ten and one, two, three, four, five, six, seven, eight, or nine ones.
Instructional Days:	4	
Coherence -Links from:	GK–M4	Number Pairs, Addition and Subtraction to 10
-Links to:	G2–M3	Place Value, Counting, and Comparison of Numbers to 1,000
	G2–M5	Addition and Subtraction Within 1,000 with Word Problems to 100

Topic D closes the module with students renaming ten as a unit: *a ten* (**1.NBT.2a**). This is the very first time students are introduced to this language of ten as a unit, so this is exciting! The unit of ten is the foundation for our whole number system wherein all units are composed of ten of the adjacent unit on the place value chart.

In Lesson 26, students revisit representations of 10 ones that they have worked with in the past. They rename their Rekenrek bracelet, the ten-frame, the fingers on two hands, and two 5-groups as 1 ten. They connect teen numbers to the unit form (e.g., 1 ten and 1 one, 1 ten and 2 ones), and represent the numbers with Hide Zero cards. They now analyze the digit 1 in the tens place as representing both 10 ones and 1 unit of ten, further setting the foundation for later work with place value in Module 4. Students use their very own Magic Counting Sticks (i.e., their fingers) to help them to compose 1 ten. By bundling 1 ten, students realize that some ones are left over which clarifies the meaning of the ones unit (**1.NBT.2b**).

9 + 6 =
1 5
9 + 1 = 10
10 + 5 = 15

In Lesson 27, students solve both abstract and contextualized *result-unknown* problems (**1.OA.1**). The lesson takes them through a progression from problems with teens decomposed or composed using 1 ten and some ones to problems wherein they find the hidden ten (e.g., 8 + 6 or 12 – 5).

In Lesson 28, students solve familiar problems such as, "Maria had 8 snowballs on the ground and 5 in her arms. How many snowballs did Maria have?" As students write their solutions, they break apart an addend to make a ten with another addend and write two equations leading to the solution (see the bond and equations to the right). This movement forward in their ability to record the two steps allows them to own the structure of the ten they have been using for the entire module, on a new level (MP.7).

13 − 8
10 3
10 − 8 = 2
2 + 3 = 5

Topic D closes with Lesson 29, where students solve *add to with change unknown* and *take apart/put together with addend unknown* problems. As in Lesson 28, students write both equations leading to the solution as they take from the ten (see bond and equation to the top right).

A Teaching Sequence Toward Mastery of Solving Varied Problems with Decompositions of Teen Numbers as 1 Ten and Some Ones	
Objective 1:	**Identify 1 ten as a unit by renaming representations of 10.** **(Lesson 26)**
Objective 2:	**Solve addition and subtraction problems decomposing and composing teen numbers as 1 ten and some ones.** **(Lesson 27)**
Objective 3:	**Solve addition problems using ten as a unit, and write two-step solutions.** **(Lesson 28)**
Objective 4:	**Solve subtraction problems using ten as a unit, and write two-step solutions.** **(Lesson 29)**

Lesson 26

Objective: Identify 1 ten as a unit by renaming representations of 10.

Suggested Lesson Structure

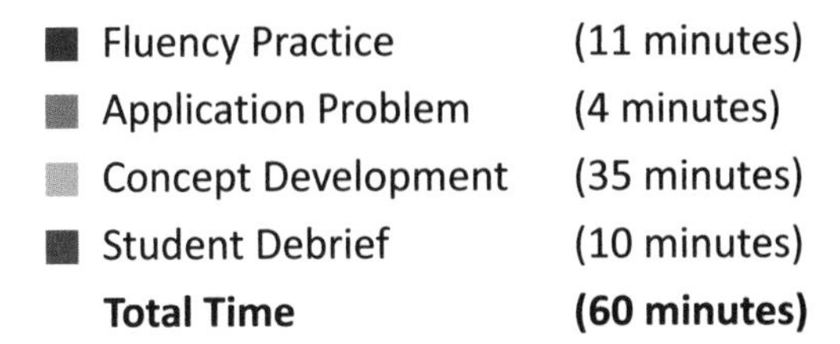

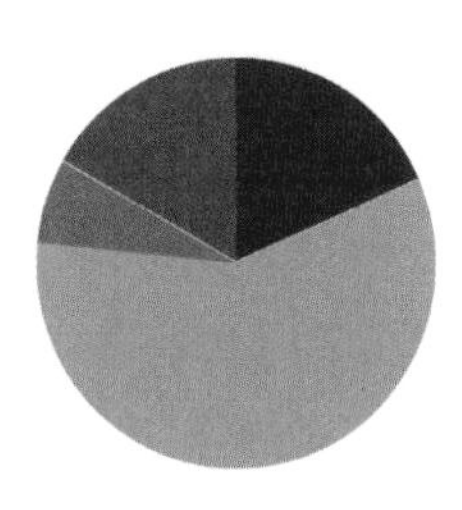

Fluency Practice (11 minutes)

- Addition with Partners **1.OA.6** (6 minutes)
- Happy Counting by Fives **1.OA.5** (2 minutes)
- 10 More/10 Less **1.NBT.5** (3 minutes)

Addition with Partners (6 minutes)

Materials: (S) Personal white board

Note: This fluency activity reviews the make ten addition strategy with addends of 7, 8, and 9. Allow students to draw 5-groups if they still need pictorial representations to solve.

Assign partners of equal ability. Partners assign each other a number from 1 to 10 (e.g., 5). On their personal white boards, they write number sentences with 9, 8, and 7 as the other addend and solve them (e.g., 9 + 5 = 14, 8 + 5 = 13, 7 + 5 = 12). Partners then exchange boards and check each other's work.

Happy Counting by Fives (2 minutes)

Note: This maintenance fluency activity reviews adding and subtracting 5.

Do the Happy Counting activity from Lesson 4, counting by fives from 0 to 40 and back. First count the Say Ten way, and then count the regular way.

10 More/10 Less (3 minutes)

Materials: (T) 20-bead Rekenrek

Note: This activity addresses the grade-level standard of finding 10 more and 10 less than a number without having to count and prepares students to see ten as a unit.

Practice identifying 10 more and 10 less on the Rekenrek.

T: (Show a number within 3 on the Rekenrek.) Say the number.

S: 3.

T: (Slide over 10 from the next row). What's 10 more than 3, the Say Ten way?
S: Ten 3.
T: What is it the regular way?
S: 13.

Repeat a few times to practice 10 more. Next, show a teen number and have students practice identifying 10 less. Then, put the Rekenrek away, and switch to cold calling students or groups of students.

T: 10 more than 5? Boys.
S: (Boys only.) 15.
T: 10 less than 14? Girls.
S: (Girls only.) 4.

Continue playing, varying the sentences: Take 10 out of 16. Add 10 to 2. 12 is 10 more than ...?

Application Problem (4 minutes)

Ruben has 18 toy cars. His car carrier holds 10 toy cars. If Ruben's carrier is full, how many cars are in the carrier, and how many cars are outside of the carrier?

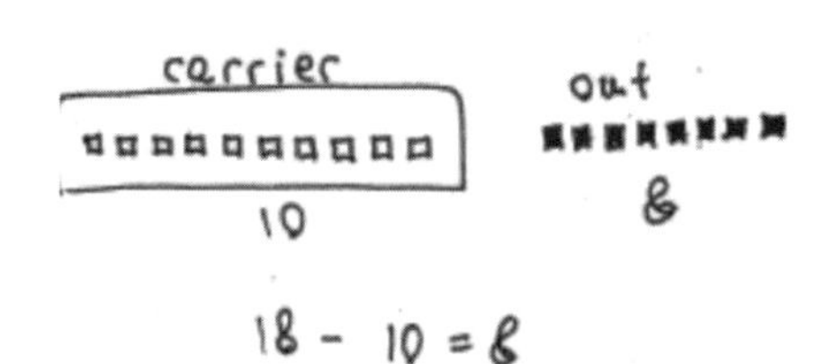

Note: This problem enables students to continue considering situations with missing parts where the context presents a grouping of 10. This grouping of 10 leads into today's lesson during which students focus on ten as a unit.

Concept Development (35 minutes)

Materials: (T) Rekenrek bracelet stretched into a straight line (first used in Grade 1 Module 1 Lesson 8), 5-group cards (Lesson 1 Fluency Template), Hide Zero cards (Lesson 18 Fluency Template 1), 9 Rekenrek beads (separated from pipe cleaner), grouping ten images (Template) (S) Personal white board

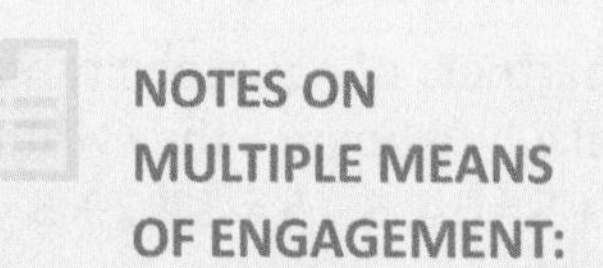

NOTES ON MULTIPLE MEANS OF ENGAGEMENT:

Students demonstrate an understanding of math concepts when they can apply them in a variety of situations. It is important for students to recognize the relationship between 1 ten and the tools used in this lesson. They see that no matter what tool they use, they still think about 1 ten.

Students sit in a semi-circle in the meeting area, next to their partners.

T: (Lay materials in front of class.) We have used many different tools during math this year. Can you name each of these models?
S: The Rekenrek bracelet! → A 5-group card of ten! → A ten-frame!
T: Talk with a partner. What do these models have in common?
S: (Discuss.) They all show ten!

T: We have another math tool that we carry around with us everywhere we go. Show me the math tools you carry everywhere.

S: (Wave hands.)

T: (Wiggle your fingers.) These fingers can help us with our math in so many ways. How many fingers do we carry around with us?

S: Ten!

T: (Pick up the Rekenrek.) We can carry around loose beads to count, but instead we use Rekenrek bracelets. Why do we like using the bracelet?

S: It keeps the beads together. → They're organized and we can count them quickly. → We can look at it and see the amount right away.

T: Right! Instead of having 10 loose beads to count one by one, we can pick up this Rekenrek bracelet and count all 10 at once. When I pick up this one bracelet, I know that I have 10 beads altogether. I can call this 1 group of ...

S: Ten!

T: We call this 1 ten.

T: Why do we frame the 10 circles when we use 5-group rows?

S: It's easier to see ten. → We don't have to recount them, because we know there are ten. → Then, we can just count on the extras and quickly know how many there.

T: Just like we called our Rekenrek bracelet 1 ten, when we frame 10 circles in the 5-group rows, we have 1 frame of ten, or ...

S: 1 ten!

T: Let's see if we can make 1 ten with our fingers. Let's bundle them up into a set of 10. First, show me all 10 of your fingers.

S: (Raise hands, palms out.)

T: Count with me.

S/T: (Count on fingers from left to right, starting with the pinky.) 1, 2, 3, 4, 5, 6, 7, 8, 9, 10. (As you say 10, clasp two hands together.)

T: With our hands bundled like this, we have taken our 10 fingers and put them together to show 1 set of ten, or 1 ten.

T: Let's make 12 with our fingers, including pretend fingers now. Put out all of your fingers. How many pretend ones can you see to make 12?

S: Two!

T: Let's bundle the 10 fingers on our hands. 1, 2, 3, 4, 5, 6, 7, 8, 9, 10. (Clasp hands.) We have 1 ten, and, hmm, how many more fingers?

S: Two more fingers!

T: Let's make more with a partner. (Show the number 19 with Hide Zero cards.) Use your fingers to show the number 19.

S: (Show 10 on one partner's hands and 9 on the other partner's hands.)

T: (Pull apart Hide Zero cards.) Right now, you are showing 10 fingers (hold up the 10 card) and 9 fingers (hold up the 9 card).

T: If you are showing 10 fingers, bundle them together to make 1 ten.

S: (One student in partnership bundles to form clasped hands, illustrating 1 ten.)

T: Do you still have 19 fingers?

S: Yes!

T: How many tens do you have?

S: 1 ten.

T: How many extra ones do you have out?

S: Nine!

T: We call these 9 ones, since they are all apart and we can count them one by one. (Touch each extended finger of a student's hand who is holding out 9 fingers.)

T: So, our 10 fingers and our 9 fingers become how many tens and how many ones? (Hold up the 10 and 9 Hide Zero cards.)

S: 1 ten and 9 ones!

T: (Slip the Hide Zero cards together to show 19.)

Repeat the process with the following numbers: 18, 15, and 14.

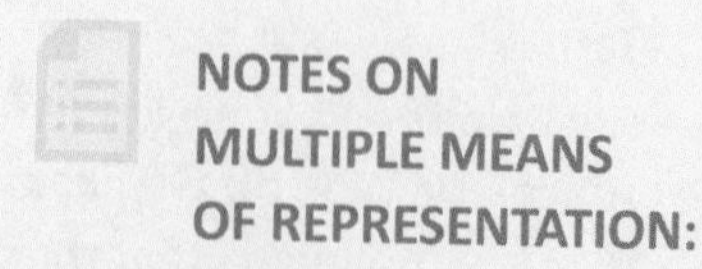

NOTES ON MULTIPLE MEANS OF REPRESENTATION:

When using the new format of drawing 5-groups as a stick and some circles, it is important to be sure that students grasp the meaning and are able to connect this new representation to ways of drawing they already know. Allow for some time to transition from drawing them horizontally to vertically.

T: (Place 1 Rekenrek bracelet and 4 separate beads in front of the class.) How many beads are here?

S: Fourteen!

T: How did you know that so quickly?

S: There are 10 on the Rekenrek, and I can see 4 more.

T: So, I have how many tens and how many ones?

S: 1 ten and 4 ones!

T: When we draw our 5-groups, let's draw a stick through our circles, like the beads, whenever it is 1 ten. (Draw a vertical line, and add 10 circles to it. Then, draw 4 circles in a vertical formation without the line.) We can call this a 5-group column. Can you pick out the ten from the ones? Draw 14 in 5-group columns like mine.

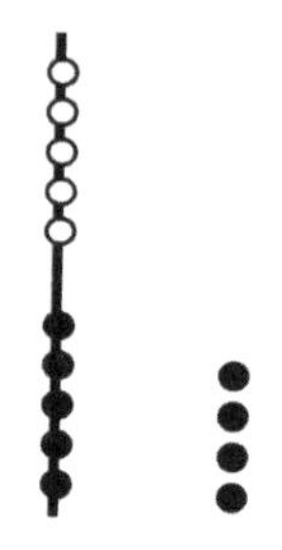

5-group column

MP.4

S: (Draw the same picture.)

T: Put your finger on your 1 ten.

S: (Touch 1 ten in 5-group column.)

T: Put your finger on your 4 ones.

S: (Touch 4 ones in 5-group drawing.)

Project images, one at a time. Have students draw 5-group columns as above and state the number of tens and ones for each picture. Use the Hide Zero cards to show students the ten and ones separated as well as together.

Problem Set (10 minutes)

Students should do their personal best to complete the Problem Set within the allotted 10 minutes. For some classes, it may be appropriate to modify the assignment by specifying which problems they work on first. Some problems do not specify a method for solving. Students should solve these problems using the RDW approach used for Application Problems.

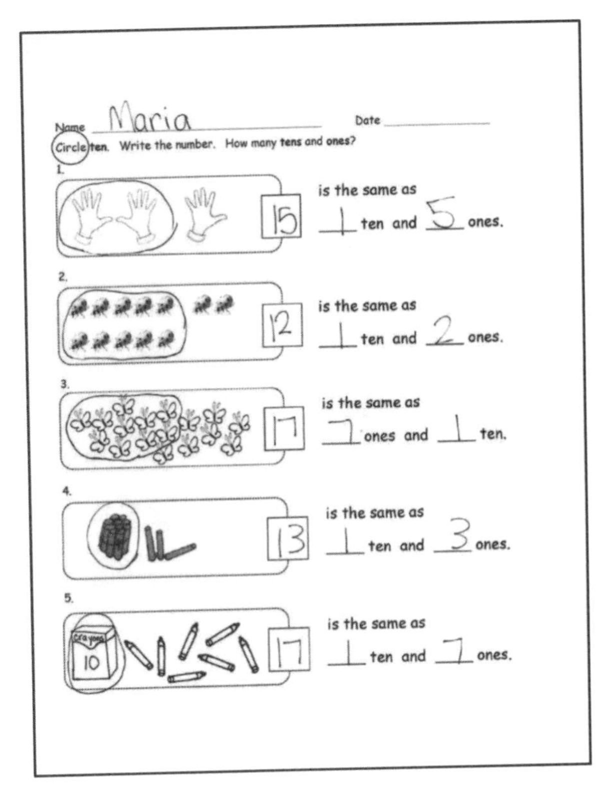
Name Maria Date

Circle ten. Write the number. How many tens and ones?

1. 15 is the same as 1 ten and 5 ones.
2. 12 is the same as 1 ten and 2 ones.
3. 17 is the same as 7 ones and 1 ten.
4. 13 is the same as 1 ten and 3 ones.
5. 17 is the same as 1 ten and 7 ones.

Student Debrief (10 minutes)

Lesson Objective: Identify 1 ten as a unit by renaming representations of 10.

The Student Debrief is intended to invite reflection and active processing of the total lesson experience.

Invite students to review their solutions for the Problem Set. They should check work by comparing answers with a partner before going over answers as a class. Look for misconceptions or misunderstandings that can be addressed in the Debrief. Guide students in a conversation to debrief the Problem Set and process the lesson.

Any combination of the questions below may be used to lead the discussion.

- Look at Problems 1–5. Which were you able to answer most quickly? Why?
- The cards we used today are called Hide Zero cards. Why do you think they have that name? Explain how they work.
- Look at Problems 6 and 7. What is the same about them? What is different?
- Talk with a partner. How do you know 9 ones and 1 **ten** is the same as 1 ten and 9 ones? How is this like other addition rules we have learned?
- (Hold up a 5-group row next to a **5-group column**.) How are these different? How are they the same? How can the 5-group column help us *see* the ten better than with the 5-group row?
- Today, we talked about 1 ten. How is 1 ten the same as having 10 ones? How is it different?
- How did the Application Problem connect to today's lesson?

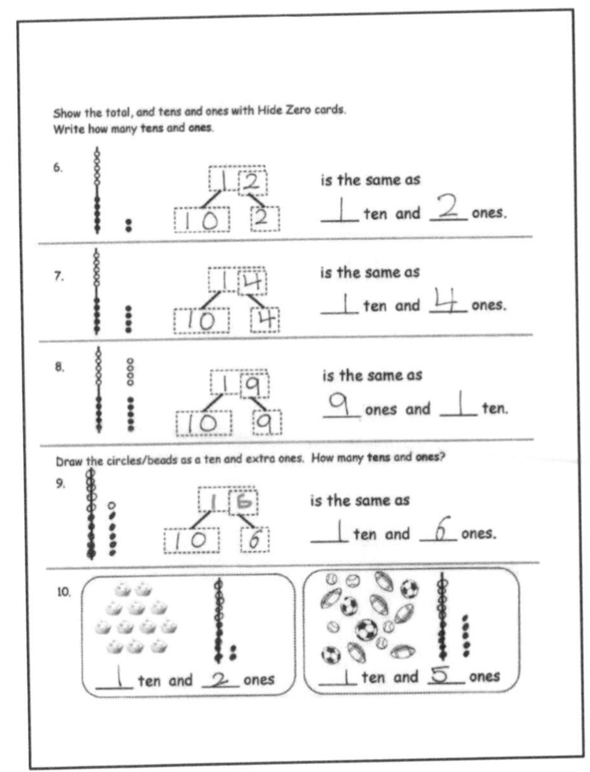
Show the total, and tens and ones with Hide Zero cards.
Write how many tens and ones.

6. 12 → 10, 2 is the same as 1 ten and 2 ones.
7. 14 → 10, 4 is the same as 1 ten and 4 ones.
8. 19 → 10, 9 is the same as 9 ones and 1 ten.

Draw the circles/beads as a ten and extra ones. How many tens and ones?

9. 16 → 10, 6 is the same as 1 ten and 6 ones.
10. 1 ten and 2 ones; 1 ten and 5 ones

Exit Ticket (3 minutes)

After the Student Debrief, instruct students to complete the Exit Ticket. A review of their work will help with assessing students' understanding of the concepts that were presented in today's lesson and planning more effectively for future lessons. The questions may be read aloud to the students.

Name ______________________________ Date ________________

Circle **ten**. Write the number. How many **tens** and **ones**?

1.

☐ is the same as

____ ten and ____ ones.

2.

☐ is the same as

____ ten and ____ ones.

3.

☐ is the same as

____ ones and ____ ten.

4.

☐ is the same as

____ ten and ____ ones.

5.

☐ is the same as

____ ten and ____ ones.

Show the total and tens and ones with Hide Zero cards.
Write how many **tens** and **ones**.

6.

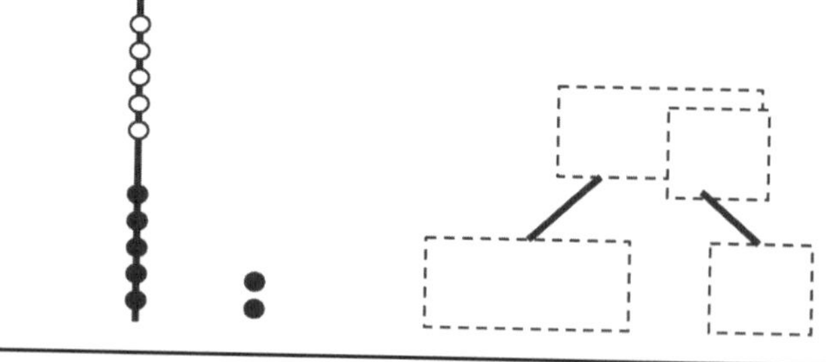

is the same as

_____ ten and _____ ones.

7.

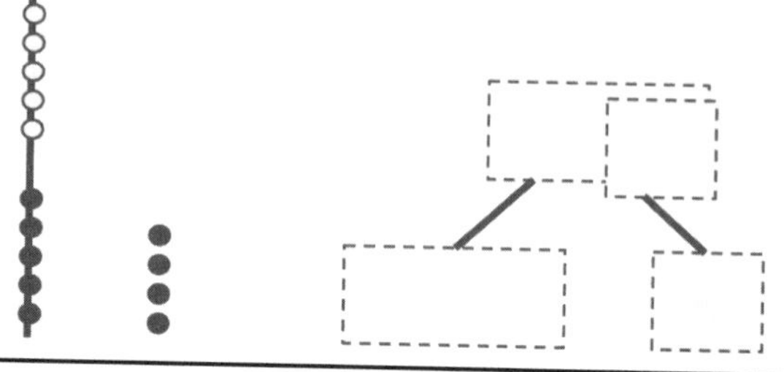

is the same as

_____ ten and _____ ones.

8.

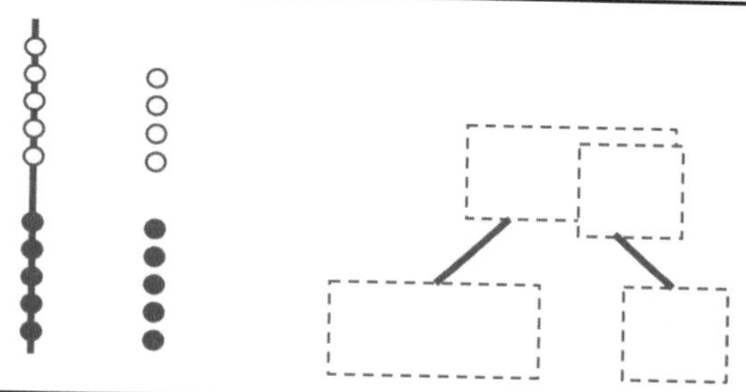

is the same as

_____ ones and _____ ten.

Draw the circles as a ten and extra ones. How many **tens** and **ones**?

9.

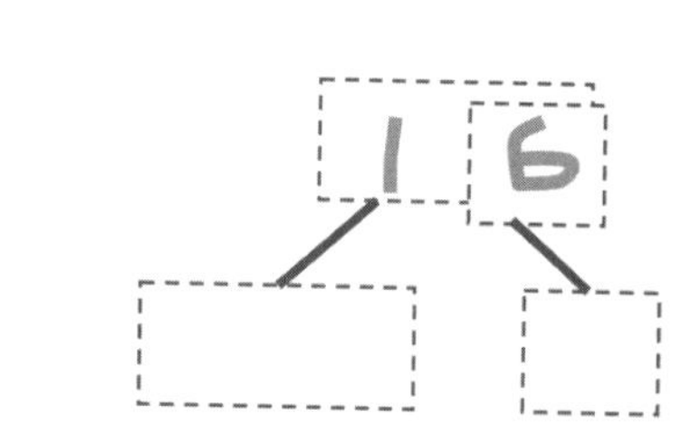

is the same as

_____ ten and _____ ones.

10.

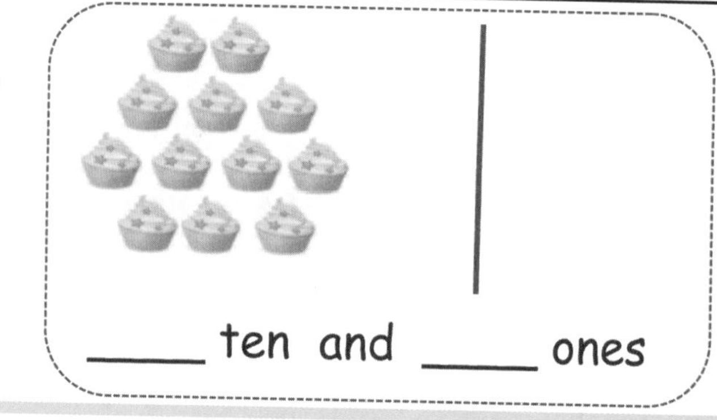

_____ ten and _____ ones

_____ ten and _____ ones

Name ______________________________ Date ________________

Match the pictures of tens and ones to the Hide Zero cards. How many tens and ones?

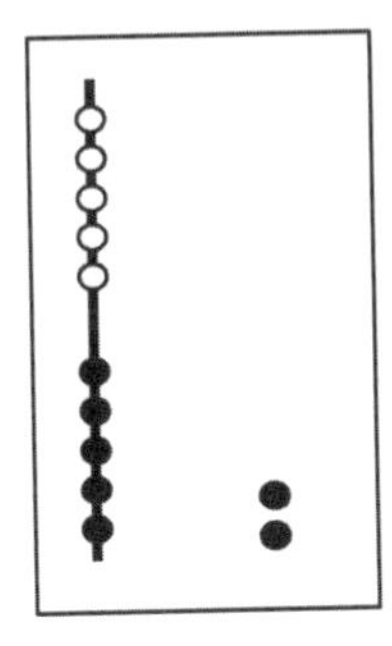

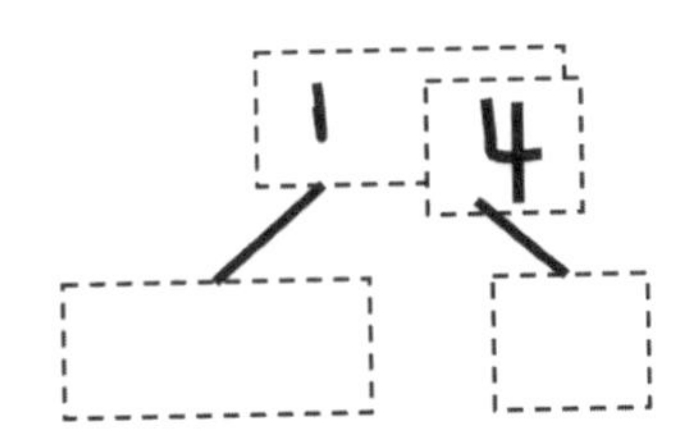

is the same as

_____ ten and _____ ones.

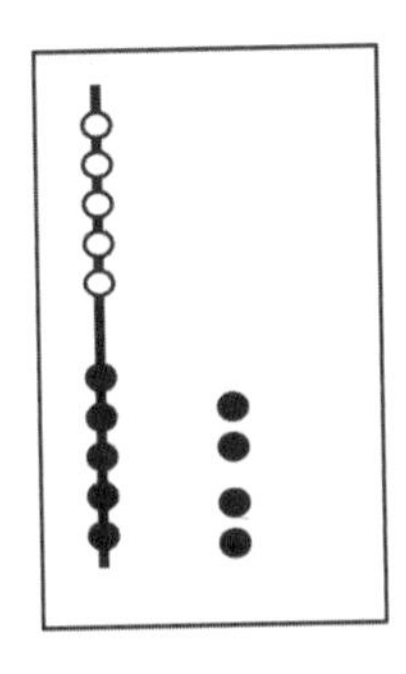

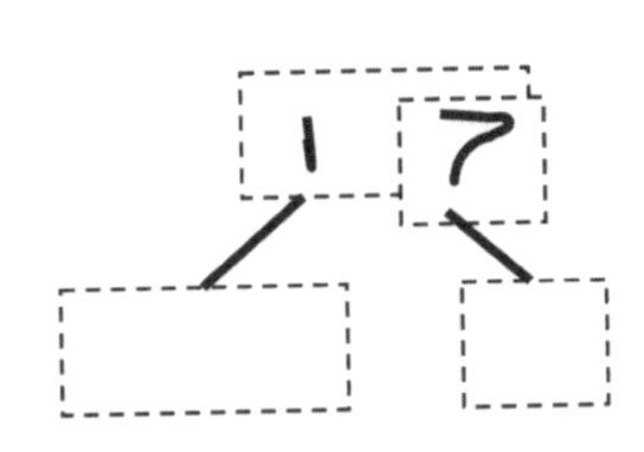

is the same as

_____ ten and _____ ones.

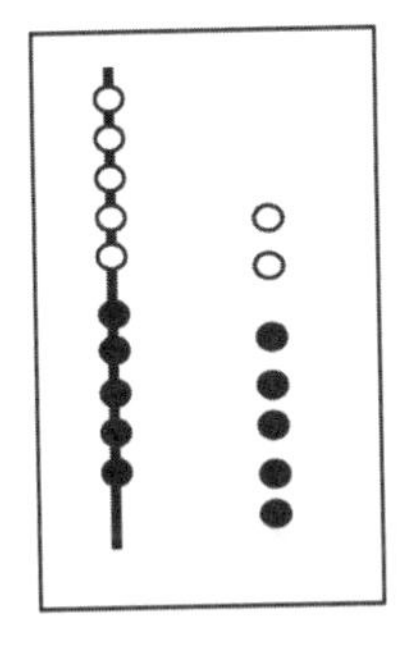

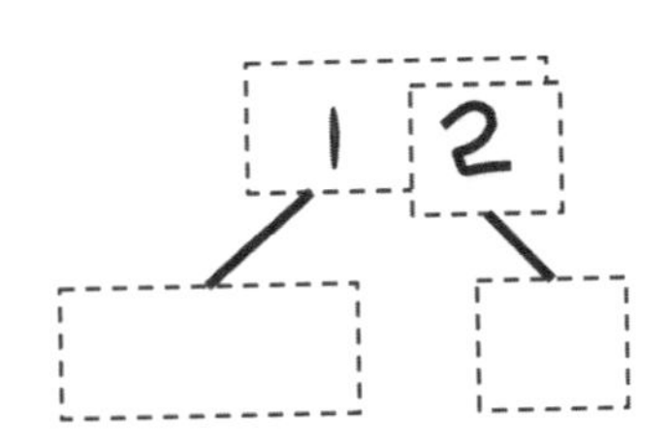

is the same as

_____ ten and _____ ones.

Name ______________________________ Date ______________

Circle **ten**. Write the number. How many **tens** and **ones**?

1.

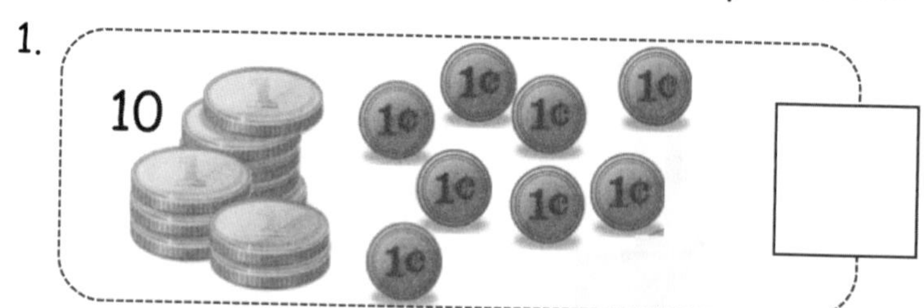

☐ is the same as

_____ ten and _____ ones.

2.

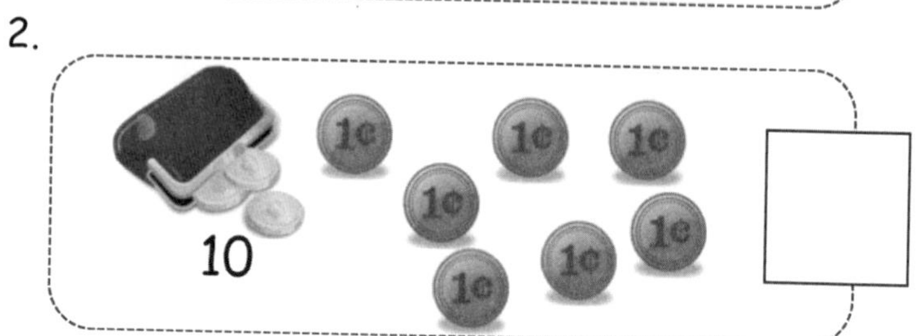

☐ is the same as

_____ ones and _____ ten.

Use the Hide Zero pictures to draw the ten and ones shown on the cards.

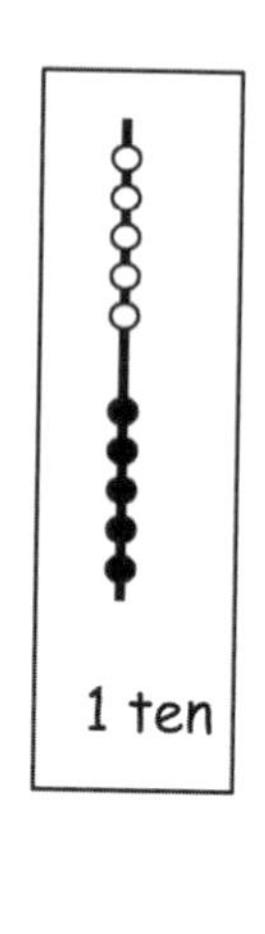

3.

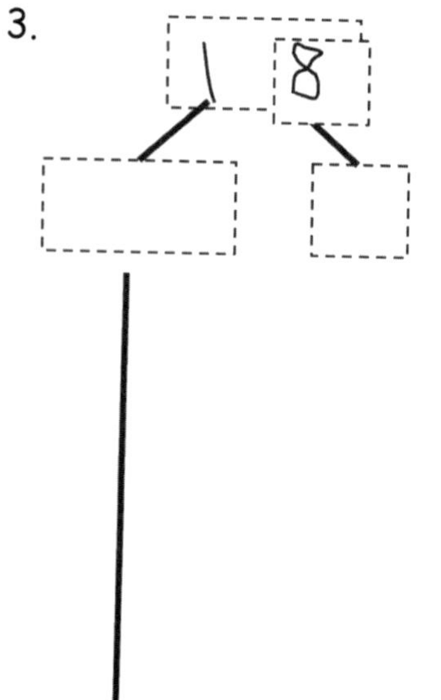

_____ ten and _____ ones

4.

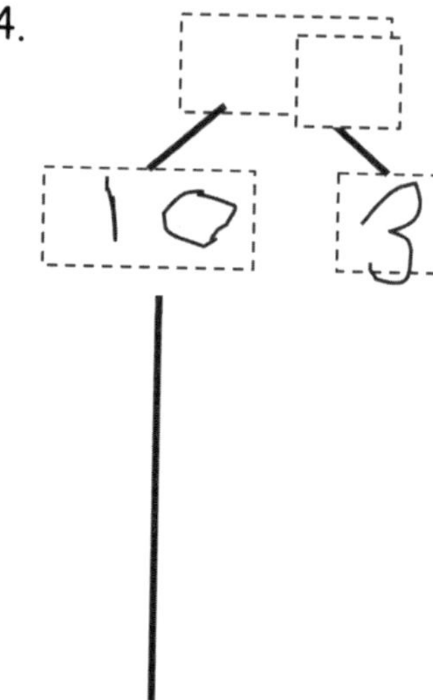

_____ ten and _____ ones

EUREKA MATH

Draw using 5-groups columns to show the tens and ones.

5.

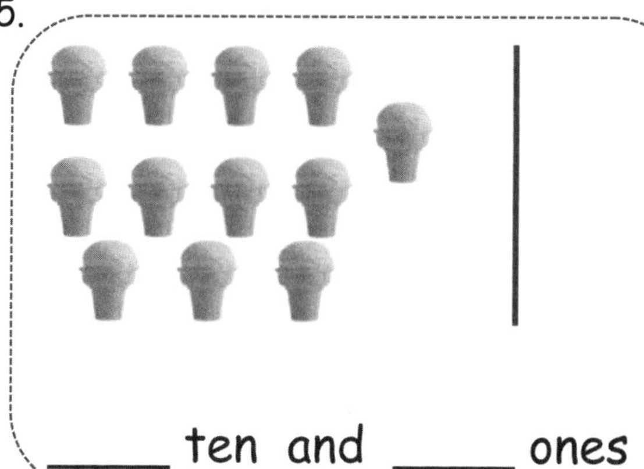

_____ ten and _____ ones

6.

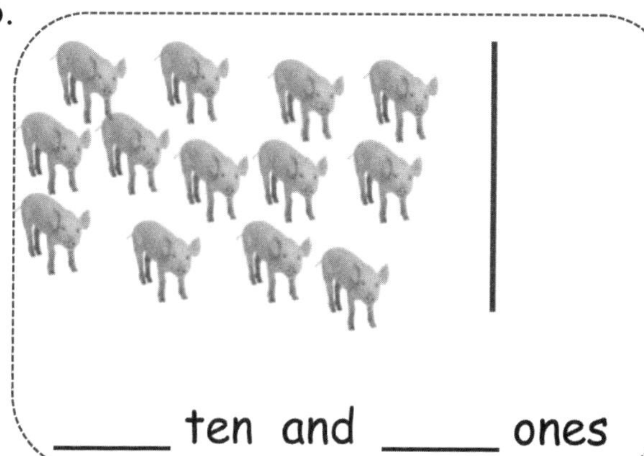

_____ ten and _____ ones

Draw your own examples using 5-groups columns to show the tens and ones.

7. 16

16 is the same as

____ ten and _____ ones.

8. 19

19 is the same as

_____ ones and _____ ten.

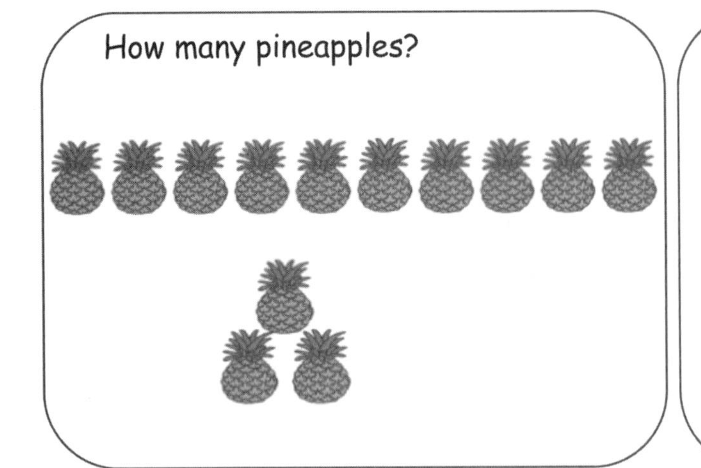

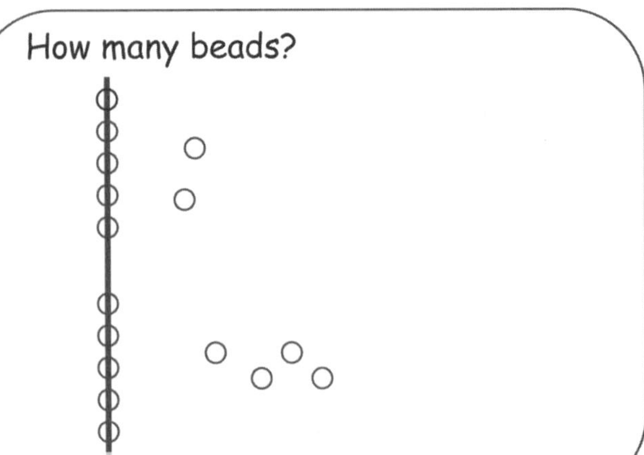

grouping ten images

Lesson 27

Objective: Solve addition and subtraction problems decomposing and composing teen numbers as 1 ten and some ones.

Suggested Lesson Structure

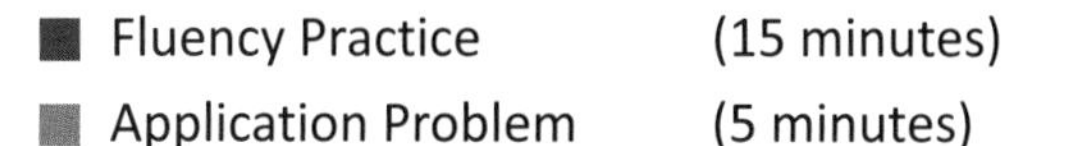

- Fluency Practice (15 minutes)
- Application Problem (5 minutes)
- Concept Development (30 minutes)
- Student Debrief (10 minutes)

Total Time (60 minutes)

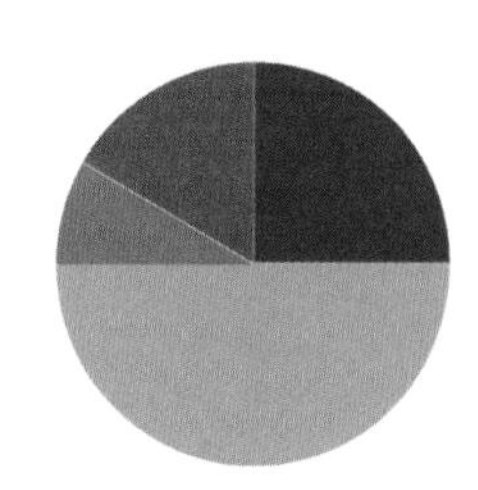

Fluency Practice (15 minutes)

- Say Ten: 5-Group Column **1.NBT.2** (2 minutes)
- Sprint: 10 More and 10 Less **1.NBT.5** (10 minutes)
- Magic Counting Sticks **1.NBT.2** (3 minutes)

Say Ten: 5-Group Column (2 minutes)

Materials: (T) 5-group column cards (Fluency Template)

Note: This fluency activity reviews the unit of 1 ten as a 5-group column, which was introduced in yesterday's lesson.

T: (Hold up the card showing 14.) Tell me how many the Say Ten way.
S: Ten 4.
T: How many tens?
S: 1 ten.
T: How many ones?
S: 4 ones.

Repeat this process and alternate between requesting that students respond the Say Ten way and saying the number of tens and ones.

Sprint: 10 More and 10 Less (10 minutes)

Materials: (S) 10 More and 10 Less Sprint

Note: This activity addresses the grade-level standard of mentally finding 10 more and 10 less than a number.

Magic Counting Sticks (3 minutes)

Materials: (T) Hide Zero cards (Lesson 18 Fluency Template 1)

Note: This activity reviews the concept of ten as a unit and prepares students for today's lesson.

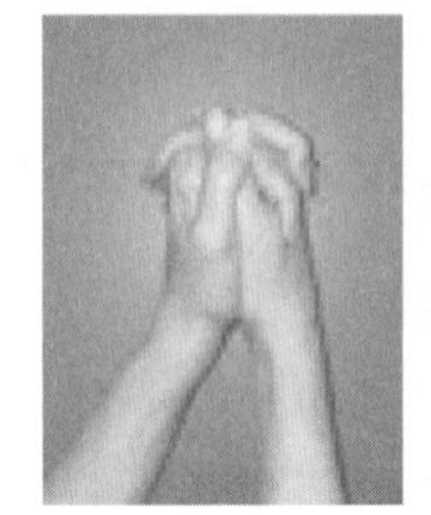

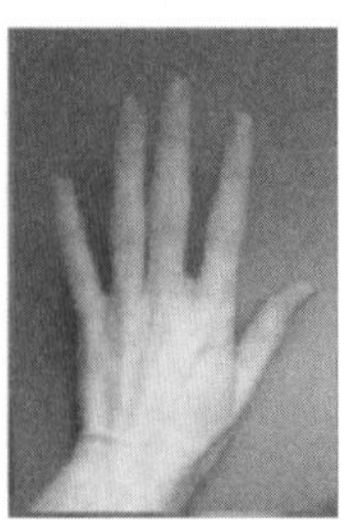

Assign partners. Show a teen number with Hide Zero cards (e.g., 15). Partner A uses his *magic counting sticks* (fingers) to show a bundle of ten, and Partner B shows 5 ones. Ask students to identify how many tens and ones they made. Repeat with other teen numbers, alternating the roles of Partners A and B. Extend the game by calling out a teen number and letting one partner choose whether to show the ten or the ones. Then, ask the other partner to show the missing part.

Application Problem (5 minutes)

Ruben was putting away his 14 toy cars. He filled his car carrier and had 4 cars left that could not fit. How many cars fit in his car carrier?

Note: This problem continues to consider contexts where 10 is grouped together within a unit. During the Debrief, the unitization of ten is discussed.

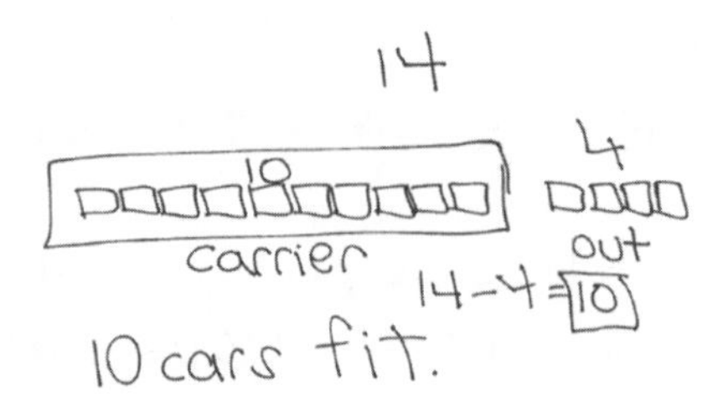

Concept Development (30 minutes)

Materials: (T) Hide Zero cards (Lesson 18 Fluency Template 1) (S) Personal white board, Hide Zero cards (Lesson 18 Fluency Template 1)

Students sit in a semicircle next to their partner in the meeting area with their personal white boards.

T: Get out your magic counting sticks! With your partner, show 13.

S: (One student bundles 10 fingers by clasping her hands together; the other student shows 3 fingers.)

T: Good! Now, make 13 with your Hide Zero cards. You can talk with your partner if you are stuck.

S: (Layer 3 on top of 10 to make 13.)

T: How many tens do you have in 13?

S: 1 ten!

T: (Hold up the 10 Hide Zero card.) How many extra ones do you have in 13?

S: 3 ones!

T: (Hold up the 3 Hide Zero card.) Yes, 13 is made of 1 ten (hold the 10 card out) and 3 ones (hold the 3 card out). (Layer the Hide Zero cards again to show 13.)

T: (Project 13 – 3.) How can you use your Hide Zero cards to solve this?

S: Just take away 3.

T: And, how many are left?

S: Ten!

T: We can also call that ...

S: 1 ten.

Repeat this process as needed with the following suggested sequence: 15 – 5, 16 – 4, and 18 – 7, asking, "How many tens and ones are left?"

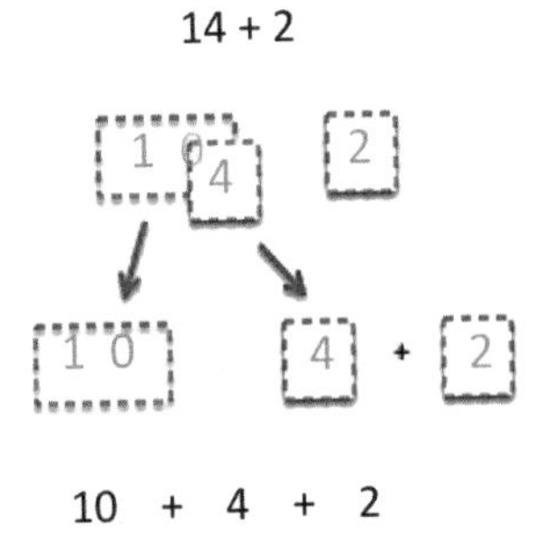

T: Work with your partner to show 14 with your magic counting sticks and your Hide Zero cards.

S: (One student bundles 10 fingers by clasping hands together; the other student shows 4 fingers. They put 14 in front of them.)

T: (Project 14 + 2.) How can you do this? Will you add to the ten or the ones?

S: Just add more to the ones! → Count 2 more! → Use your fingers to count 2 more!

T: So, we don't have to add to the ten in order to figure this out, we can just add to the ones?

S: Yes! It's sixteen!

T: How many tens and ones make up 16?

S: 1 ten and 6 extra ones.

Project 14 + 2 and ask students to model it with their Hide Zero cards. As they take apart the 4 from 14, they add 4 ones and 2 ones together first to make 6. Now, with a ten and a 6, they layer these to make their total. Discuss the tens and ones that comprise the total each time. Repeat this process as needed with the following suggested sequence: 15 + 3, 17 + 2, and 13 + 7. Be sure to discuss the significance of 2 tens.

> **NOTES ON MULTIPLE MEANS OF REPRESENTATION:**
>
> While some students are experts at solving these number sentences, others may need more support with their Hide Zero cards. Students should use the Hide Zero cards as much as necessary. Remove the scaffold for students who are able to do more mental calculations. Others can use the cards to help develop their ability to calculate mentally.

T: (Project 8 + 5.) Work with your partner. Partner A, use your personal board to show how to make 1 ten. Partner B, when she is done, use your Hide Zero cards to show the solution.

S: (Partner A models the addition with the number sentence and number bond. Partner B shows 13 with Hide Zero cards.)

T: Point to the card that tells how many tens are in your answer, and say the number of tens. If you are not sure, you can check!

S: (Point to the 1 in 13.) 1 ten.

T: Point to the card on your Hide Zero cards that tells how many ones are in your answer, and say how many ones.

S: 3 ones.

If students need more practice with this process, switch the partners, and repeat the same process with the following suggested sequence: 8 + 6, 7 + 5, and 6 + 9.

T: Hmm. I wonder how we can use our Hide Zero cards and personal white boards to help us solve 13 – 4?

S: Take from the ten! → Count back! → Count on!

MP.7

T: Let's try taking from the ten, just like [Student 1] said. Let's make our total of 13 with our cards.

S: (Make 13 with Hide Zero cards.)

T: How can we take from the ten here?

S: Take apart the 10 and take 4 away from the ten!

T: (Draw a matching illustration on the board, showing 10 and 3 separated. Touch the 10.) And, how many are left?

S: 6.

T: (Write 10 – 4 = 6 on the board.) How many do we have altogether? (Touch the 6 and the remaining 3.)

S: 9. (Write 6 + 3 = 9 on the board when students answer.)

T: 9 tens or 9 ones?

S: 9 ones!

T: How many tens are left?

S: 0 tens!

NOTES ON MULTIPLE MEANS OF ENGAGEMENT:

Remember to challenge advanced learners. Students enjoy working with larger numbers, so extend their knowledge of place value. Give them a larger two-digit number, and they can articulate how many ones and tens are in that number. Find interactive games online by searching with the keywords *place value games*. These games can be played with numbers appropriate for the students in class.

Repeat this process as needed with the following suggested sequence: 12 – 5, 14 – 8, and 15 – 7.

Problem Set (10 minutes)

Students should do their personal best to complete the Problem Set within the allotted 10 minutes. For some classes, it may be appropriate to modify the assignment by specifying which problems they work on first. Some problems do not specify a method for solving. Students should solve these problems using the RDW approach used for Application Problems.

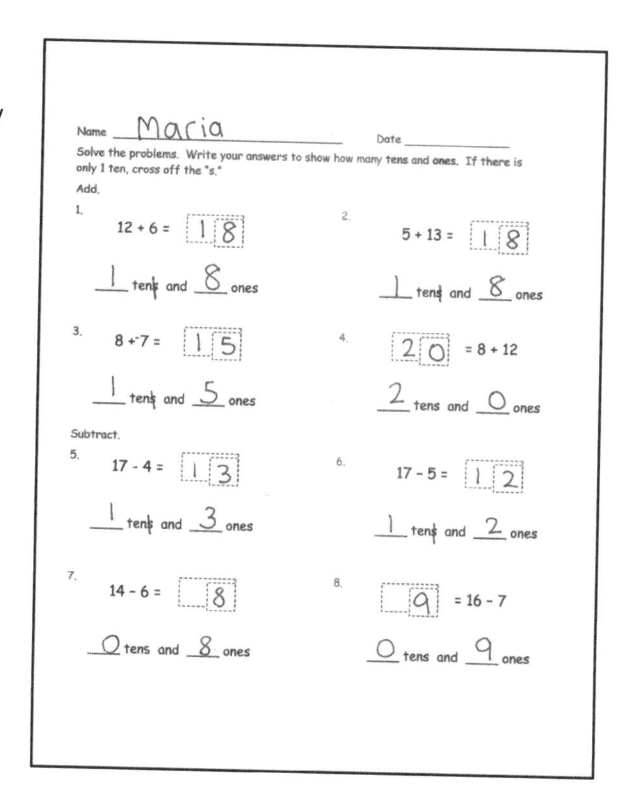

Name Maria Date

Solve the problems. Write your answers to show how many tens and ones. If there is only 1 ten, cross off the "s."

Add.

1. 12 + 6 = 18
   1 tens and 8 ones
2. 5 + 13 = 18
   1 tens and 8 ones
3. 8 + 7 = 15
   1 tens and 5 ones
4. 20 = 8 + 12
   2 tens and 0 ones

Subtract.

5. 17 - 4 = 13
   1 tens and 3 ones
6. 17 - 5 = 12
   1 tens and 2 ones
7. 14 - 6 = 8
   0 tens and 8 ones
8. 9 = 16 - 7
   0 tens and 9 ones

Student Debrief (10 minutes)

Lesson Objective: Solve addition and subtraction problems decomposing and composing teen numbers as 1 ten and some ones.

The Student Debrief is intended to invite reflection and active processing of the total lesson experience.

Invite students to review their solutions for the Problem Set. They should check work by comparing answers with a partner before going over answers as a class. Look for misconceptions or misunderstandings that can be addressed in the Debrief. Guide students in a conversation to debrief the Problem Set and process the lesson.

Any combination of the questions below may be used to lead the discussion.

- How did you use what we learned during the lesson to help you solve the word problems in the Problem Set?
- How was Problem 3 helpful in solving Problem 4?
- Look at Problem 4. How many tens are there altogether? Explain how you solved this.
- What do you notice about the problems that have 0 tens in the answer? What is similar about them?
- What do you notice about the problems that have 1 ten in the answer? How are they similar and different?
- Look at your work from the Application Problem. What is another way to say the answer using tens and ones? If Ruben and his friend played with a total of 6 cars, how many tens and ones would be left in the carrier?

Read the word problem. Draw and label. Write a number sentence and statement that matches the story. Rewrite your answer to show its tens and ones. If there is only 1 ten or 1 one, cross off the "s."

9. Frankie and Maya made 4 big sandcastles at the beach. If they made 10 small sandcastles, how many total sandcastles did they make?

$10 + 4 = 14$

They made 14 sandcastles. 1 ten~~s~~ and 4 ones

10. Ronnie has 8 stickers that are stars. Her friend Sina gives her 7 more. How many stickers does Ronnie have now?

$8 + 7 = 15$

Ronnie has 15 stickers. 1 ten~~s~~ and 5 ones

11. We tied 14 balloons to the tables for a party, but 3 floated away! How many balloons were still tied to the tables?

$14 - 3 = 11$

11 balloons were tied. 1 ten~~s~~ and 1 one~~s~~

12. I ate 5 of the 16 strawberries that I picked. How many did I have left over?

$16 - 5 = 11$

I had 11 strawberries left. 1 ten~~s~~ and 1 one~~s~~

Exit Ticket (3 minutes)

After the Student Debrief, instruct students to complete the Exit Ticket. A review of their work will help with assessing students' understanding of the concepts that were presented in today's lesson and planning more effectively for future lessons. The questions may be read aloud to the students.

A

Name ______________________ Number Correct: ☆

Date ____________

*Write the missing number.

1.	10 + 3 = □		16.	10 + □ = 11	
2.	10 + 2 = □		17.	10 + □ = 12	
3.	10 + 1 = □		18.	5 + □ = 15	
4.	1 + 10 = □		19.	4 + □ = 14	
5.	4 + 10 = □		20.	□ + 10 = 17	
6.	6 + 10 = □		21.	17 - □ = 7	
7.	10 + 7 = □		22.	16 - □ = 6	
8.	8 + 10 = □		23.	18 - □ = 8	
9.	12 - 10 = □		24.	□ - 10 = 8	
10.	11 - 10 = □		25.	□ - 10 = 9	
11.	10 - 10 = □		26.	1 + 1 + 10 = □	
12.	13 - 10 = □		27.	2 + 2 + 10 = □	
13.	14 - 10 = □		28.	2 + 3 + 10 = □	
14.	15 - 10 = □		29.	4 + □ + 3 = 17	
15.	18 - 10 = □		30.	□ + 5 + 10 = 18	

B

Number Correct:

Name ____________________ Date ____________

*Write the missing number.

1.	10 + 1 = □		16.	10 + □ = 10	
2.	10 + 2 = □		17.	10 + □ = 11	
3.	10 + 3 = □		18.	2 + □ = 12	
4.	4 + 10 = □		19.	3 + □ = 13	
5.	5 + 10 = □		20.	□ + 10 = 13	
6.	6 + 10 = □		21.	13 - □ = 3	
7.	10 + 8 = □		22.	14 - □ = 4	
8.	8 + 10 = □		23.	16 - □ = 6	
9.	10 - 10 = □		24.	□ - 10 = 6	
10.	11 - 10 = □		25.	□ - 10 = 8	
11.	12 - 10 = □		26.	2 + 1 + 10 = □	
12.	13 - 10 = □		27.	3 + 2 + 10 = □	
13.	15 - 10 = □		28.	2 + 3 + 10 = □	
14.	17 - 10 = □		29.	4 + □ + 4 = 18	
15.	19 - 10 = □		30.	□ + 6 + 10 = 19	

Name ______________________________ Date ______________

Solve the problems. Write your answers to show how many **tens** and **ones**. If there is only 1 ten, cross off the "s."

Add.

1. 12 + 6 = [ ][ ]

_____ tens and _____ ones

2. 5 + 13 = [ ][ ]

_____ tens and _____ ones

3. 8 + 7 = [ ][ ]

_____ tens and _____ ones

4. [ ][ ] = 8 + 12

_____ tens and _____ ones

Substract.

5. 17 - 4 = [ ][ ]

_____ tens and _____ ones

6. 17 - 5 = [ ][ ]

_____ tens and _____ ones

7. 14 - 6 = [ ][ ]

_____ tens and _____ ones

8. [ ][ ] = 16 - 7

_____ tens and _____ ones

Read the word problem. Draw and label. Write a number sentence and statement that matches the story. Rewrite your answer to show its tens and ones. If there is only 1 ten or 1 one, cross off the "s."

9. Frankie and Maya made 4 big sandcastles at the beach. If they made 10 small sandcastles, how many total sandcastles did they make?

_____ tens and _____ ones

10. Ronnie has 8 stickers that are stars. Her friend Sina gives her 7 more. How many stickers does Ronnie have now?

_____ tens and _____ ones

11. We tied 14 balloons to the tables for a party, but 3 floated away! How many balloons were still tied to the tables?

_____ tens and _____ ones

12. I ate 5 of the 16 strawberries that I picked. How many did I have left over?

_____ tens and _____ ones

Name ______________________________ Date ______________

Solve the problems. Write the answers to show how many tens and ones. If there is only one ten, cross off the "s."

1.

13 + 6 = ☐☐

_____ tens and _____ ones

2.

7 + 6 = ☐☐

_____ tens and _____ ones

Read the word problem. **D**raw and label. **W**rite a number sentence and statement that matches the story. Rewrite your answer to show its tens and ones.

3. Kendrick went bowling. He knocked down 16 pins in the first two frames. If he knocked down 9 in the first frame, how many pins did he knock down in the second frame?

_____ tens and _____ ones

Name ______________________ Date ____________

Solve the problems. Write the answers to show how many tens and ones. If there is only one ten, cross off the "s."

1. 8 + 5 = ☐☐

____ tens and ____ ones

2. 12 - 4 = ☐☐

____ tens and ____ ones

3. 15 - 6 = ☐☐

____ tens and ____ ones

4. 14 + 5 = ☐☐

____ tens and ____ ones

5. 13 + 5 = ☐☐

____ tens and ____ ones

6. 17 - 8 = ☐☐

____ tens and ____ ones

Read the word problem. Draw and label. Write a number sentence and statement that matches the story. Rewrite your answer to show its tens and ones. If there is only 1 ten, cross off the "s."

7. Mike has some red cars and 8 blue cars. If Mike has 9 red cars, how many cars does he have in all?

_____ tens and _____ ones

8. Yani and Han had 14 golf balls. They lost some balls. They had 8 golf balls left. How many balls did they lose?

_____ tens and _____ ones

9. Nick rides his bike for 6 miles over the weekend. He rides 14 miles during the week. How many total miles does Nick ride?

_____ tens and _____ ones

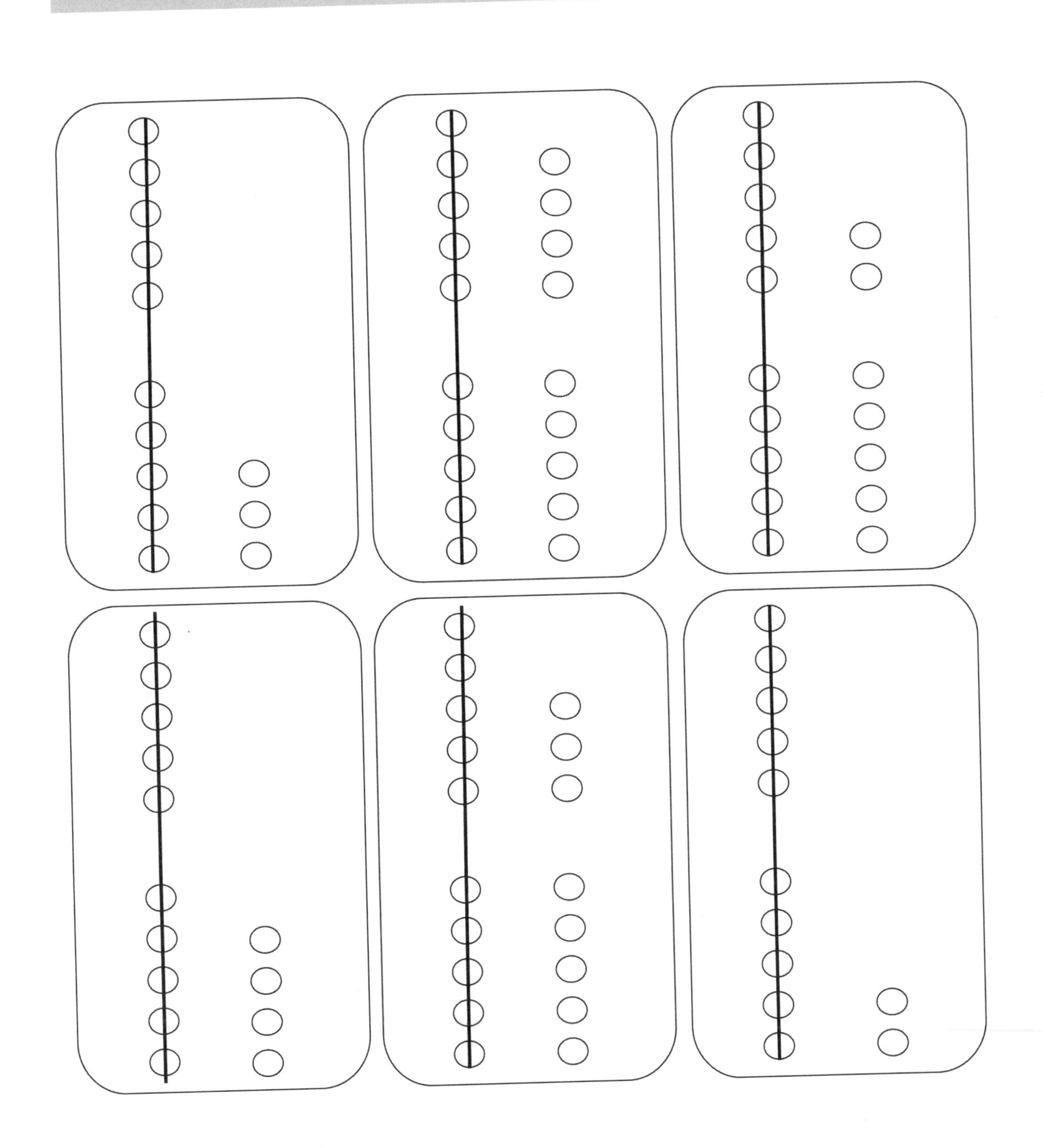

5-group column cards

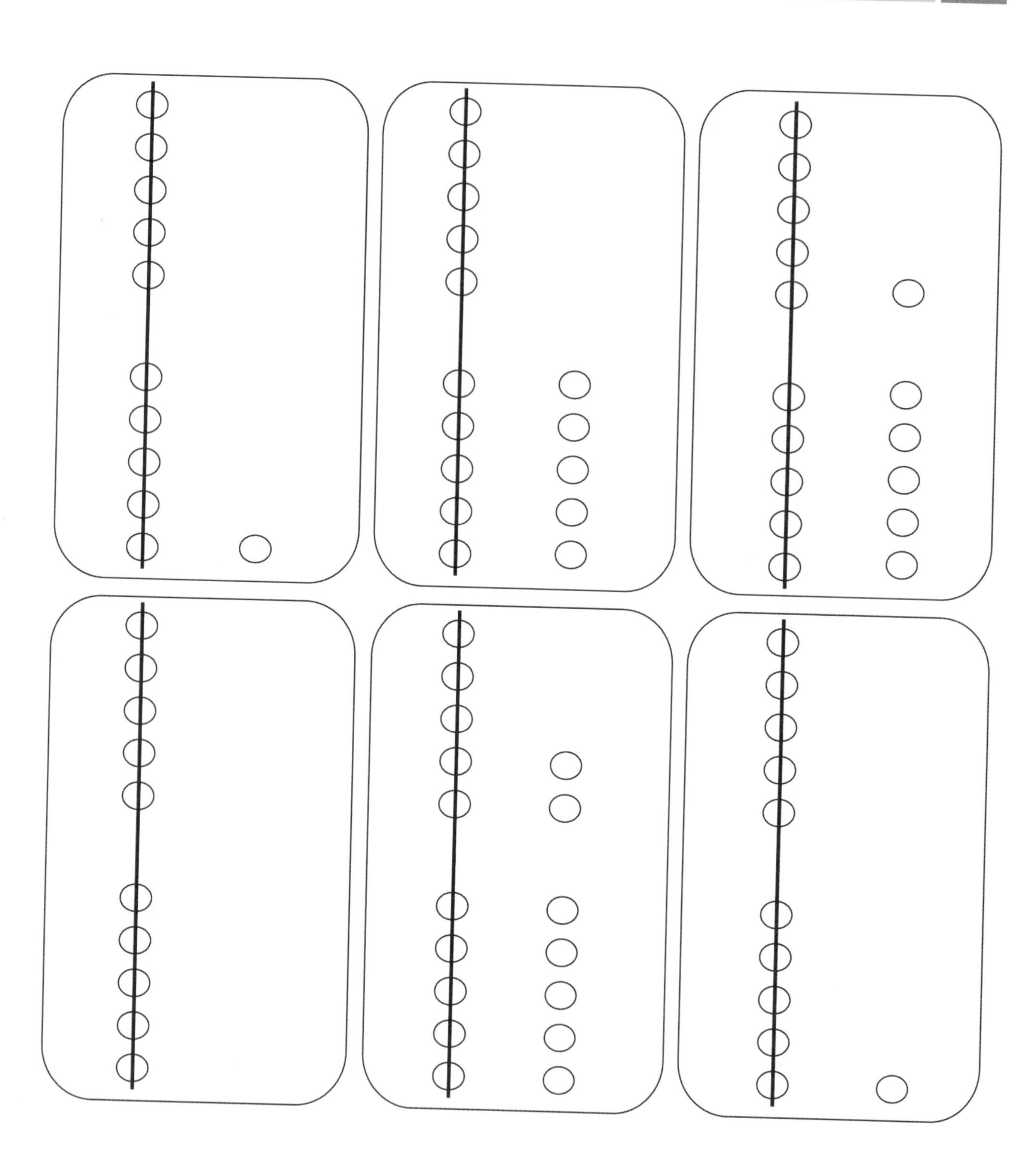

5-group column cards

Lesson 28

Objective: Solve addition problems using ten as a unit, and write two-step solutions.

Suggested Lesson Structure

■ Fluency Practice	(13 minutes)
■ Application Problem	(5 minutes)
■ Concept Development	(32 minutes)
■ Student Debrief	(10 minutes)
Total Time	**(60 minutes)**

Fluency Practice (13 minutes)

- Magic Counting Sticks **1.NBT.2** (3 minutes)
- Sprint: Adding by Decomposing Teen Numbers **1.OA.6** (10 minutes)

Magic Counting Sticks (3 minutes)

Materials: (T) Hide Zero cards (Lesson 18 Fluency Template 1)

Note: This activity reviews the concept of ten as a unit and prepares students for today's lesson.

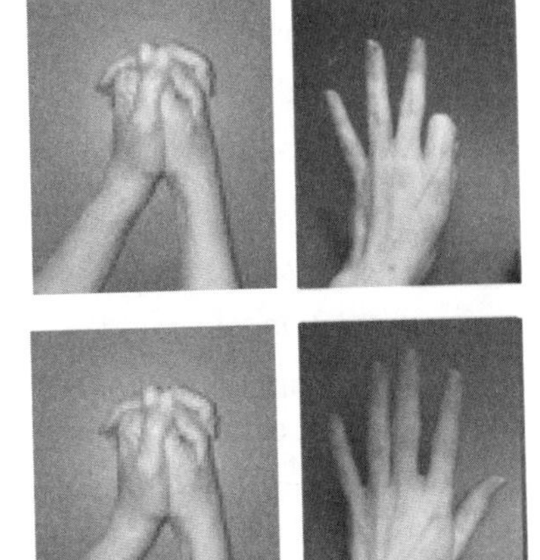

T: (Divide students into partners. Show 13 with Hide Zero cards.) Partner A, show the ones. Partner B, show the tens. How many tens are in 13?

S: 1.

T: How many ones?

S: 3.

T: If I wanted to add 2, which partner could do it?

S: Partner A.

T: Yes. Add 2 to 13. What number do you see?

S: 15.

Alternate partners and continue with the suggested sequence: 12 + 2, 14 + 1, 15 + 3, 14 + 2, 15 + 3, 16 + 3. All sums should be between 11 and 19.

Sprint: Adding by Decomposing Teen Numbers (10 minutes)

Materials: (S) Adding by Decomposing Teen Numbers Sprint

Note: This Sprint addresses the Grade 1 core fluency objective of adding and subtracting within 20.

Application Problem (5 minutes)

Ruben has 7 blue cars and 6 red cars. If Ruben puts all of the blue cars in his car carrier that carries 10 cars, how many red cars will fit in the carrier, and how many will be left out of the carrier?

Note: This Application Problem serves multiple purposes. Some students may respond to the problem with an answer of 13 cars, anticipating the question as, "How many cars does Ruben have?" Look for such misinterpretations as an opportunity to reinforce the importance of reading the question carefully. In addition, the problem gives students a chance to focus on the decomposition of the second addend when creating a unit of ten. This leads into today's lesson in which students write number sentences to show the two steps in the Level 3 strategy of making ten.

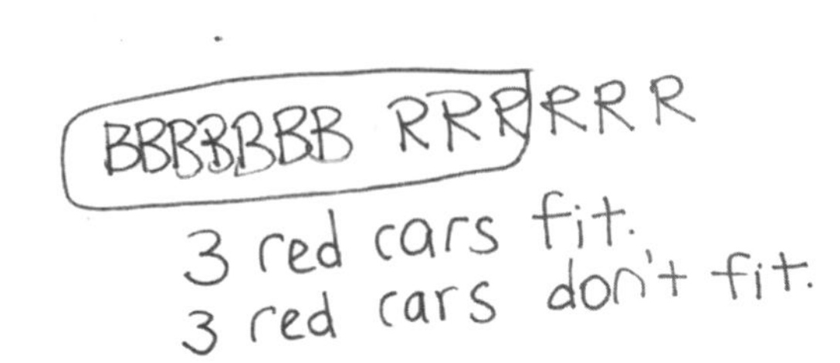

Concept Development (32 minutes)

Materials: (T) Hide Zero cards (Lesson 18 Fluency Template 1) (S) Personal white board

Have students gather in a semicircle in the meeting area with their personal white boards.

T: (Project 8 + 4.) Solve this problem with a partner.
S: (Discuss with partners.)
T: How many is 8 + 4?
S: Twelve!
T: In the number 12, do we have any tens? How many tens do we have?
S: Yes! 1 ten!
T: Along with 1 ten, do we have any extra ones? How many?
S: Yes! 2 ones!
T: (Hold up the number 12 with Hide Zero cards.) Right, the number 12 is made of 1 ten and 2 ones. (Pull apart the two cards to show 10 card and 2 card separately.)
T: How many tens in the number 8?
S: None!
T: How many tens in the number 4?
S: None!
T: Then, how did we take 8 and 4, which didn't have any tens, to make a number that has 1 ten and 2 ones? Talk with your partner.
S: (Discuss with partners.)
T: (Listen for students to articulate the making of 1 ten and extra ones when adding 8 and 4.)
T: How did we add 8 and 4 to make 12, which has 1 ten and 2 ones?

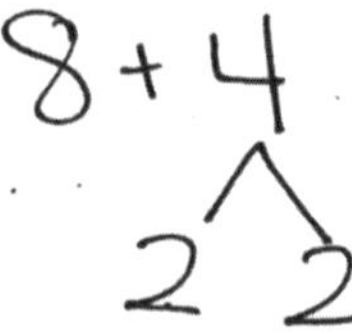

As students share, the goal is to create two number sentences. The first shows the addition that makes ten, and the second shows the addition of the ten and extra ones to make the final total as pictured.

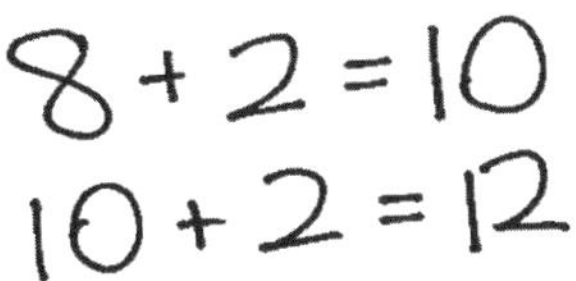

S: Break apart 4 into 2 and 2. Add 8 and 2 to make 1 ten, and then add 2 more ones. → If you start at 8 and count on, you get to ten after 2 counts. That's 1 ten. Then, you still have 2 more. That makes 12. 1 ten and 2 ones.

T: While you were sharing, I wrote your explanations as number sentences. You said that to solve 8 + 4, you started with 8 and added 2 out of the 4. That made 1 ten. (Point to 8 + 2 = 10 in the first number sentence.)

T: Then, we have 2 more left from the 4, so you added your 1 ten and 2 ones to make 12. (Point to 10 + 2 = 12 in the second number sentence.) Did I explain that correctly?

S: Yes!

T: Write down the two number sentences I have on the board, and talk with your partner to explain how it shows the way we made 1 ten and 2 ones when adding 8 + 4.

S: (Discuss with partners.)

T: (Listen for students who are using accurate language. If students are not explaining 1 ten, emphasize the creation of 1 ten in upcoming examples.)

T: Today, let's write two number sentences each time we solve a problem like this, so we can see how we made 1 ten first and then added the ones.

Repeat the process, having students write two number sentences to show making 1 ten and adding the extra ones, using the following sequence: 8 + 5, 8 + 6, 9 + 6, 7 + 5, and 7 + 9. If students appear to require more support at the onset, complete the first problem or two as a class.

NOTES ON MULTIPLE MEANS OF ACTION AND EXPRESSION:

In the beginning, students may get confused about which numbers go where when writing their two number sentences. Emphasize the importance of the addition of the ten for the first number sentence. Some students may need number sentences more concretely framed out. Have an example of a completed problem where they can easily see it to reference if they get confused.

NOTES ON MULTIPLE MEANS OF REPRESENTATION:

For those students who have difficulty writing, provide the sentence frame when doing word problems. This helps students focus on their math without worrying about the writing.

Problem Set (10 minutes)

Students should do their personal best to complete the Problem Set within the allotted 10 minutes. For some classes, it may be appropriate to modify the assignment by specifying which problems they work on first. Some problems do not specify a method for solving. Students should solve these problems using the RDW approach used for Application Problems.

Student Debrief (10 minutes)

Lesson Objective: Solve addition problems using ten as a unit, and write two-step solutions.

The Student Debrief is intended to invite reflection and active processing of the total lesson experience.

Invite students to review their solutions for the Problem Set. They should check work by comparing answers with a partner before going over answers as a class. Look for misconceptions or misunderstandings that can be addressed in the Debrief. Guide students in a conversation to debrief the Problem Set and process the lesson.

Any combination of the questions below may be used to lead the discussion.

- Look at Problem 1. How many tens and how many ones are there? Use a yellow crayon and find all of the places 1 ten is hiding within the Problem Set. (Color the 1 in all numbers from 10 through 19 within the Problem Set. They may also color the two-digit representation of 10.)
- Look at Problems 1 and 3. What do the number sentences have in common? (The first number sentence is the same.) Do you have any other problems on the Problem Set that have 9 + 1 = 10 as the first number sentence? What is similar about the problems that caused them to have the same number sentence as part of the solution?
- Look at Problems 1 and 2. How are they the same, and how are they different?
- Look at the Application Problem. Ruben has a carrier that fits 10 cars. How is Ruben's 1 carrier like 1 ten?
- How many cars does Ruben have? Use two number sentences to show how we can make 1 ten and then add the extra ones.

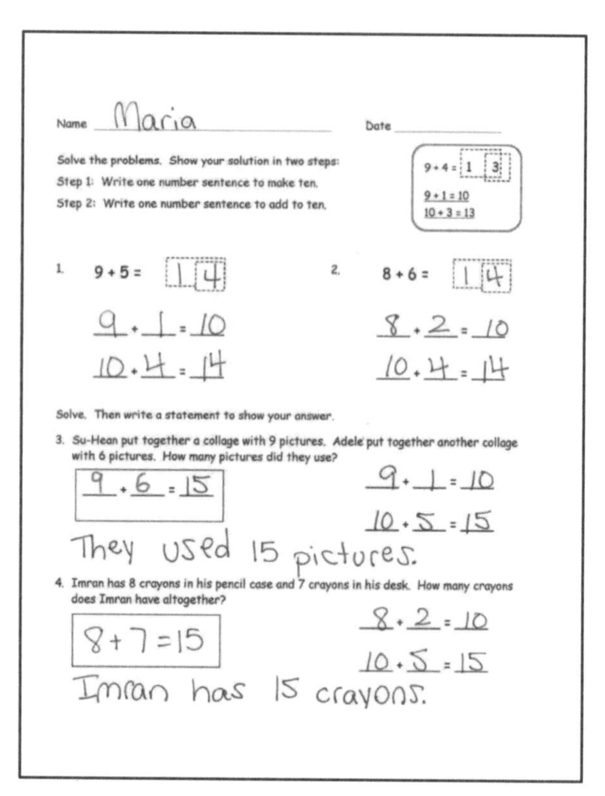
Name Maria Date

Solve the problems. Show your solution in two steps:
Step 1: Write one number sentence to make ten.
Step 2: Write one number sentence to add to ten.

9 + 4 = 1 3
9 + 1 = 10
10 + 3 = 13

1. 9 + 5 = 1 4
9 + 1 = 10
10 + 4 = 14

2. 8 + 6 = 1 4
8 + 2 = 10
10 + 4 = 14

Solve. Then write a statement to show your answer.

3. Su-Hean put together a collage with 9 pictures. Adele put together another collage with 6 pictures. How many pictures did they use?
9 + 6 = 15
9 + 1 = 10
10 + 5 = 15
They used 15 pictures.

4. Imran has 8 crayons in his pencil case and 7 crayons in his desk. How many crayons does Imran have altogether?
8 + 7 = 15
8 + 2 = 10
10 + 5 = 15
Imran has 15 crayons.

5. At the park, there were 4 ducks swimming in the pond. If there were 9 ducks resting on the grass, how many ducks were at the park in all?
4 + 9 = 13
9 + 1 = 10
10 + 3 = 13
There are 13 ducks at the park.

6. Cece made 7 frosted cookies and 8 cookies with sprinkles. How many cookies did Cece make?
7 + 8 = 15
8 + 2 = 10
10 + 5 = 15
Cece made 15 cookies.

7. Payton read 8 books about dolphins and whales. She read 9 books about dogs and cats. How many books did she read about animals altogether?
8 + 9 = 17
9 + 1 = 10
10 + 7 = 17
Payton read 17 books.

Exit Ticket (3 minutes)

After the Student Debrief, instruct students to complete the Exit Ticket. A review of their work will help with assessing students' understanding of the concepts that were presented in today's lesson and planning more effectively for future lessons. The questions may be read aloud to the students.

A

Name ______________________

Number Correct: 

Date ______________

*Write the missing number.

1.	10 + 2 = □		16.	12 + 3 = □	
2.	2 + 1 = □		17.	13 + 3 = □	
3.	10 + 3 = □		18.	14 + 3 = □	
4.	4 + 10 = □		19.	13 + 5 = □	
5.	4 + 2 = □		20.	14 + 5 = □	
6.	6 + 10 = □		21.	15 + 5 = □	
7.	10 + 3 = □		22.	4 + 14 = □	
8.	3 + 3 = □		23.	4 + 15 = □	
9.	10 + 6 = □		24.	12 + □ = 14	
10.	2 + 1 = □		25.	12 + □ = 15	
11.	12 + 1 = □		26.	12 + □ = 16	
12.	2 + 2 = □		27.	□ + 4 = 16	
13.	12 + 2 = □		28.	5 + □ = 16	
14.	3 + 3 = □		29.	5 + □ = 26	
15.	13 + 3 = □		30.	4 + □ = 36	

B

Name ______________________

Number Correct:

Date ______________

*Write the missing number.

1.	10 + 1 = □		16.	12 + 2 = □	
2.	1 + 1 = □		17.	13 + 2 = □	
3.	10 + 2 = □		18.	14 + 2 = □	
4.	3 + 10 = □		19.	13 + 4 = □	
5.	3 + 2 = □		20.	14 + 4 = □	
6.	5 + 10 = □		21.	15 + 4 = □	
7.	10 + 2 = □		22.	5 + 14 = □	
8.	2 + 2 = □		23.	5 + 15 = □	
9.	10 + 4 = □		24.	11 + □ = 12	
10.	2 + 1 = □		25.	11 + □ = 13	
11.	12 + 1 = □		26.	11 + □ = 14	
12.	1 + 1 = □		27.	□ + 3 = 14	
13.	11 + 1 = □		28.	6 + □ = 19	
14.	3 + 2 = □		29.	6 + □ = 29	
15.	13 + 2 = □		30.	5 + □ = 39	

Name ______________________________ Date ______________

Solve the problems. Show your solution in two steps:

Step 1: Write one number sentence to make ten.
Step 2: Write one number sentence to add to ten.

$9 + 4 = \boxed{1}\boxed{3}$

9 + 1 = 10

10 + 3 = 13

1. 9 + 5 = ☐☐

_____ + _____ = _____

_____ + _____ = _____

2. 8 + 6 = ☐☐

_____ + _____ = _____

_____ + _____ = _____

Solve. Then, write a statement to show your answer.

3. Su-Hean put together a collage with 9 pictures. Adele put together another collage with 6 pictures. How many pictures did they use?

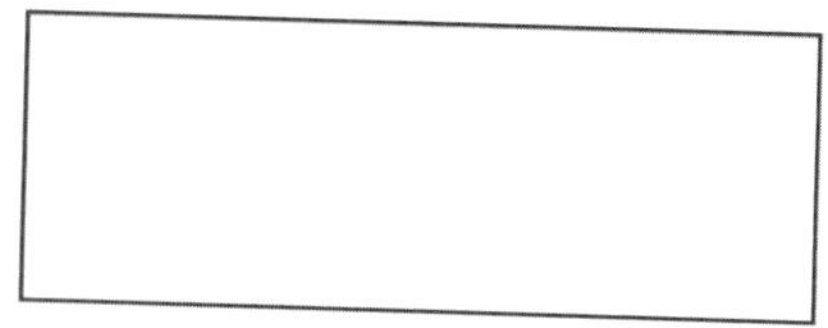

_____ + _____ = _____

_____ + _____ = _____

4. Imran has 8 crayons in his pencil case and 7 crayons in his desk. How many crayons does Imran have altogether?

_____ + _____ = _____

_____ + _____ = _____

5. At the park, there were 4 ducks swimming in the pond. If there were 9 ducks resting on the grass, how many ducks were at the park in all?

_____ + _____ = _____

_____ + _____ = _____

6. Cece made 7 frosted cookies and 8 cookies with sprinkles. How many cookies did Cece make?

7. Payton read 8 books about dolphins and whales. She read 9 books about dogs and cats. How many books did she read about animals altogether?

Name ______________________ Date __________

Solve the problems. Write your answers to show how many **tens** and **ones**.

$9 + 7 = $ 1 6

9 + 1 = 10

10 + 6 = 16

1. $9 + 4 =$ ☐☐

____ + ____ = ____

____ + ____ = ____

2. $8 + 7 =$ ☐☐

____ + ____ = ____

____ + ____ = ____

Name ______________________ Date ____________

Solve the problems. Write your answers to show how many **tens** and **ones**.

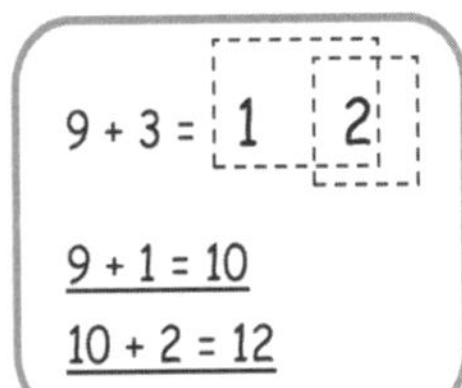

1. 9 + 7 =

____ + ____ = ____

____ + ____ = ____

2. 8 + 5 =

____ + ____ = ____

____ + ____ = ____

Solve. Write the two number sentences for each step to show how you make **a ten**.

3. Boris has 9 board games on his shelf and 8 board games in his closet. How many board games does Boris have altogether?

____ + ____ = ____

____ + ____ = ____

---

4. Sabra built a tower with 8 blocks. Yuri put together another tower with 7 blocks. How many blocks did they use?

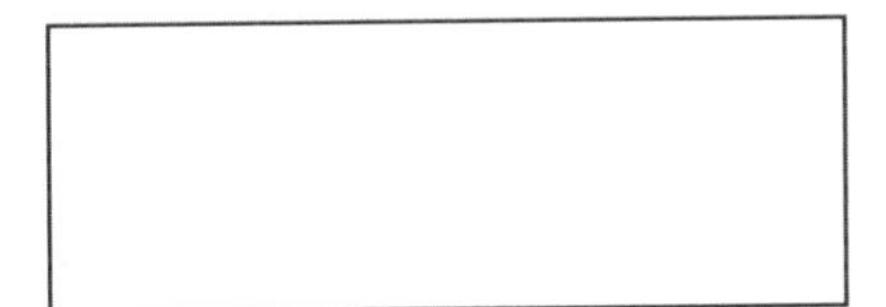

5. Camden solved 6 addition word problems. She also solved 9 subtraction word problems. How many word problems did she solve altogether?

6. Minna made 4 bracelets and 8 necklaces with her beads. How many pieces of jewelry did Minna make?

7. I put 5 peaches into my bag at the farmer's market. If I already had 7 apples in my bag, how many pieces of fruit did I have in all?

Lesson 29

Objective: Solve subtraction problems using ten as a unit, and write two-step solutions.

Suggested Lesson Structure

Fluency Practice	(15 minutes)
Application Problem	(5 minutes)
Concept Development	(30 minutes)
Student Debrief	(10 minutes)
Total Time	**(60 minutes)**

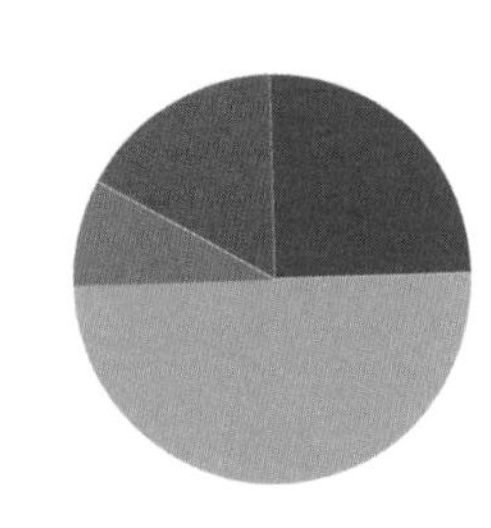

Fluency Practice (15 minutes)

- Say Ten: 5-group Columns **1.NBT.2** (3 minutes)
- Magic Counting Sticks **1.OA.6** (4 minutes)
- Happy Counting by Fives **1.OA.5** (3 minutes)
- Take from Ten Subtraction **1.OA.6** (5 minutes)

Say Ten: 5-group Columns (3 minutes)

Materials: (T) 5-group column cards (Lesson 27 Fluency Template)

Note: This fluency activity reviews the unit of 1 ten as a 5-group column, which was introduced in the last lesson.

T: (Hold up the card showing 13.) Tell me how many, the Say Ten way.
S: Ten 3.
T: How many tens?
S: 1 ten.
T: How many ones?
S: 3 ones.

Repeat this process and alternate between requesting that students respond the Say Ten way and saying the number of tens and ones.

Magic Counting Sticks (4 minutes)

Materials: (T) Hide Zero cards (Lesson 18 Fluency Template 1)

Note: This activity reviews decomposing teen numbers in order to subtract.

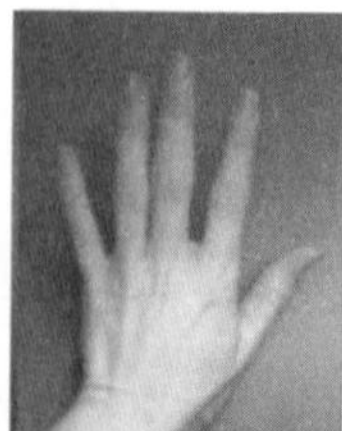

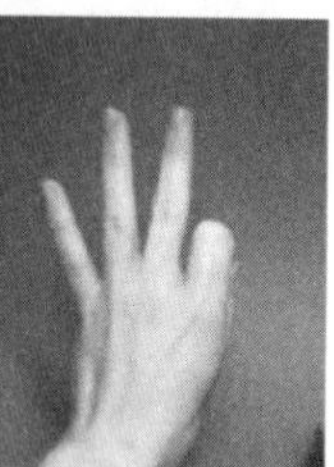

T: (Assign partners. Show 15 with Hide Zero cards.) Partner A, show the ones. Partner B, show the tens. How many ones are in 15?

S: 5.

T: How many tens?

S: 1.

T: If we wanted to subtract 2, which partner should do it?

S: Partner A.

T: Yes. Subtract 2 from 15. What number do you see?

S: 13.

Alternate partners and continue with the suggested sequence: 12 – 2, 13 – 1, 14 – 2, 14 – 3, 15 – 3, 16 – 4. Differences should be between 10 and 19.

Happy Counting by Fives (3 minutes)

Note: This maintenance fluency activity reviews adding and subtracting 5.

Repeat the Happy Counting activity from Lesson 4, counting by fives from 0 to 40 and back. First, count the Say Ten way, and then count the regular way.

Take from Ten Subtraction (5 minutes)

Materials: (T) Subtract 9 flash cards (Lesson 17 Fluency Template), subtract 8 flash cards (Lesson 20 Fluency Template), subtract 7 and 6 flash cards (Fluency Template)

Note: This activity reviews the take from ten subtraction strategy.

Show a flash card (e.g., 12 – 8 = ____). Cold call a student or group of students to answer. If students need additional help subtracting 8, use the following vignette.

T: Say 12 the Say Ten way.

S: Ten 2.

T: 10 – 8 = ____. (Snap.)

S: 2.

T: 2 + 2 = ____. (Point to the 2 on the flash card, and snap.)

S: 4.

T: So, 12 – 8 = ____. (Snap.)

S: 4.

Repeat the process using subtract 9, 8, 7, and 6 flash cards.

Application Problem (5 minutes)

Hae Jung had 13 markers, and she gave some to Lily. If Hae Jung then had 5 markers, how many markers did she give to Lily?

13 − 5 = 8

SHe gave Lily 8 markers.

Note: Students continue to consider *take apart with addend unknown* problem types in this problem. During the Debrief, students have the opportunity to apply today's objective to the problem, writing number sentences to show the two steps in the Level 3 strategy of taking from ten.

Concept Development (30 minutes)

Materials: (T) Hide Zero cards (Lesson 18 Fluency Template 1) (S) Personal white board

Have students gather in a semi-circle in the meeting area with their personal white boards.

T: (Project and read.) Suhani has some presents left to open. If she received 13 presents and already opened 8 of them, how many presents does Suhani still need to open? Solve this problem with your partner.

T: I see that many of you used a subtraction sentence, 13 – 8, to solve this problem. What is 13 – 8? How many presents does Suhani need to open?

S: 5 presents!

T: In the number 13, do we have any tens? How many tens do we have?

S: Yes! 1 ten!

T: Along with 1 ten, do we have any extra ones? How many?

S: Yes! 3 ones!

T: (Hold up the number 13 with Hide Zero cards.) The number 13 is made of 1 ten and 3 ones. (Pull apart the two cards to show the 10 card and the 3 card separately.)

NOTES ON MULTIPLE MEANS OF REPRESENTATION:

Some students may benefit from connecting the abstract number bonds and equations with concrete materials. Linking cubes in sticks of 10 and separated ones or Rekenreks can be used along with the numbers. Using concrete and abstract representations simultaneously develops stronger mental images. Moving to the use of the abstract while visualizing the concrete materials can increase students' confidence and math fluency.

MP.7

T: Where should I take 8 from? The 1 ten or the 3 ones?

S: From the ten.

T: How many ones are left over when we take 8 from the ten?

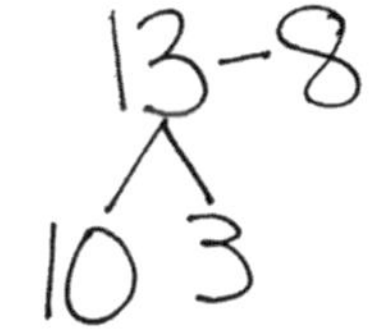

S: 2 ones.

T: Write down the number sentence to show how we just subtracted 8.

S: 10 – 8 = 2.

T: (Put down the 10 card and hold up 2 fingers next to the 3 card.)

MP.7

T: Did we have any extra ones from the starting number?

S: Yes. We had 3 ones.

T: Let's put the ones together. (Continue to hold up 2 fingers and the 3 card.) Write down the number sentence that tells how many ones we have altogether.

S: 2 ones + 3 ones = 5 ones.

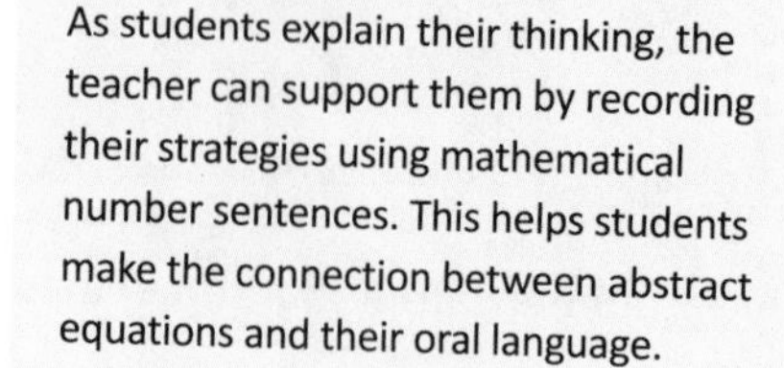

T: So, when we solved 13 – 8 and got 5, we started with 1 ten and 3 ones. We ended with no tens and 5 ones. Where did the ten go? Turn and talk to your partner. Point to the number sentence that shows how we ended with 0 tens.

S: We don't have a ten anymore because we used it to take away 8. (Point to 10 – 8.)

T: Explain to your partner how we then ended with 5 ones. (Point to 2 + 3 = 5.)

S: We had 2 ones left from 10 – 8, and we still had 3 extra ones, so we added 2 and 3 to get 5.

T: Today, let's write two number sentences each time we solve a problem like this, so we can see how we took away from the ten first and then added the extra ones.

NOTES ON MULTIPLE MEANS OF EXPRESSION:

As students explain their thinking, the teacher can support them by recording their strategies using mathematical number sentences. This helps students make the connection between abstract equations and their oral language.

Repeat the process, having students write two number sentences to show taking away from the ten and adding the extra ones, using the following sequence with *add to with change unknown* and *take apart/put together with addend unknown* problem types: 11 – 5, 12 – 9, 14 – 6, 17 – 8, and 16 – 7. If students appear to require more support at the onset, complete the first problem or two as a class.

Note: Some students may find it easier to count back when subtracting. When solving 13 – 8, students may subtract 3 first to make a ten, then subtract 5 more, writing 13 – 3 = 10 and 10 – 5 = 5 as their number sentences. This is another efficient Level 3 strategy that uses two steps. Today, however, students are asked to solve the problems by taking from ten first.

Problem Set (10 minutes)

Students should do their personal best to complete the Problem Set within the allotted 10 minutes. For some classes, it may be appropriate to modify the assignment by specifying which problems they work on first. Some problems do not specify a method for solving. Students should solve these problems using the RDW approach used for Application Problems.

Student Debrief (10 minutes)

Lesson Objective: Solve subtraction problems using ten as a unit, and write two-step solutions.

The Student Debrief is intended to invite reflection and active processing of the total lesson experience.

Invite students to review their solutions for the Problem Set. They should check work by comparing answers with a partner before going over answers as a class. Look for misconceptions or misunderstandings that can be addressed in the Debrief. Guide students in a conversation to debrief the Problem Set and process the lesson.

Any combination of the questions below may be used to lead the discussion.

- Look at Problem 7 in your Problem Set. How many tens do you have left? Explain how we started with 1 ten and some ones and ended with 0 tens and some ones.
- How is Problem 6 different from the rest of the problems in your Problem Set? How did you solve Problem 6 using two number sentences? Explain why we still have 1 ten as a part of your answer.
- In what new way did we solve subtraction problems today?
- How can you solve today's Application Problem using two number sentences so we can see how we took away from the ten first and then added the extra ones?

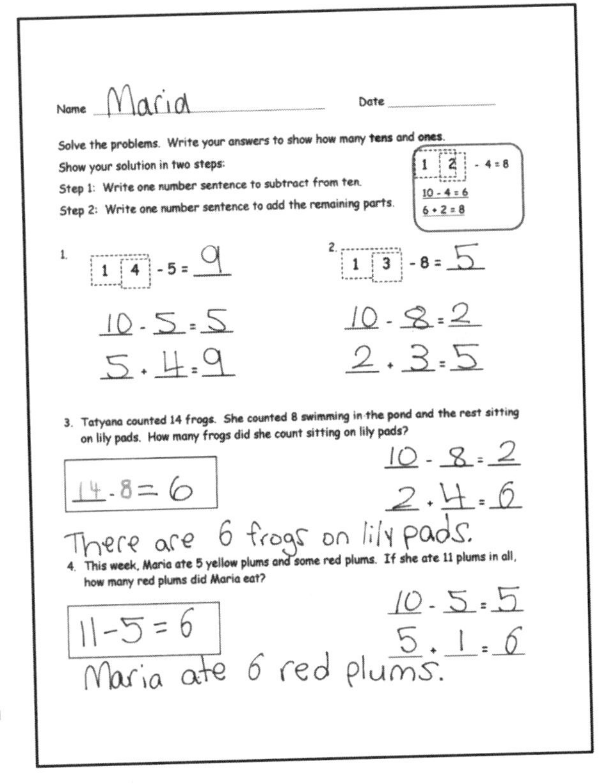

Name Maria Date

Solve the problems. Write your answers to show how many **tens** and **ones**.
Show your solution in two steps:
Step 1: Write one number sentence to subtract from ten.
Step 2: Write one number sentence to add the remaining parts.

1 2 - 4 = 8
10 - 4 = 6
6 + 2 = 8

1. 1 4 - 5 = 9
10 - 5 = 5
5 + 4 = 9

2. 1 3 - 8 = 5
10 - 8 = 2
2 + 3 = 5

3. Tatyana counted 14 frogs. She counted 8 swimming in the pond and the rest sitting on lily pads. How many frogs did she count sitting on lily pads?

14 - 8 = 6
10 - 8 = 2
2 + 4 = 6
There are 6 frogs on lily pads.

4. This week, Maria ate 5 yellow plums and some red plums. If she ate 11 plums in all, how many red plums did Maria eat?

11 - 5 = 6
10 - 5 = 5
5 + 1 = 6
Maria ate 6 red plums.

5. Some children are on the playground playing tag. Eight are on the swings. If there are 16 children on the playground in all, how many children are playing tag?

16 - 8 = 8
10 - 8 = 2
2 + 6 = 8
There are 8 children playing tag.

6. Oziah read some nonfiction books. Then he read 6 fiction books. If he read 18 books altogether, how many nonfiction books did Oziah read?

18 - 6 = 12
8 - 6 = 2
10 + 2 = 12
He read 12 non-fiction books.

7. Hadley has 9 buttons on her jacket. She has some more buttons on her shirt. Hadley has a total of 17 buttons on her jacket and shirt. How many buttons does she have on her shirt?

17 - 9 = 8
10 - 9 = 1
1 + 7 = 8
She has 8 buttons on her shirt.

Exit Ticket (3 minutes)

After the Student Debrief, instruct students to complete the Exit Ticket. A review of their work will help with assessing students' understanding of the concepts that were presented in today's lesson and planning more effectively for future lessons. The questions may be read aloud to the students.

Name ______________________________ Date ______________

Solve the problems. Write your answers to show how many **tens** and **ones**. Show your solution in two steps:

Step 1: Write one number sentence to subtract from ten.
Step 2: Write one number sentence to add the remaining parts.

1 2 - 4 = 8
10 - 4 = 6
6 + 2 = 8

1. 1 4 - 5 = _____

_____ - _____ = _____

_____ + _____ = _____

2. 1 3 - 8 = _____

_____ - _____ = _____

_____ + _____ = _____

3. Tatyana counted 14 frogs. She counted 8 swimming in the pond and the rest sitting on lily pads. How many frogs did she count sitting on lily pads?

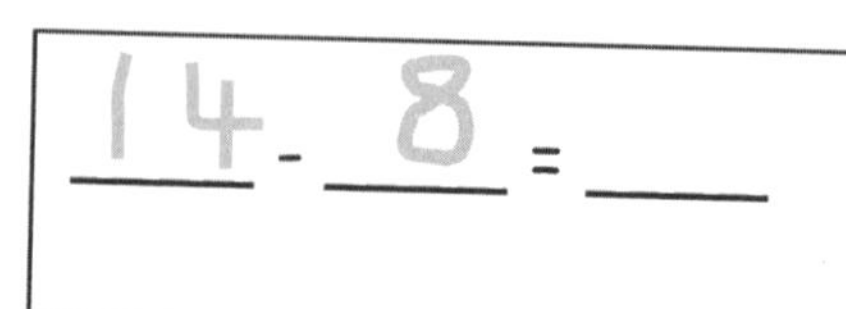

_____ - _____ = _____

_____ + _____ = _____

4. This week, Maria ate 5 yellow plums and some red plums. If she ate 11 plums in all, how many red plums did Maria eat?

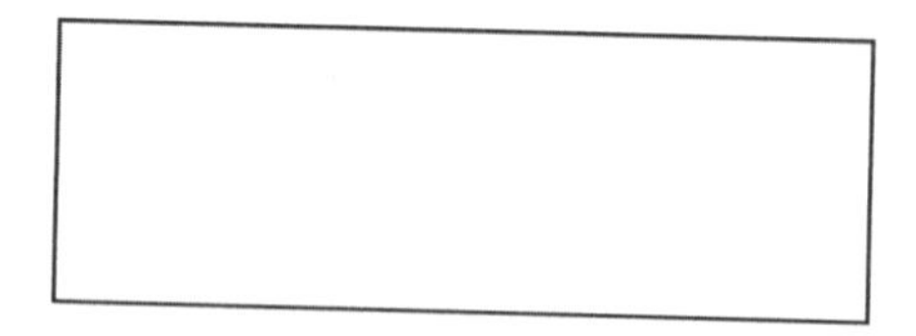

_____ - _____ = _____

_____ + _____ = _____

EUREKA MATH

5. Some children are on the playground playing tag. Eight are on the swings. If there are 16 children on the playground in all, how many children are playing tag?

____ - ____ = ____

____ + ____ = ____

6. Oziah read some nonfiction books. Then, he read 6 fiction books. If he read 18 books altogether, how many nonfiction books did Oziah read?

7. Hadley has 9 buttons on her jacket. She has some more buttons on her shirt. Hadley has a total of 17 buttons on her jacket and shirt. How many buttons does she have on her shirt?

Name ______________________________ Date ______________

Solve the problems. Write your answers to show how many **tens** and **ones**.

1	2	- 5 = 7

10 - 5 = 5

5 + 2 = 7

1. | 1 | 5 | - 6 = ____

____ - ____ = ____

____ + ____ = ____

2. | 1 | 4 | - 8 = ____

____ - ____ = ____

____ + ____ = ____

Name ______________________ Date ____________

Solve the problems. Write your answers to show how many **tens** and **ones**.

1 2 - 5 = 7

10 - 5 = 5

5 + 2 = 7

1. 1 7 - 8 = ____

____ - ____ = ____

____ + ____ = ____

2. 1 6 - 7 = ____

____ - ____ = ____

____ + ____ = ____

Solve. Write the two number sentences for each step to show how you take from **ten**. Remember to put a box around your solution and write a statement.

3. Yvette counted 12 kids at the park. She counted 3 on the playground and the rest playing in the sand. How many kids did she count playing in the sand?

12 - 3 = ____

____ - ____ = ____

____ + ____ = ____

---

4. Eli read some science magazines. Then, he read 9 sports magazines. If he read 18 magazines altogether, how many science magazines did Eli read?

____ - ____ = ____

____ + ____ = ____

5. On Monday, Paulina checked out 6 whale books and some turtle books from the library. If she checked out 13 books in all, how many turtle books did Paulina check out?

_____ - _____ = _____

_____ + _____ = _____

---

6. Some children are at the park playing soccer. Seven are wearing white shirts. If there are 14 children playing soccer in all, how many children are not wearing white shirts?

_____ - _____ = _____

_____ + _____ = _____

---

7. Dante has 9 stuffed animals in his room. The rest of his stuffed animals are in the TV room. Dante has 15 stuffed animals. How many of Dante's stuffed animals are in the TV room?

_____ - _____ = _____

_____ + _____ = _____

10 - 7	11 - 7
12 - 7	13 - 7
14 - 7	15 - 7
16 - 7	17 - 7

subtract 7 and 6 flash cards

10 - 6	11 - 6
12 - 6	13 - 6
14 - 6	15 - 6
16 - 6	

subtract 7 and 6 flash cards

Name ______________________________ Date ____________

1. Mr. Baggy owns a pet store.
   He counted 10 goldfish in a big tank and 5 goldfish in a small tank. He sold 8 goldfish out of the big tank. How many goldfish did he have left in all? Explain your answer using a labeled math drawing and a number sentence.

   Mr. Baggy had ______ goldfish.

2. Write the numbers that make the number sentences true.

   a. $12 - 9 =$ ______ $11 - 8 =$ ______ $15 - 6 =$ ______

   b. $9 +$ ______ $= 13$ $8 +$ ______ $= 12$ $12 =$ ______ $+ 7$

   c. Write a related subtraction fact for each of the three problems in the last row in the spaces below.

   ______________ ______________ ______________

3. Write a number bond in each number sentence to show how to use ten to subtract. Draw 5-groups and some ones to show each subtraction sentence.

a. 13 − 9 = 4	b. 12 − 8 = 4

c. Use your pictures and numbers to explain how both subtraction problems equal 4.

4. Mr. Baggy also has 9 birds, 15 snakes, and 12 turtles.

a. Show the number of snakes as a ten and some ones with a number bond, a 5-group drawing, and a number sentence.

b. Mr. Baggy sold some snakes. Now, he has 5. How many snakes did he sell? Explain your solution using a number bond or a math drawing. Write a number sentence. Complete the statement.

Mr. Baggy sold _______ snakes.

c. Mr. Baggy sold 8 turtles. How many turtles does he have left? Explain your solution using a number bond or a math drawing. Write a number sentence. Complete the statement.

Mr. Baggy has _______ turtles left.

d. Mr. Baggy's daughter says she can find the number of turtles Mr. Baggy has left using subtraction or addition. Show two ways Mr. Baggy's daughter can solve this problem.

e. As Mr. Baggy gets ready to close his pet store for the day, he needs to know how many animals he has altogether. How many birds, snakes, and turtles does Mr. Baggy have left in his store altogether? Explain your solution using number bonds or math drawings. Write a number sentence. Complete the statement.

Mr. Baggy has _______ animals left.

f. True or false: You will get a different answer if you add 9 and 5 first, then add 4, than if you add 9 and 4 first, then add 5. (Circle one.) **True** **False**
Use pictures or words to show how you know.

End-of-Module Assessment Task Standards Addressed	Topics A–D

Represent and solve problems involving addition and subtraction.

1.OA.1 Use addition and subtraction within 20 to solve word problems involving situations of adding to, taking from, putting together, taking apart, and comparing, with unknowns in all positions, e.g., by using objects, drawings, and equations with a symbol for the unknown number to represent the problem.

1.OA.2 Solve word problems that call for addition of three whole numbers whose sum is less than or equal to 20, e.g., by using objects, drawings, and equations with a symbol for the unknown number to represent the problem.

Understand and apply properties of operations and the relationship between addition and subtraction.

1.OA.3 Apply properties of operations as strategies to add and subtract. (Students need not use formal terms for these properties.) *Examples: If 8 + 3 = 11 is known, then 3 + 8 = 11 is also known. (Commutative property of addition.) To add 2 + 6 + 4, the second two numbers can be added to make a ten, so 2 + 6 + 4 = 2 + 10 = 12. (Associative property of addition.)*

1.OA.4 Understand subtraction as an unknown-addend problem. *For example, subtract 10 – 8 by finding the number that makes 10 when added to 8.*

Add and subtract within 20.

1.OA.6 Add and subtract within 20, demonstrating fluency for addition and subtraction within 10. Use strategies such as counting on; making ten (e.g., 8 + 6 = 8 + 2 + 4 = 10 + 4 = 14); decomposing a number leading to a ten (e.g., 13 – 4 = 13 – 3 – 1 = 10 – 1 = 9); using the relationship between addition and subtraction (e.g., knowing that 8 + 4 = 12, one knows 12 – 8 = 4); and creating equivalent but easier or known sums (e.g., adding 6 + 7 by creating the known equivalent 6 + 6 + 1 = 12 + 1 = 13).

Understand place value.

1.NBT.2 Understand that the two digits of a two-digit number represent amounts of tens and ones. Understand the following as special cases:

a. 10 can be thought of as a bundle of ten ones—called a "ten."

b. The numbers from 11 to 19 are composed of a ten and one, two, three, four, five, six, seven, eight, or nine ones.

Evaluating Student Learning Outcomes

A Progression Toward Mastery is provided to describe steps that illuminate the gradually increasing understandings that students develop *on their way to proficiency*. In this chart, this progress is presented from left (Step 1) to right (Step 4). The learning goal for students is to achieve Step 4 mastery. These steps are meant to help teachers and students identify and celebrate what the students CAN do now and what they need to work on next.

A Progression Toward Mastery				
Assessment Task Item	**STEP 1 Little evidence of reasoning without a correct answer. (1 Point)**	**STEP 2 Evidence of some reasoning without a correct answer. (2 Points)**	**STEP 3 Evidence of some reasoning with a correct answer or evidence of solid reasoning with an incorrect answer. (3 Points)**	**STEP 4 Evidence of solid reasoning with a correct answer. (4 Points)**
1 **1.OA.1**	Student's drawing and number sentence are completely unrelated to the problem, showing no understanding of the problem.	Student has the incorrect answer but shows some understanding through drawings or number sentences.	Student answers correctly (7) but is missing the drawing or the number sentence. OR Student draws a picture or number sentences to show her thinking but has an incorrect answer.	Student correctly: ▪ Answers 7. ▪ Explains using a drawing and any number sentence that matches her work (e.g., 15 – 8 = 7 or 2 + 5 = 7).
2 **1.OA.3** **1.OA.4** **1.OA.6**	Student answers one to two problems correctly, demonstrating a limited understanding of the problems.	For each problem, student: ▪ Subtracts from a teen number, ▪ Finds the missing addend, ▪ Writes the corresponding subtraction sentences, with three or four calculation errors.	For each problem, student: ▪ Subtracts from a teen number, ▪ Finds the missing addend, ▪ Writes the corresponding subtraction sentences, with one or two calculation errors.	For each problem, student correctly: a. Subtracts from a teen number: 3, 3, 9. b. Finds the missing addend: 4, 4, 5. c. Writes the corresponding subtraction sentences: ▪ 13 – 9 = 4 ▪ 12 – 8 = 4 ▪ 12 – 7 = 5
3 **1.OA.3** **1.OA.6**	Student is not able to correctly accomplish any component of the task, demonstrating a lack of understanding of the problems.	Student may show some understanding and skill with 5-group drawings but is unable to execute the bonds or explain his thinking. OR Student is able to show the bonds but is unable to draw the 5-groups or explain appropriately.	Student draws the bonds and 5-groups but is unable to explain how both have an answer of 4. OR Student explains well, and draws 5-groups well, but does not execute the bonds accurately.	Student correctly: ▪ Models the number bonds and 5-group drawings. ▪ Explains how both problems equal 4 using pictures or numbers (i.e., 1 + 3 = 2 + 2).

A Progression Toward Mastery				
4 **1.OA.1** **1.OA.2** **1.OA.3** **1.OA.4** **1.OA.6** **1.NBT.2a** **1.NBT.2b**	Student answers one or fewer questions correctly and is unable to show work, thus demonstrating a lack of understanding of the concepts.	Student answers two of the questions correctly with all accompanying models but demonstrates inconsistent understanding of the take from ten strategy, the connection between addition and subtraction, or the associative property.	Student answers three of the four questions correctly and with all requested models and number sentences. OR Student computes and explains the final question but may have errors in previous computations that impact accuracy (i.e., 1 or 2 off).	Student correctly: ▪ Represents 15 with a number bond, 5-group drawing, and number sentence. ▪ Explains that 10 snakes were sold. ▪ Explains that 4 turtles are left. ▪ Writes both an addition and subtraction equation 12 – 8 = 4 and 8 + 4 = 12. ▪ Explains that 18 animals are left altogether. ▪ Identifies the statement as false and explains why, citing the associative property with pictures or words (no formal terms necessary).

Name ______________________________ Date ______________

1. Mr. Baggy owns a pet store.

   He counted 10 goldfish in a big tank and 5 goldfish in a small tank. He sold 8 goldfish out of the big tank. How many goldfish did he have left in all? Explain your answer using a labeled math drawing and a number sentence.

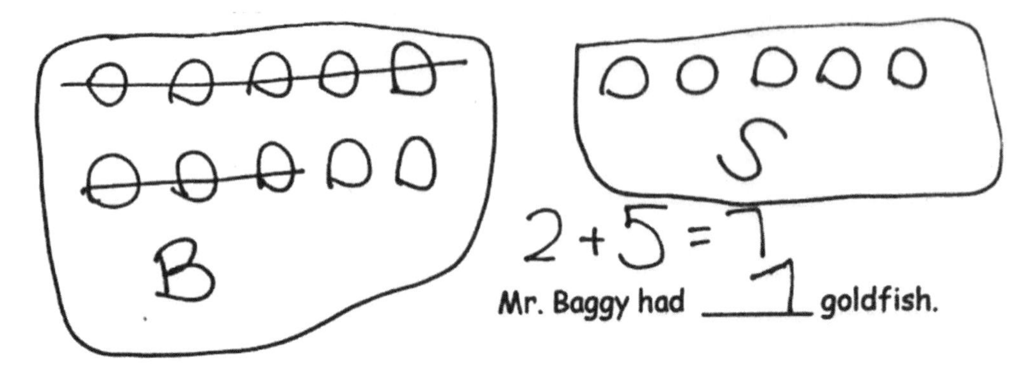

2. Write the numbers that make the number sentences true.

a. $12 - 9 =$ 3 (10, 2; 1+2) $11 - 8 =$ 3 (10, 1; 2+1) $15 - 6 =$ 9 (10, 5; 4+5)

b. $9 +$ 4 $= 13$ $8 +$ 4 $= 12$ $12 =$ 5 $+ 7$

c. Write a related subtraction fact for each of the three problems in the last row in the spaces below.

13-9=4 12-8=4 12-7=5

3. Write a number bond in each number sentence to show how to use ten to subtract. Draw 5-groups and some ones to show each subtraction sentence.

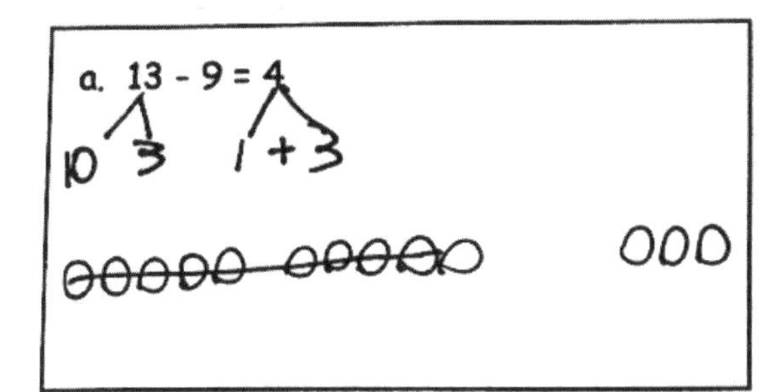

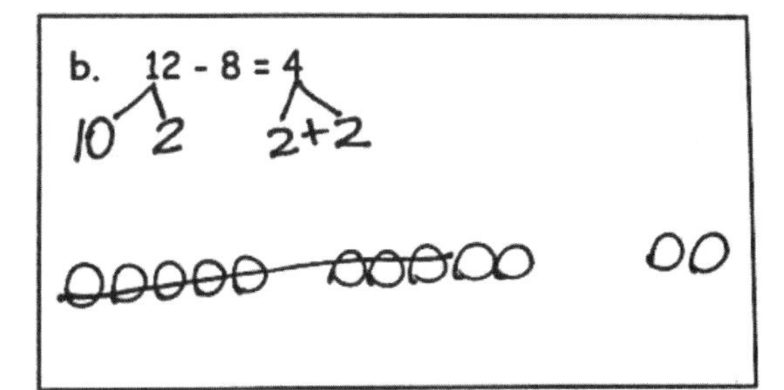

c. Use your pictures and numbers to explain how both subtraction problems equal 4.

1 + 3 = 2 + 2

10-9 10-8

4 = 4

4. Mr. Baggy also has 9 birds, 15 snakes, and 12 turtles.

a. Show the number of snakes as a ten and some ones with a number bond, a 5-group drawing, and a number sentence.

15 = 10 + 5

15

10 5

00000 00000 00000

15

b. Mr. Baggy sold some snakes. Now, he has 5. How many snakes did he sell? Explain your solution using a number bond or a math drawing. Write a number sentence. Complete the statement.

15 - □ = 5

15 - 5 = 10

15
10 5

Mr. Baggy sold 10 snakes.

c. Mr. Baggy sold 8 turtles. How many turtles does he have left? Explain your solution using a number bond or a math drawing. Write a number sentence. Complete the statement.

12 - 8 = 4

10 2   2 + 2

Mr. Baggy has 4 turtles left.

d. Mr. Baggy's daughter says she can find the number of turtles Mr. Baggy has left using subtraction or addition. Show two ways Mr. Baggy's daughter can solve this problem.

8 + □ = 12   12 - 8 = □

e. As Mr. Baggy gets ready to close his pet store for the day, he needs to know how many animals he has altogether. How many birds, snakes, and turtles does Mr. Baggy have left in his store altogether? Explain your solution using number bonds or math drawings. Write a number sentence. Complete the statement.

B S t
9 + 5 + 4
1 3
10 + 8 = 18

Mr. Baggy has 18 animals left.

f. True or false: You will get a different answer if you add 9 and 5 first, then add 4, than if you add 9 and 4 first, then add 5. (Circle one.) True (False)

Use pictures or words to show how you know.

00000 0000● ●●●●●0 000
9 + 5 = 14 14 + 4 = (18)

00000 0000● ●●●00 000
9 + 4 = 13 13 + 5 = (18)

You are just starting with a different number but you are just adding them all.

Answer Key

Eureka Math®
Grade 1
Module 2

Special thanks go to the Gordon A. Cain Center and to the Department of Mathematics at Louisiana State University for their support in the development of *Eureka Math*.

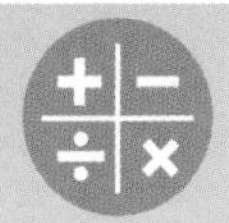

Answer Key

GRADE 1 • MODULE 2

Introduction to Place Value Through Addition and subtraction Within 20

Lesson 1

Problem Set

1. Partial answer provided; 16; 6, 16; 16
2. Drawing with 10 circled; 4, 7, 3, 14; 4, 14; 14
3. Drawing with 10 circled; 8, 3, 2, 13; 3, 13; 13
4. Drawing with 10 circled; 4, 2, 6, 12; 2, 12; 12

Exit Ticket

Drawing with 10 circled; 2, 4, 8, 14; 4, 14; 14

Homework

1. Drawing with 10 circled; 5, 6, 4, 15; 5, 15; 15
2. Drawing with 10 circled; 5, 7, 5, 17; 7, 17; 17
3. Drawing with 10 circled; 7, 6, 3, 16; 6, 16; 16
4. Drawing with 10 circled; 4, 9, 1, 14; 10, 4, 14; 14

Lesson 2

Problem Set

1. Partial answer provided; 4, 14
2. 9 and 1 circled; picture drawn; 9, 1, 4; 4, 14
3. 5 and 5 circled; picture drawn; 5, 5, 6; 6, 16
4. 7 and 3 circled; picture drawn; 3, 7, 4; 4, 14
5. 2 and 8 circled; picture drawn; 2, 8, 7; 7, 17
6. Partial answer provided; 15
7. 8 and 2 circled into a number bond; 14
8. 5 and 5 circled into a number bond; 13
9. 3 and 7 circled into a number bond; 16

Exit Ticket

a. 8 and 2 circled; picture drawn; 13; 2, 8, 10; 3, 13
b. 7 and 3 circled; picture drawn; 14; 3, 7, 10; 4, 14

Homework

1. Partial answer provided; 4; 2, 12
2. 5 and 5 circled; picture drawn; 10, 5, 5, 3; 3, 13
3. 2 and 8 circled; picture drawn; 10, 5, 2, 8; 5, 15
4. 7 and 3 circled; picture drawn; 10, 2, 3, 7; 2, 12
5. Partial answer provided; 15; 10, 5, 15
6. 8 and 2 circled into a number bond of 10; 14; 10, 4, 14

   Challenge:

   a. Sentence circled
   b. 4, 6 circled; sentence circled
   c. 3, 7 circled
   d. 2, 8 circled; sentence circled

Lesson 3

Problem Set

1. 10 circled; 6, 15; 5, 15; 15
2. 10 circled in drawing; 4, 13; 3, 13; 13
3. 10 circled in drawing; 3, 12; 2, 12; 12
4. 10 circled in drawing; 7, 16; 6, 16; 16

Exit Ticket

10 circled in drawing; 9, 4, 13; 10, 3, 13; 13

Homework

1. 10 circled in drawing; 4, 13; 3, 13; 13
2. 10 circled in drawing; 5, 14; 4, 14; 14
3. 10 circled in drawing; 6, 15; 5, 15; 15
4. 10 circled in drawing; 7, 16; 6, 16; 16
5. 10 circled in drawing; 8, 17; 7, 17; 17

Lesson 4

Sprint

Side A

1.	13	11.	17	21.	13
2.	15	12.	13	22.	16
3.	15	13.	16	23.	14
4.	11	14.	17	24.	4
5.	14	15.	17	25.	1
6.	16	16.	15	26.	6
7.	15	17.	16	27.	8
8.	11	18.	16	28.	9
9.	13	19.	15	29.	5
10.	17	20.	15	30.	10

Side B

1.	14	11.	17	21.	14
2.	16	12.	17	22.	18
3.	15	13.	18	23.	17
4.	11	14.	19	24.	1
5.	12	15.	19	25.	2
6.	15	16.	12	26.	7
7.	15	17.	13	27.	10
8.	16	18.	13	28.	2
9.	14	19.	16	29.	7
10.	17	20.	17	30.	10

Problem Set

1. Partial answer provided; 1, 4, 14; 4, 14
2. 10 circled; 1, 2; 2, 12
3. 10 circled; 3, 1; 3, 13
4. 1, 4; 14; 10, 4, 14; picture drawn
5. 5, 1; 15; 10, 5, 15; picture drawn
6. 7, 1; 17; 10, 7, 17; picture drawn
7. Numbers circled into a number bond; 4, 1; 14
8. Numbers circled into a number bond; 16, 1, 6

Exit Ticket

1. 15; 5, 15; picture drawn
2. 13; 10, 3, 13; picture drawn

Homework

1. 1, 2; 12; 10, 2, 12; picture drawn
2. 1, 5; 15; 10, 5, 15; picture drawn
3. 6, 1; 16; 10, 6, 16; picture drawn
4. a. 17; matched with 1, 7
   b. 15; matched with 1, 5
   c. 16; matched with 6, 1
5. a. 1, 1
   b. 1, 3; picture drawn
   c. 5, 1; picture drawn

Lesson 5

Problem Set

1. 1, 2, 12; 2, 12; 12
2. 1, 3; 13; 3, 13
3. 1, 1; 11; 10, 1, 11
4. 1, 4; 14; 10, 4, 14
5. 1, 3; 13; 10, 3, 13
6. 1, 6; 16; 10, 6, 16
7. 8, 1, 7; 17; 17
8. a. 11; 10, 1
   b. 11; 9, 2
9. a. 18; 10, 8
   b. 18; 9, 9
10. a. 17; 10, 7, 17
    b. 17; 9, 8, 17
11. a. 15; 15, 10, 5
    b. 15; 15, 6, 9
12. a. 16; 10, 6, 16
    b. 16; 7, 9, 16

Exit Ticket

1. Most efficient solution shown; 11
2. Most efficient solution shown; 16
3. Most efficient solution shown; 14

Homework

1. 15; 5, 15; number bond shows 5, 10, 15
2. 17; 10, 7, 17; number bond shows 7, 10, 17
3. 14; 10, 4, 14; number bond shows 4, 10, 14
4. 16; 10, 6, 16; number bond shows 6, 10, 16
5. a. 14; matched with 10, 4, 14
   b. 15; matched with 5, 10, 15
   c. 17; matched with 7, 10, 17
6. Most efficient strategy shown; 16
7. Most efficient strategy shown; 11
8. Most efficient strategy shown; 10
9. Most efficient strategy shown; 17
10. Most efficient strategy shown; 13
11. Most efficient strategy shown; 18

Lesson 6

Problem Set

1. Answer provided
2. 1, 5; 15; 5, 1; 15; 10, 5, 15
3. 6, 1; 16; 1, 6; 16; 10, 6, 16
4. 13; 10, 3, 13
5. 12; 10, 2, 12
6. 14; 10, 4, 14
7. a. Answer provided
   b. 10 + 4
   c. 10 + 5
   d. 10 + 7
   e. 10 + 6
   f. 10 + 0
8. a. 12
   b. 16
   c. 4
   d. 12
   e. 13
   f. 5
   g. 19
   h. 17
   i. 10
   j. 14
   k. 8
   l. 8
   m. 16
   n. 7

Exit Ticket

1. 1, 4; 14; 4, 1; 14; 4, 14
2. a. 16; 16; answer provided
   b. 15; 9 + 6 = 15; 10 + 5 = 15
   c. 17; 17 = 9 + 8; 10 + 7 = 17

Homework

1. a. 15; 6 + 9 = 15, answer provided
   b. 12; 9 + 3 = 12; 10 + 2 = 12
   c. 14; 5 + 9 = 14; 10 + 4 = 14
   d. 17; 17 = 9 + 8; 10 + 7 = 17
   e. 16; 16 = 7 + 9; 10 + 6 = 16
2. a. 13
   b. 13
   c. 15
   d. 15
   e. 17
   f. 16
   g. 8
   h. 17
   i. 10
   j. 14
3. a. 9 + 4 colored; answer provided; 9 + 4
   b. 7 + 9 colored; 10 + 6 = 7 + 9
   c. 8 + 9 colored; 10 + 7 = 8 + 9
   d. 9 + 9 colored; 10 + 8 = 9 + 9

Lesson 7

Friendly Fact Go Around: Make It Equal

0	2	4
3	6	5
2	1	7
4	3	8
3	5	4
1	6	8
6	8	9
4	5	7

Problem Set

1. 10 circled; 5, 13; 3, 13; 13
2. Picture drawn showing 8 and 4; 10 circled; 4, 12; 2, 12; 12
3. Picture drawn showing 3 and 8; 10 circled; 3, 11; 1, 11; 11
4. Picture drawn showing 7 and 8; 10 circled; 7, 15; 5, 15; 15

Exit Ticket

Picture drawn and labeled showing 5 and 8; 10 circled; 5, 13; 3, 13; 13

Homework

1. Picture drawn and labeled showing 4 and 8; 10 circled; 12; 2, 12; 12
2. Picture drawn and labeled showing 6 and 8; 10 circled; 6, 8, 14; 10, 4, 14; 14
3. Picture drawn and labeled showing 7 and 8; 10 circled; 7, 8, 15; 10, 5, 15; 15
4. Picture drawn and labeled showing 9 and 8; 10 circled; 9, 8, 17; 10, 7, 17; 17

Lesson 8

Sprint

Side A

1. 10
2. 11
3. 11
4. 10
5. 12
6. 12
7. 10
8. 14
9. 14
10. 10
11. 16
12. 16
13. 10
14. 18
15. 18
16. 14
17. 15
18. 15
19. 13
20. 13
21. 17
22. 18
23. 9
24. 9
25. 10
26. 4
27. 9
28. 3
29. 4
30. 10

Side B

1. 10
2. 12
3. 12
4. 10
5. 11
6. 11
7. 10
8. 13
9. 13
10. 10
11. 15
12. 15
13. 10
14. 14
15. 14
16. 14
17. 15
18. 15
19. 16
20. 16
21. 17
22. 18
23. 8
24. 9
25. 10
26. 6
27. 9
28. 5
29. 5
30. 10

Problem Set

1. Partial answer provided; 2, 3; 13; 3, 13
2. 10 circled; 2, 1; 11; 1, 11
3. 10 circled; 2, 2; 12; 2, 12
4. Picture drawn; 2, 2; 12; 10, 2, 12
5. Picture drawn; 4, 2; 14; 10, 4, 14
6. Picture drawn; 2, 3; 13; 10, 3, 13
7. 3, 2; 13
8. 15; 2, 5

Exit Ticket

1. Picture drawn; 14; 4, 14
2. Picture drawn; 12; 10, 2, 12

Homework

1. Picture drawn; 2, 2; 12; 10, 2, 12
2. Picture drawn; 2, 4; 14; 10, 4, 14
3. Picture drawn; 5, 2; 15; 10, 5, 15
4. a. Picture drawn; 2, 2; circled
   b. Picture drawn; 2, 6; circled
   c. Picture drawn; 5, 2; crossed out
   d. Picture drawn; 3, 2; crossed out
   e. Picture drawn; 2, 1; crossed out
   f. Picture drawn; 7, 1; circled

Lesson 9

Problem Set

1. 2, 1; 11; 1, 11; 11
2. 2, 2; 12; 2, 12
3. 2, 3; 13; 10, 3, 13
4. 2, 5; 15; 10, 5, 15
5. 2, 2; 12; 10, 2, 12
6. 5, 2; 15; 10, 5, 15
7. 7, 1; 9; 10, 7, 17
8. a. 11; 10, 1
   b. 11; 8, 3
9. a. 15; 10, 5
   b. 15; 8, 7
10. a. 16; 10, 6, 16
    b. 16; 8, 8, 16
11. a. 12; 12, 2, 10
    b. 12; 12, 4, 8
12. a. 14; 4, 10, 14
    b. 14; 6, 8, 14

Exit Ticket

1. 1, 2; 11; 1, 11
2. a. 14; 14, 8, 6
   b. 4; 14, 10, 4

Homework

1. 2, 1; 11; 1, 11
2. 4, 2; 14; 4, 14
3. 16; 2, 6; 16, 6
4. 13; 3, 2; 13, 3
5. a. 15; 7, 8
   b. 15; 15, 10, 5
6. a. 8; 16, 8, 8
   b. 16; 16, 10, 6
7. a. 17; 17, 9, 8
   b. 17; 17, 10, 7
8. 10 + 1 = 11
9. 13 = 10 + 3
10. 8 + 6 = 14

Lesson 10

Problem Set

1. 13; 3, 13
2. 14; 4, 14
3. 11; 1, 11
4. a. Answer provided
   b. 10 + 3
   c. 10 + 5
   d. 10 + 7
   e. 10 + 1
   f. 10 + 4
5. a. 11
   b. 12
   c. 12
6. a. 14
   b. 11
   c. 13
7. a. 15
   b. 14
   c. 11
8. a. 16
   b. 13
   c. 14
9. a. 8
   b. 8
   c. 9
10. a. 6
    b. 7
    c. 10

Exit Ticket

a. 14; 4, 14
b. 12; 2, 12
c. 13; 3, 13

Homework

1. 14; matched with 14-10-4 bond; 10 + 4 = 14
2. 12; matched with 12-10-2 bond; 10 + 2 = 12
3. 13; matched with 13-10-3 bond; 10 + 3 = 13
4. 11; matched with 11-10-1 bond; 10 + 1 = 11
5. 15; matched with 15-10-5 bond; 10 + 5 = 15
6. 12; 3; 4; 5
7. 4; 5; 6
8. 5; 6; 7
9. 15; 6; 15, 7; 15, 8
10. 7; 16, 8; 9, 16
11. 17; 17; 17, 9; 17, 10

Lesson 11

Sprint

Side A

1.	11	11.	14	21.	13
2.	12	12.	14	22.	18
3.	14	13.	15	23.	9
4.	13	14.	15	24.	9
5.	10	15.	15	25.	8
6.	11	16.	12	26.	9
7.	13	17.	12	27.	8
8.	12	18.	11	28.	5
9.	13	19.	12	29.	7
10.	13	20.	13	30.	7

Side B

1.	10	11.	14	21.	12
2.	11	12.	15	22.	16
3.	13	13.	16	23.	8
4.	12	14.	16	24.	9
5.	10	15.	16	25.	8
6.	11	16.	11	26.	9
7.	13	17.	11	27.	7
8.	12	18.	10	28.	4
9.	13	19.	11	29.	7
10.	13	20.	12	30.	6

G1-M2-TE-BK2-1.3.1-01.2016

Problem Set

1. Student work (a), (b), (d), (e), (f) circled
2. Answers will vary.
3. Drawings will vary; 4 + 8 = 12
4. Drawings will vary; 5 + 7 = 12

Exit Ticket

Problem solved and most efficient solution circled; 5 + 8 = 13

Homework

1. Joe's work corrected to show number bond of 1, 6; 9 + 7 = 16
2. Lori's work corrected to show number bond of 5, 3; 8 + 5 = 13
3. Mary's work corrected to show drawing and number sentence of 13 = 7 + 6
4. Frank's work corrected to show number bond of 1, 7; 9 + 8 = 17

Lesson 12

Problem Set

1. 7
2. Groups of 10 and 2 drawn; 9 crossed off; 3
3. Groups of 10 and 4 drawn; 9 crossed off; 5
4. 15; groups of 10 and 5 drawn; 9 crossed off; 6
5. 17, 10, 7; groups of 10 and 7 drawn; 9 crossed off from the table; 8
6. 18, 10, 8; groups of 10 and 8 drawn; 9 crossed off from the tray; 9

Exit Ticket

16, 10, 6; groups of 10 and 6 drawn; 9 crossed off from dinosaur books; 7

Homework

1. 15, 10, 5; groups of 10 and 5 drawn; 9 crossed off from eating nuts; 6
2. 17, 10, 7; groups of 10 and 7 drawn; 9 crossed off from the leaf; 8
3. Groups of 10 and 3 drawn; 9 crossed off from the sleeping ants; 4
4. Groups of 10 and 4 drawn; answers will vary.

Lesson 13

Problem Set

1. 9 crossed off; 4
2. 5-group rows showing 15; 9 crossed off; number bond shows 10, 5; 6
3. 5-group rows showing 17; 9 crossed off; number bond shows 17, 10, 7; 8
4. 7; number bond showing 16, 10, 6; 5-group rows showing 16; 9 crossed off; 16 – 9 = 7
5. 3; number bond showing 12, 10, 2; 5-group rows showing 12; 9 crossed off; 12 – 9 = 3
6. 10; number bond showing 19, 10, 9; 5-group rows showing 19; 9 crossed off; 19 – 9 = 10

Exit Ticket

Number bond showing 14, 10, 4; 5-group rows showing 14; 9 crossed off; 5

Homework

1. 11; 9 crossed off; 11 – 9 = 2; 2
2. 5-group rows showing 19; 9 crossed off; number bond showing 19, 10, 9; 19 – 9 = 10; 10
3. 5-group rows showing 18; 9 crossed off; number bond showing 18, 10, 8; 18 – 9 = 9; 9
4. 14; 5-group rows showing 14; 9 crossed off; 14, 10; 14 – 10 = 4; 4

Lesson 14

Sprint

Side A

1. 1
2. 2
3. 4
4. 3
5. 4
6. 5
7. 4
8. 6
9. 7
10. 3
11. 2
12. 8
13. 9
14. 1
15. 0
16. 5
17. 4
18. 3
19. 7
20. 6
21. 5
22. 10
23. 9
24. 8
25. 3
26. 4
27. 1
28. 2
29. 3
30. 8

Side B

1. 2
2. 1
3. 2
4. 1
5. 3
6. 1
7. 2
8. 3
9. 7
10. 3
11. 4
12. 6
13. 7
14. 3
15. 5
16. 10
17. 9
18. 8
19. 9
20. 8
21. 7
22. 10
23. 9
24. 8
25. 7
26. 8
27. 0
28. 1
29. 2
30. 8

Problem Set

1. Number sentences on the left appropriately matched with pictures
2. Group of 10 circled; 3
3. Group of 10 circled; 5
4. Group of 10 circled; 6
5. Group of 10 circled; 4
6. Group of 10 circled; 7
7. Group of 10 circled; 8
8. Group of 10 drawn and circled; 3
9. Group of 10 drawn and circled; 4
10. Group of 10 drawn and circled; 5
11. Group of 10 drawn and circled; 6

Exit Ticket

1. Group of 10 drawn and circled; number bond showing 17, 9, 8; 8
2. Group of 10 drawn and circled; number bond showing 14, 9, 5; 5
3. Group of 10 drawn and circled; number bond showing 15, 9, 6; 6
4. Group of 10 drawn and circled; number bond showing 18, 9, 9; 9

Homework

1. Group of 10 circled; number bond showing 15, 9, 6; 6
2. Group of 10 drawn and circled; number bond showing 14, 9, 5; 5
3. Group of 10 drawn and circled; number bond showing 12, 9, 3; 3
4. Group of 10 drawn and circled; number bond showing 13, 9, 4; 4
5. Group of 10 drawn and circled; number bond showing 16, 9, 7; 7
6. a. 4; $1 + 3 = 4$
   b. 5; $1 + 4 = 5$
   c. 6; $1 + 5 = 6$
   d. 7; $1 + 6 = 7$
7. Number bond drawn showing 17, 9, 8; $9 + 8 = 17$

Lesson 15

Problem Set

1. Number sentences appropriately matched to pictures
2. 5-group rows drawn; 9 crossed off; 2
3. 5-group rows drawn; 9 crossed off; 4
4. 5-group rows drawn; 9 crossed off; 7
5. 5-group rows drawn; 9 crossed off; 8
6. 5-group rows drawn; 9 crossed off; 5
7. 5-group rows drawn; 9 crossed off; 4
8. 5-group rows drawn; 9 crossed off; 3
9. 5-group rows drawn; 9 crossed off; 6
10. a. 5-group rows drawn showing 10 and 4; 14

    b. 5-group rows drawn; 9 crossed off; 5
11. 14, 9, 5; 9 + 5 = 14; 14 – 5 = 9

Exit Ticket

1. 5-group rows drawn; 9 crossed off; 8
2. 5-group rows drawn; 9 crossed off; 10

Homework

1. Answer provided

   16 – 9 = 7

   19 – 9 = 10

   17 – 9 = 8

   18 – 9 = 9

   14 – 9 = 5
2. 5-groups drawn; 6; 15 – 9 = 6
3. 5-groups drawn; 8; 17 – 9 = 8
4. 5-groups drawn; 7; 16 – 9 = 7
5. 5-groups drawn; 17; 17, 9, 8; 9 + 8 = 17; 17 – 9 = 8
6. 5-groups drawn; 8; 17, 9, 8; 17 – 8 = 9; 8 + 9 = 17

Lesson 16

Problem Set

1. a. 3; addition number sentence or drawing
   b. Number bond showing 2, 10; 3; 3
2. a. 6; addition number sentence or drawing
   b. Number bond showing 5, 10; 6; 6
3. a. 2; addition number sentence or drawing
   b. Number bond showing 1, 10; 2
   Strategy will vary.
4. a. 9; addition number sentence or drawing
   b. Number bond showing 8, 10; 9
   Strategy will vary.
5. Answers will vary.

Exit Ticket

1. a. 4; addition number sentence or drawing
   b. Number bond showing 3, 10; 4
2. a. 8; addition number sentence or drawing
   b. Number bond showing 7, 10; 8

Homework

1. 8; work based on strategy used to solve
2. 3; work based on strategy used to solve
3. 7; work based on strategy used to solve
4. 2; work based on strategy used to solve
5. 5; work will vary.
6. 8; work will vary.
7. Paul: 3, Lisa: 9; work will vary.
8. Answers will vary.

Lesson 17

Sprint

Side A

1. 1	11. 4	21. 4
2. 3	12. 4	22. 8
3. 1	13. 1	23. 9
4. 4	14. 6	24. 4
5. 1	15. 6	25. 5
6. 2	16. 1	26. 7
7. 1	17. 2	27. 6
8. 3	18. 3	28. 8
9. 3	19. 6	29. 9
10. 1	20. 5	30. 10

Side B

1. 1	11. 4	21. 3
2. 2	12. 4	22. 6
3. 1	13. 1	23. 7
4. 3	14. 3	24. 3
5. 1	15. 3	25. 4
6. 4	16. 1	26. 6
7. 1	17. 2	27. 5
8. 5	18. 4	28. 6
9. 5	19. 5	29. 8
10. 1	20. 4	30. 7

Problem Set

1. Number sentences appropriately matched to pictures
2. Group of 10 circled; 8 crossed off; 5
3. Group of 10 circled; 8 crossed off; 3
4. Group of 10 circled; 8 crossed off; 7
5. Group of 10 circled; 8 crossed off; 11
6. Group of 10 circled; 8 crossed off; 8
7. Group of 10 circled; 8 crossed off; 9
8. 10 drawn and circled or number bond drawn; 4
9. 10 drawn and circled or number bond drawn; 5
10. 10 drawn and circled or number bond drawn; 6
11. 10 drawn and circled or number bond drawn; 7

Exit Ticket

1. a. 10 drawn and circled; 8 crossed off; 4
   b. 10 drawn and circled; 8 crossed off; 6
2. Number bond shows 5, 10; 7

Homework

1. a. 6; matched to first picture
   b. 8; matched to hearts
   c. 3; matched to stars
   d. 5; matched to third picture
2. Number bond or drawing to match number sentence; 6
3. 10 circled; 17 – 8 = 9; 9
4. 10 circled; 8 crossed off; 12 – 8 = 4; 4
5. Number bond or drawing; 15 – 8 = 7; 7
6. 18, 8; 18 – 8 = 10; number bond showing 18, 10, 8

Lesson 18

Problem Set

1. a. Matched to third picture
   b. Matched to second picture
   c. Matched to first picture
   d. Matched to fifth picture
   e. Matched to fourth picture
2. 5-group row and ones drawn; 8 crossed off; 3; 2 + 1 = 3
3. 5-group row and ones drawn; 8 crossed off; 4; 2 + 2 = 4
4. 5-group row and ones drawn; 8 crossed off; 7; 2 + 5 = 7
5. 5-group row and ones drawn; 8 crossed off; 11; 2 + 9 = 11
6. 5-group row and ones drawn; 8 crossed off; 8; 2 + 6 = 8
7. 5-group row and ones drawn; 9 crossed off; 7; 1 + 6 = 7
8. 5-group row and ones drawn; 9 crossed off; 5; 1 + 4 = 5
9. a. 14; number bond drawn showing 4, 2
   b. 6; number bond drawn showing 10, 4

Exit Ticket

1. 5-group row and ones drawn; 8 crossed off; 6; 4, 6
2. 5-group row and ones drawn; 8 crossed off; 9; 7, 9

Homework

1. 5-group row and ones drawn; 8 crossed off; 2 + 3 = 5; 13 – 8 = 5; 5
2. 5-group row and ones drawn; 8 crossed off; 2 + 5 = 7; 15 – 8 = 7; 7
3. 5-group row and ones drawn; 8 crossed off; 2 + 9 = 11; 19 – 8 = 11; 11
4. 15; number bond drawn showing 5, 2
5. 7; number bond drawn showing 10, 5
6. 5-group row and ones drawn; 9 crossed off; 2
7. 5-group row and ones drawn; 9 crossed off; 5
8. 5-group row and ones drawn; 8 crossed off or number bond drawn; 14 – 8 = 6; 6

Lesson 19

Problem Set

1. 6; number bond provided; 6
2. 9; number bond showing 10, 7
3. 10; number bond showing 10, 8
4. 5
5. 7
6. a. 4; number bond showing 10, 2
   b. 4; strategy will vary.
7. a. 3; number bond showing 10, 1
   b. 3; strategy will vary.
8. a. 8; number bond showing 10, 6
   b. 8; strategy will vary.
9. a. 11; number bond showing 10, 9
   b. 11; strategy will vary.

Exit Ticket

1. a. Number bond showing 10, 1; 3
   b. 3
2. a. Number bond showing 10, 5; 7
   b. 7

Homework

1. a. Number bond showing 10, 2; 4
   b. 4
2. a. Number bond showing 10, 5; 7
   b. 7
3. Strategy will vary; 3
4. Strategy will vary; 9
5. 8; 8
6. 13 – 8 = 5; number bond showing 10, 3; 5
7. a. 4; $8 + \underline{4} = 12$
   b. 7; $8 + \underline{7} = 15$
   c. 10; $8 + \underline{10} = 18$
   d. 3; $8 + \underline{3} = 11$

Lesson 20

Sprint

Side A

1. 2	11. 4	21. 5
2. 4	12. 4	22. 9
3. 2	13. 2	23. 10
4. 5	14. 7	24. 3
5. 2	15. 7	25. 4
6. 6	16. 2	26. 7
7. 2	17. 3	27. 6
8. 3	18. 4	28. 8
9. 3	19. 7	29. 9
10. 2	20. 6	30. 10

Side B

1. 2	11. 5	21. 4
2. 3	12. 5	22. 7
3. 2	13. 2	23. 8
4. 4	14. 4	24. 2
5. 2	15. 4	25. 3
6. 5	16. 2	26. 5
7. 2	17. 3	27. 4
8. 4	18. 5	28. 5
9. 4	19. 6	29. 7
10. 2	20. 5	30. 8

Problem Set

1. Drawing or number bond shown; 2
2. Drawing or number bond shown; 3
3. Drawing or number bond shown; 4
4. Drawing or number bond shown; 5
5. Drawing or number bond shown; 6
6. Drawing or number bond shown; 5
7. a. Answer provided
   b. 18 – 8
   c. 13 – 9
   d. 15 – 9
8. a. 3
   b. 4
   c. 5
9. a. 4
   b. 5
   c. 6
10. a. 4
    b. 5
    c. 6
11. a. 7
    b. 9
    c. 8
12. a. 7
    b. 6
    c. 8
13. a. 9
    b. 8
    c. 9

Exit Ticket

Appropriate drawings or number bond included for each

a. 5
b. 7
c. 6
d. 9
e. 7
f. 8

Homework

1. 6
2. 7
3. 8
4. 8
5. 9
6. 10
7. 7
8. 8
9. 9
10. 10
11. 11
12. 12
13. a. 18 – 8
    b. 12 – 7
14. a. Drawing or number bond drawn showing Elsie is right since both expressions are equal
    b. Drawing or number bond drawn showing John is right since both expressions are not equal
    c. Drawing or number bond drawn showing Elsie made a mistake and John is correct since 17 – 9 and 16 – 10 are not equal
    d. 7; other answers will vary.

Lesson 21

Sprint

Side A

1.	1	11.	4	21.	8
2.	2	12.	5	22.	10
3.	4	13.	4	23.	9
4.	2	14.	5	24.	10
5.	3	15.	6	25.	7
6.	5	16.	5	26.	8
7.	3	17.	6	27.	9
8.	4	18.	7	28.	8
9.	6	19.	6	29.	9
10.	3	20.	7	30.	16

Side B

1.	1	11.	3	21.	8
2.	2	12.	6	22.	7
3.	3	13.	3	23.	8
4.	2	14.	4	24.	9
5.	3	15.	7	25.	7
6.	4	16.	4	26.	8
7.	3	17.	5	27.	9
8.	4	18.	8	28.	7
9.	5	19.	6	29.	9
10.	2	20.	7	30.	16

Problem Set

1. Student work (a), (b), (c), (e), (f) circled
2. New drawing shown for (d) starting at 7; 16 – 7 = 9
3. Work shown; 12 – 5 = 7; 7
4. Work shown; 17 – 9 = 8; 8

Exit Ticket

Work shown for each strategy; explanations will vary; 14 – 6 = 8

Homework

1. a. Take from ten; 13 – 6 = 7; strategy shown correctly solving problem
   b. Make ten; different strategy shown correctly solving problem
   c. Explanations may vary.
2. a. Take from ten; different strategy shown correctly solving problem
   b. Make ten; 9 + 8 = 17; strategy shown correctly solving problem
   c. Explanations may vary.

Lesson 22

Sprint

Side A

1. 1
2. 2
3. 2
4. 2
5. 1
6. 3
7. 4
8. 1
9. 2
10. 3
11. 4
12. 6
13. 6
14. 2
15. 3
16. 6
17. 4
18. 2
19. 5
20. 6
21. 7
22. 8
23. 5
24. 5
25. 4
26. 3
27. 4
28. 4
29. 2
30. 5

Side B

1. 2
2. 3
3. 0
4. 2
5. 1
6. 0
7. 1
8. 4
9. 3
10. 2
11. 4
12. 3
13. 4
14. 3
15. 2
16. 5
17. 6
18. 7
19. 4
20. 7
21. 5
22. 4
23. 3
24. 3
25. 3
26. 1
27. 2
28. 3
29. 5
30. 5

Problem Set

1. Labeled drawing; 5 + 6 = 11 or 11 – 5 = 6
2. Labeled drawing; 8 + 6 = 14 or 14 – 8 = 6
3. Labeled drawing; 8 + 7 = 15 or 15 – 8 = 7
4. Labeled drawing; 7 + 9 = 16 or 16 – 7 = 9

Exit Ticket

1. Labeled drawing; 8 + 9 = 17 or 17 – 8 = 9
2. Labeled drawing; 8 + 5 = 13 or 13 – 8 = 5

Homework

1. Labeled drawing and statement; 6 + 8 = 14 or 14 – 6 = 8
2. Labeled drawing and statement; 6 + 9 = 15 or 15 – 6 = 9
3. Labeled drawing and statement; 9 + 9 = 18 or 18 – 9 = 9
4. Labeled drawing and statement; 8 + 9 = 17 or 17 – 8 = 9
5. Labeled drawing and statement; 9 + 8 = 17 or 17 – 9 = 8
6. Labeled drawing and statement; 9 + 7 = 16 or 16 – 9 = 7

Lesson 23

Sprint

Side A

1. 1
2. 2
3. 2
4. 2
5. 1
6. 3
7. 4
8. 1
9. 2
10. 3
11. 4
12. 6
13. 6
14. 2
15. 3
16. 6
17. 4
18. 2
19. 5
20. 6
21. 7
22. 8
23. 5
24. 5
25. 4
26. 3
27. 4
28. 4
29. 2
30. 5

Side B

1. 2
2. 3
3. 0
4. 2
5. 1
6. 0
7. 1
8. 4
9. 3
10. 2
11. 4
12. 3
13. 4
14. 3
15. 2
16. 5
17. 6
18. 7
19. 4
20. 7
21. 5
22. 4
23. 3
24. 3
25. 3
26. 1
27. 2
28. 3
29. 5
30. 5

Problem Set

1. Labeled drawing; 8 + 4 = 12 or 12 − 8 = 4; 4
2. Labeled drawing; 5 + 8 = 13 or 13 − 5 = 8; 8
3. Labeled drawing; 8 + 7 = 15 or 15 − 8 = 7; 7
4. Labeled drawing; 9 + 10 = 19 or 19 − 9 = 10; 10

Exit Ticket

Labeled drawing; 7 + 6 = 13 or 13 − 7 = 6; 6

Homework

1. Labeled drawing; 9 + 5 = 14 or 14 − 9 = 5; 5
2. Labeled drawing; 8 + 9 = 17 or 17 − 8 = 9; 9
3. Labeled drawing; 5 + 8 = 13 or 13 − 5 = 8; 8
4. Labeled drawing; 12 + 6 = 18 or 18 − 12 = 6; 6

Lesson 24

Sprint

Side A

1. 1
2. 0
3. 2
4. 1
5. 2
6. 3
7. 0
8. 0
9. 1
10. 2
11. 3
12. 5
13. 4
14. 3
15. 2
16. 4
17. 3
18. 2
19. 4
20. 5
21. 6
22. 6
23. 5
24. 4
25. 7
26. 6
27. 5
28. 2
29. 4
30. 5

Side B

1. 0
2. 1
3. 2
4. 0
5. 1
6. 2
7. 3
8. 0
9. 1
10. 2
11. 3
12. 0
13. 1
14. 2
15. 3
16. 3
17. 2
18. 1
19. 3
20. 4
21. 5
22. 5
23. 4
24. 3
25. 6
26. 5
27. 4
28. 3
29. 5
30. 4

Problem Set

1. Labeled drawing and statement; 11 – 3 = 8 or 8 + 3 = 11; 3
2. Labeled drawing and statement; 15 – 9 = 6 or 9 + 6 = 15; 6
3. Labeled drawing and statement; 16 – 8 = 8 or 8 + 8 = 16; 8
4. Labeled drawing and statement; 18 – 9 = 9 or 9 + 9 = 18; 9

Exit Ticket

Labeled drawing and statement; 18 – 9 = 9 or 9 + 9 = 18; 9

Homework

1. Labeled drawing and statement; 12 – 9 = 3 or 9 + 3 = 12; 3
2. Labeled drawing and statement; 11 – 7 = 4 or 7 + 4 = 11; 4
3. Labeled drawing and statement; 17 – 8 = 9 or 8 + 9 = 17; 9
4. Labeled drawing and statement; 12 – 9 = 3 or 9 + 3 = 12; 3

Lesson 25

Sprint

Side A

1. 5
2. 6
3. 7
4. 6
5. 7
6. 8
7. 7
8. 1
9. 1
10. 8
11. 0
12. 7
13. 8
14. 2
15. 2
16. 6
17. 5
18. 4
19. 7
20. 4
21. 4
22. 6
23. 4
24. 3
25. 1
26. 5
27. 5
28. 6
29. 6
30. 6

Side B

1. 4
2. 5
3. 6
4. 5
5. 6
6. 7
7. 6
8. 7
9. 8
10. 1
11. 0
12. 6
13. 7
14. 1
15. 1
16. 6
17. 3
18. 2
19. 6
20. 3
21. 4
22. 6
23. 5
24. 3
25. 1
26. 6
27. 6
28. 7
29. 5
30. 7

Problem Set

Answers will vary.

Exit Ticket

Answers will vary.

Homework

1. a. False
   b. True
   c. True
   d. False
   e. True
   f. True
2. a. Lola; pictures drawn to explain thinking
   b. Lola; pictures drawn to explain thinking
   c. Charlie; pictures drawn to explain thinking
3. a. Answers will vary.
   b. Answers will vary.
   c. Answers will vary.

Lesson 26

Problem Set

1. 10 circled; 15, 1, 5
2. 10 circled; 12, 1, 2
3. 10 circled; 17, 7, 1
4. 10 circled; 13, 1, 3
5. 10 circled; 17, 1, 7
6. 1, 2, 10, 2 shown on cards; 1, 2
7. 1, 4, 10, 4 shown on cards; 1, 4
8. 1, 9, 10, 9 shown on cards; 9, 1
9. Circles drawn to show 16; 10, 6; 1, 6
10. Circles drawn to show 12; 1, 2
    Circles drawn to show 15; 1, 5

Exit Ticket

12 circles matched to cards that read 12, 10, 2; 1, 2

14 circles matched to cards that read 14, 10, 4; 1, 4

17 circles matched to cards that read 17, 10, 7; 1, 7

Homework

1. Pile of 10 coins circled; 18; 1, 8
2. Purse of coins circled; 17; 7, 1
3. Circles drawn to show 18; 10, 8; 1, 8
4. Circles drawn to show 13; 1, 3; 1, 3
5. Circles drawn to show 12; 1, 2
6. Circles drawn to show 13; 1, 3
7. 5-group column, circles drawn to show 16; 1, 6
8. 5-group column, circles drawn to show 19; 9, 1

Lesson 27

Sprint

Side A

1. 13	11. 0	21. 10
2. 12	12. 3	22. 10
3. 11	13. 4	23. 10
4. 11	14. 5	24. 18
5. 14	15. 8	25. 19
6. 16	16. 1	26. 12
7. 17	17. 2	27. 14
8. 18	18. 10	28. 15
9. 2	19. 10	29. 10
10. 1	20. 7	30. 3

Side B

1. 11	11. 2	21. 10
2. 12	12. 3	22. 10
3. 13	13. 5	23. 10
4. 14	14. 7	24. 16
5. 15	15. 9	25. 18
6. 16	16. 0	26. 13
7. 18	17. 1	27. 15
8. 18	18. 10	28. 15
9. 0	19. 10	29. 10
10. 1	20. 3	30. 3

Problem Set

1. 1, 8; 1, 8; *s* crossed off appropriately
2. 1, 8; 1, 8; *s* crossed off appropriately
3. 1, 5; 1, 5; *s* crossed off appropriately
4. 2, 0; 2, 0
5. 1, 3; 1, 3; *s* crossed off appropriately
6. 1, 2; 1, 2; *s* crossed off appropriately
7. 0, 8; 0, 8
8. 0, 9; 0, 9
9. Labeled drawing; 4 + 10 = 14; 1, 4; *s* crossed off appropriately
10. Labeled drawing; 8 + 7 = 15; 1, 5; *s* crossed off appropriately
11. Labeled drawing; 14 – 3 = 11; 1, 1; *s* crossed off appropriately
12. Labeled drawing; 16 – 5 = 11; 1, 1; *s* crossed off appropriately

Exit Ticket

1. 1, 9; 1, 9; *s* crossed off appropriately
2. 1, 3; 1, 3; *s* crossed off appropriately
3. Labeled drawing; 16 – 9 = 7; 7; 0, 7

Homework

1. 1, 3; 1, 3; *s* crossed off appropriately
2. 0, 8; 0, 8
3. 0, 9; 0, 9
4. 1, 9; 1, 9; *s* crossed off appropriately
5. 1, 8; 1, 8; *s* crossed off appropriately
6. 0, 9; 0, 9
7. Labeled drawing; 8 + 9 = 17; 1, 7; *s* crossed off appropriately
8. Labeled drawing; 14 – 8 = 6; 0, 6
9. Labeled drawing; 6 + 14 = 20; 2, 0

Lesson 28

Sprint

Side A

1. 12
2. 3
3. 13
4. 14
5. 6
6. 16
7. 13
8. 6
9. 16
10. 3
11. 13
12. 4
13. 14
14. 6
15. 16
16. 15
17. 16
18. 17
19. 18
20. 19
21. 20
22. 18
23. 19
24. 2
25. 3
26. 4
27. 12
28. 11
29. 21
30. 32

Side B

1. 11
2. 2
3. 12
4. 13
5. 5
6. 15
7. 12
8. 4
9. 14
10. 3
11. 13
12. 2
13. 12
14. 5
15. 15
16. 14
17. 15
18. 16
19. 17
20. 18
21. 19
22. 19
23. 20
24. 1
25. 2
26. 3
27. 11
28. 13
29. 23
30. 34

Problem Set

1. 1, 4; 9 + 1 = 10; 10 + 4 = 14
2. 1, 4; 8 + 2 = 10; 10 + 4 = 14
3. 15; 9 + 1 = 10; 10 + 5 = 15
4. 8 + 7 = 15; 8 + 2 = 10; 10 + 5 = 15
5. 4 + 9 = 13; 9 + 1 = 10; 10 + 3 = 13
6. 7 + 8 = 15; 8 + 2 = 10; 10 + 5 = 15
7. 8 + 9 = 17; 9 + 1 = 10; 10 + 7 = 17

Exit Ticket

1. 1, 3; 9 + 1 = 10; 10 + 3 = 13
2. 1, 5; 8 + 2 = 10; 10 + 5 = 15

Homework

1. 1, 6; 9 + 1 = 10; 10 + 6 = 16
2. 1, 3; 8 + 2 = 10; 10 + 3 = 13
3. 17; 9 + 1 = 10; 10 + 7 = 17
4. 8 + 7 = 15; 8 + 2 = 10; 10 + 5 = 15
5. 6 + 9 = 15; 9 + 1 = 10; 10 + 5 = 15
6. 4 + 8 = 12; 8 + 2 = 10; 10 + 2 = 12
7. 5 + 7 = 12; 7 + 3 = 10; 10 + 2 = 12

EUREKA MATH